THE SELF
AS PROJECT

Politics and the Human Sciences

EDITED BY

*Greg Eghigian, Andreas Killen,
and Christine Leuenberger*

OSIRIS | 22

A Research Journal Devoted to the
History of Science and Its Cultural Influences

Osiris

Series editor, 2002–2012

KATHRYN OLESKO, *Georgetown University*

Volumes 17 to 27 in this series are designed to dissolve boundaries between history and the history of science. They cast science in the framework of larger issues prominent in the historical discipline but infrequently treated in the history of science, such as the development of civil society, urbanization, and the evolution of international affairs. They aim to open up new categories of analysis, to stimulate fresh areas of investigation, and to explore novel ways of synthesizing major historical problems that demand consideration of the role science has played in them. They are written not only for historians of science, but also for historians and other scholars who wish to integrate issues concerning science into courses on broader themes, as well as for readers interested in viewing science from a general historical perspective. Special attention is paid to the international dimensions of each volume's topic.

Cover Illustration:

"The Pupil becomes an Individual." From *The American School Board Journal,* Volume LXIV, Number 4, 4 April 1922. (Image is in the public domain.) *Courtesy of Cornell University Library.*

THE SELF AS PROJECT:
POLITICS AND THE HUMAN SCIENCES

Introduction:

The Self as Project: Politics and the Human Sciences in the Twentieth Century

By Greg Eghigian, Andreas Killen, and Christine Leuenberger[*]

On May 8, 1945, Germany formally surrendered to Allied forces, ending the Second World War in Europe. Soon after Soviet, American, British, and French troops occupied German territory, something unique in the annals of military history took place. Scores of researchers and practitioners from clinical medicine, law, psychiatry, and the social sciences descended upon Germany at the invitation of occupying authorities. They were charged with two main tasks: help determine who was responsible for the murderous actions committed in the name of National Socialism, and contribute to the reconstruction of a democratic Germany. To these ends, legal scholars, forensic pathologists, physicians, psychiatrists, psychologists, and psychometrists were recruited to assist military tribunals in western occupation zones in analyzing evidence and interrogating defendants and witnesses in criminal cases against Nazi officials, administrators, scientists, and doctors.[1] At the same time, social scientists were brought in to revamp the German educational system and to track public opinion using a relatively new tool, the mass survey.[2]

[*] Greg Eghigian, Department of History, 108 Weaver Building, Penn State University, University Park, PA 16802; gae2@psu.edu. Andreas Killen, City College of New York, CUNY, North Academic Center, Room 5/144, Convent Ave. at 138th Street, New York, NY 10031; akillen@ccny.cuny.edu. Christine Leuenberger, Cornell University, Department of Science and Technology Studies, 301 Rockefeller Hall, Ithaca, NY 14853 7601; cal22@cornell.edu.

This project began as a conference, "The Self as Scientific and Political Project in the Twentieth Century: The Human Sciences between Utopia and Reform," held at Pennsylvania State University in October 2003. From the early planning for that conference to the publication of this volume, numerous organizations and individuals have offered their support, advice, and encouragement. We would like to thank the National Science Foundation as well as the College of Liberal Arts; the Rock Ethics Institute; the Department of History; the Science, Medicine, and Technology in Culture Program; and the Social Thought Program at Pennsylvania State University for their generous funding. In addition, we would like to extend our gratitude to all those who worked hard to make the original conference possible: Amy Bennett, Luc Berlivet, John Carson, John Christman, Moritz Foellmer, Sarah Goodfellow, Rüdiger Graf, David Horn, Anna Krylova, Elizabeth Lunbeck, Rose Niman, Daniel Purdy, Sophie de Schaepdrijver, Londa Schiebinger, Cathie Stong, Mark Sullivan, Nancy Tuana, Adrian Wanner, and Karin Weaver. Finally, we are enormously grateful to Kathy Olesko, the editorial board of Osiris, and two anonymous readers for their thoughtful and constructive comments.

[1] Ulf Schmidt, *Justice at Nuremberg: Leo Alexander and the Nazi Doctors' Trial* (New York, 2004); Paul Julian Weindling, *Nazi Medicine and the Nuremberg Trials: From Medical War Crimes to Informed Consent* (New York, 2005).

[2] Christopher P. Loss, "Party School: Education and Political Ideology in Cold War Eastern Europe," *Journal of Policy History* 16 (2004): 99–116; James F. Tent, *Mission on the Rhine: Reeducation and Denazification in American-Occupied Germany* (Chicago, 1982); Anna J. Merritt and Richard L. Merritt, eds., *Public Opinion in Occupied Germany: The OMGUS Surveys, 1945–1949* (Urbana, Ill., 1970).

Here, then, at one of the most defining moments of modern political history, were the human sciences, accepted by civil and military authorities as essential to understanding and overcoming the destruction associated with a war that had claimed some 55 million lives. Yet this represents only one aspect of the involvement of the human sciences in that conflict. For as historians have since noted, the genocidal domestic, foreign, and military policies of the Third Reich were themselves partly products of the work of demographers, sociologists, physicians, engineers, economists, architects, anthropologists, criminologists, and psychologists. The history of the Holocaust and its legacy is, at once, a chapter in the political as well as the scientific, medical, and technological history of the Western world.[3]

How did this particular marriage of politics and science come to be? It is apparent that by the middle of the twentieth century, political parties, states, and the human sciences, by and large, shared a common belief: the belief that societies and the human beings that inhabit them could be known, changed, and managed. The first three-quarters of the twentieth century, especially, were stamped by a widespread enthusiasm for public projects, ranging from urban renewal to macroeconomic planning to space exploration. Grand achievements were expected of governments and public institutions, and bold new futures were charted for societies. These ideals were projected not only *onto* society but also *into* individuals. Never before had states invested so much time, energy, and resources into making sense and exploiting the potential of the human being. It was here, at this juncture of public planning and human engineering, that policy makers, researchers, and clinicians envisioned a formative role for the human sciences.

The purpose of the following essays is to explore this pivotal development, to discuss the ways in which the human sciences and politics together turned the behavior and inner workings of individuals—in short, the self—into sites for ambitious scientific, medical, and political projects. In doing so, the studies here build on some promising lines of recent scholarship. Over the past decade and a half, notable scholars such as Norbert Elias, George Herbert Mead, and Michel Foucault, drawing on various disciplinary traditions ranging from cultural studies, cultural history, sociological theory, and symbolic interactionism to the Frankfurt school of critical theory, have inspired observers from the humanities and social sciences to revisit the topics of individuality, identity, personhood, and subjectivity.[4] They have focused on such issues as the distinctive characteristics of the modern individual, the manner in which indi-

[3] Dieter Kuntz and Susan D. Bachrach, eds., *Deadly Medicine: Creating the Master Race* (Chapel Hill, N.C., 2004); Zygmunt Bauman, *Modernity and the Holocaust* (Ithaca, N.Y., 1991).

[4] Christine Leuenberger, "Constructions of the Berlin Wall: How Material Culture Is Used in Psychological Theory," *Social Problems* 53 (2006): 18–37; Jerrold E. Seigel, *The Idea of the Self: Thought and Experience in Western Europe since the Seventeenth Century* (New York, 2005); Kurt Danziger, *Naming the Mind: How Psychology Found Its Language* (London, 1997); Roy Porter, ed., *Rewriting the Self: Histories from the Renaissance to the Present* (London, 1997); Ian Hacking, *Rewriting the Soul: Multiple Personality and the Sciences of Memory* (Princeton, N.J., 1995); G. M. Erchak, *The Anthropology of Self and Behavior* (New Brunswick, N.J., 1992); Anthony Giddens, *Modernity and Self-Identity: Self and Society in the Late Modern Age* (Palo Alto, Calif., 1991); John Shotter and Kenneth J. Gergen, eds., *Texts of Identity* (London, 1989); Charles Taylor, *Sources of the Self: The Making of Modern Identity* (Cambridge, Mass., 1989); Michel Foucault et al., *Technologies of the Self: A Seminar with Michel Foucault* (Amherst, Mass., 1988); Ian Hacking, "Making Up People," in *Reconstructing Individualism: Autonomy, Individuality, and the Self in Western Thought,* ed. Thomas C. Heller and Christine Brooke-Rose (Palo Alto, Calif., 1986), 222–36; Michel Foucault, *Madness and Civilization: A History of Insanity in the Age of Reason* (New York, 1965).

viduals have been labeled normal and abnormal, and the changing ways in which body and mind have been represented. Research has shown that different societies at different times have pictured and experienced selfhood in disparate ways, offering their communities what psychological anthropologists have dubbed "indigenous psychologies" to make sense of human thought, feeling, and behavior.[5] In addition, historians and sociologists have noted the importance of the human sciences in establishing governing norms of human consciousness, conduct, and identity in modern society.[6]

At the same time, social scientists and historians have begun to reexamine the field of modern politics—for our purposes, defined as the realm of ideology, statecraft, political parties, public policy, nationalism, and citizenship. Cultural history, art history, gender analysis, the theory of practice, and the history of everyday life have moved historians to search for politics outside the centers of executive, legislative, and judicial power.[7] This trend has only been reinforced by the increasing interest in global approaches to modern history.[8] Meanwhile, scholars in science and technology studies have been reassessing the relationship between science and politics, seeing them as mutually responsible for producing public policies, social knowledge, and research practices.[9] What is emerging from this wave of research is, among other things, an appreciation for the peculiarity of the twentieth century. This most ideological of ages, an "age of extremes" as Eric Hobsbawm has named it, was caught up in an international zeal for experimentation and utopian and pragmatic social reform in the name of liberalisms, socialisms, conservatisms, and fascisms.[10] Both the individual and the human sciences were swept up in the political construction, destruction, and reconstruction of twentieth-century societies in historically novel ways.

It is our view that one useful way to make sense of the interpenetration of politics

[5] Paul Heelas and Andrew Lock, eds., *Indigenous Psychologies: The Anthropology of the Self* (London, 1981).

[6] Jeroen Jansz and Peter van Drunen, *A Social History of Psychology* (Malden, Mass., 2004); Arnold Davidson, *The Emergence of Sexuality: Historical Epistemology and the Formation of Concepts* (Cambridge, Mass., 2001); Roger Smith, *The Norton History of the Human Sciences* (New York, 1997); Nikolas Rose, *Inventing Our Selves: Psychology, Power, and Personhood* (New York, 1996); Hacking, *Rewriting the Soul* (cit. n. 4); Michel Foucault, *Discipline and Punish: The Birth of the Prison* (New York, 1977); Giddens, *Modernity and Self-Identity* (cit. n. 4).

[7] Theodore R. Schatzki, Karin Knorr Cetina, and Eike von Savigny, eds., *The Practice Turn in Contemporary Theory* (London, 2001); Monica Flacke, ed., *Mythen der Nationen: Ein Europäisches Panorama* (Berlin, 1998); Geoff Eley and Ronald Grigor Suny, eds., *Becoming National: A Reader* (New York, 1996); Rogers Brubaker, *Citizenship and Nationhood in France and Germany* (Cambridge, Mass., 1992); Benedict Anderson, *Imagined Communities: Reflections on the Origin and Spread of Nationalism* (London, 1991).

[8] Michael Geyer and Charles Bright, "World History in a Global Age," *American Historical Review* 100 (1995): 1034–60.

[9] Sheila Jasanoff, *Designs on Nature: Science and Democracy in Europe and the United States* (Princeton, N.J., 2005); as well as her essays, Jasanoff, "Beyond Epistemology: Relativism and Engagement in the Politics of Science," *Social Studies of Science* 26 (1996): 393–418; and Jasanoff, "Technology as a Site and Object of Politics," in *The Oxford Handbook of Contextual Political Analysis,* ed. Charles Tilly and Robert E. Goodin (Oxford, 2006), 745–63. See also Bruno Latour and Catherine Porter, *Politics of Nature: How to Bring the Sciences into Democracy* (Boston, 2004); Karin Knorr Cetina, *Epistemic Cultures: How the Sciences Make Knowledge* (Cambridge, Mass., 1999); Nikolas Rose, *Powers of Freedom: Reframing Political Thought* (Cambridge, 1999); Steven Shapin and Simon Schaffer, *Leviathan and the Air-Pump: Hobbes, Boyle, and the Experimental Life* (Princeton, N.J., 1985); John S. Nelson, Allan Megill, and Donald N. McCloskey, eds., *The Rhetoric of the Human Sciences: Language and Argument in Scholarship and Public Affairs* (Madison, Wis., 1987).

[10] Eric Hobsbawm, *The Age of Extremes: A History of the World, 1914–1991* (New York, 1994).

and the human sciences during the last century is to understand how both enterprises attempted to turn the self into a project. As the essays in this volume demonstrate, individuals across a range of regimes became, in effect, sites for public projects and, at the same time, were encouraged to treat their own lives as projects to be fulfilled. Our approach follows lines of questioning laid out most notably by Ian Hacking, Norbert Elias, and Michel Foucault, all of whom have seen historical forms of self-reflection and self-control as intimately bound up in the production of public knowledge and social control.[11] The notion of the self as project, however, allows us to better appreciate two aspects largely bracketed in the works of Elias and Foucault: politics (in the formal sense of the term) and the twentieth century. To be sure, Elias and Foucault, in particular, were instrumental in thematizing power relations in the history of the self, and the specter of the last century clearly haunted each man's writings. But both thinkers remained most interested in expanding our definitions of coercion and in tracing the genealogy of contemporary life. If in Foucault's world, as critics have often said, politics is defined so broadly as to be everywhere, then reading the self as a scientific and political project in the way we do allows us to more precisely locate the political in science and the scientific in politics. In the process, it helps us recognize not only the reach of politics and the human sciences but also their limits.

In this regard, our reliance on the rather vague term "self"—as opposed to, for instance, "individual," "personhood," "personality," "mind," or "psychology"—might seem at odds with our interest in precision. Our use of the term, however, derives from two sets of concerns. First, it is important to keep in mind that the inner workings conventionally attributed to individuals have not been the object of any one scientific or medical discipline. Not only psychologists and psychiatrists but also anthropologists, jurists, genetic engineers, coaches, filmmakers, and military planners, to name just a few, have made it their business to know and shape the inner lives of individuals. The generic term "self" captures this diversity that is so fundamental to the history of both the individual and the human sciences. Second, in the tradition of George Herbert Mead and Erving Goffman, employing the term "self" opens up the possibility of considering human beings as both objects of and subjects acting upon science and politics.[12] Rather than assume the modern individual to be a relatively powerless victim of intrinsically colonizing institutions, it is crucial that we use language that, at the very least, leaves it an open question as to whether this has been the case.

We propose, then, to bring together in conversation the history of the self, the history of politics and public policy, and the history and social study of science, medicine, and technology to better understand the peculiar significance of the last century. Although there has been no shortage of scholars exploring the relationship between the human sciences and modern politics, what has been lacking is an international perspective. Up until now, researchers have been working almost exclusively within national historiographies, little aware of the extent to which the trends they have observed were unique or part of much broader, transnational developments. This vol-

---

[11] See Ian Hacking, "The Invention of Split Personalities," in *Human Nature and Natural Knowledge*, ed. Alan Donagan, Anthony N. Perovich, and Michael V. Wedin (Boston, 1986), 63–85; as well as Hacking, *Mad Travellers: Reflections on the Reality of Transient Mental Illness* (London, 1989); Norbert Elias, *The Civilizing Process* (Oxford, 2000); Michel Foucault, *The History of Sexuality: An Introduction* (New York, 1990); and Foucault, *Discipline and Punish* (cit. n. 6).

[12] George Herbert Mead, *Mind, Self, and Society: From the Standpoint of a Social Behaviorist* (Chicago, 1967); Erving Goffman, *The Presentation of Self in Everyday Life* (Garden City, N.Y., 1959).

ume, therefore, considers the self as a scientific and political project in a number of different countries and settings. In the interest of coherence, we limit ourselves to Europe, the Soviet Union, and the United States. We fully acknowledge that the following essays leave out important developments in not only Asia, Africa, and South America but parts of Europe as well (such as the British Isles, Scandinavia, and southern Europe). Their absence here is primarily the result of mundane time, space, and financial constraints. In the end, we have not attempted to be comprehensive but instead sought to inspire broader conversation about, and new ways of framing, the reciprocal influences of politics and the human sciences on one another. We hope that by adopting a comparative approach, it makes it easier not only to reassess the history of twentieth-century science, medicine, and technology but also to reconsider the ambitions and legacies of modern liberalism, socialism, conservatism, and fascism.

POLITICS AND THE SCIENCES OF THE SELF, 1800–1900

Arguably, no period was more preoccupied with the self and with its status as an object of scientific and political design than was the twentieth century. To be sure, the lineage of this preoccupation can be traced back to the classical era. Yet not until the nineteenth century did the possibility of turning the self into a site for scientific and political intervention, experimentation, and contestation fully emerge. It is, therefore, important to begin this survey by considering the strands in nineteenth-century intellectual and political history that contributed to this appreciation of the self as a central concern of the modern era.

The advent of a notion of the self as a scientific and political issue can be traced to the profound upheavals associated with the French and industrial revolutions. One effect of the dissolution of the old regime—and with it, the guilds, corporate bodies, and feudal ties that bound individuals to that social order—was radically to problematize the nature of the self.[13] The legacy of Enlightenment discourse concerning "man" and his rights, coupled with the dawning of a rapidly industrializing world in which social relations became ever more fluid and kaleidoscopic in their complexity, invested man with a new status and made him an object of political and scientific interest.[14] Hopes for political reform became closely tied to demands for the emancipation and the rehabilitation of those on whom the Enlightenment had bestowed a new dignity, such as the "wild child" or the insane. Around such figures, scientists, doctors, and pedagogues elaborated a new science of man and a set of practices designed to integrate them into postrevolutionary society. The rights-bearing subject invented by the French Revolution became both "an object of knowledge and a target of reform."[15] The "moral treatment" of Philippe Pinel and William Tuke, like the penological reforms of Jeremy Bentham, made the reform of individuals central to the claims of the new social order.

Nonetheless, this project of reform remained partial and contradictory. The institutionalization of disciplines such as penology, psychiatry, and pedagogy became a

[13] Jan Goldstein, *The Post-Revolutionary Self: Politics and Psyche in France, 1750–1850* (Cambridge, Mass., 2005); Seigel, *The Idea of the Self* (cit. n. 4).

[14] Giddens, *Modernity and Self-Identity* (cit. n. 4); Peter Berger, Brigitte Berger, and Hansfried Keller, *The Homeless Mind: Modernization and Consciousness* (New York, 1973).

[15] Nikolas Rose, *The Psychological Complex: Psychology, Politics, and Society in England, 1869–1939* (London, 1985), 12.

crucial feature of an emerging liberal order that, while committed in principle to the "rights of all," remained in practice highly selective in the rights and the liberties it granted individuals. Benthamite liberalism, for instance, coupled the modern utilitarian ethic of the greatest happiness for the greatest number with a new emphasis on disciplinary (self) surveillance that became an integral feature of a new productivist ethos. Bentham both insisted that the criminal should be treated as a rational subject and made him the object of techniques of social control that, as Foucault has argued, were decisive for the constitution of the modern self.[16] With the waning of hopes for integration and of the revolutionary conviction in the equality of all, scientific interest increasingly shifted to the study and the classification of individual differences.[17]

This shift was reflected in another space that, along with the prison and the asylum, became a key site for the new science of man, namely, the classroom. In France, the response to the postrevolutionary problematic of selfhood took the form of a new science of psychology: the doctrine of eclecticism, associated with Victor Cousin, who sought a middle ground between sensationalism (the doctrine associated with the most radical aspects of the revolution) and an ontological belief in the self. Its establishment within the secondary school curriculum made it part of what Jan Goldstein calls a "machine for the production of selfhood." Education became part of the cultural property of middle-class French men; women and the working classes, meanwhile, remained "un-selfed," as yet not fully integrated into this new paradigm of selfhood.[18]

Along with a growing interest in cultivating human beings, the nineteenth century witnessed the first attempts to understand social and individual behavior numerically. Beginning with the influential work of Adolphe Quetelet, social scientists applied statistics and probability to the analysis of human action.[19] The discovery of regularities, even in deviations from the average, led many in the human sciences to represent a wide range of human activities and life processes along a spectrum running between the normal and the pathological.[20] With the rise of new experimental methods and laboratories by midcentury, the life sciences made it possible to make individuals and their inner lives more observable and measurable than ever before.[21]

Thus the self-reflexive turn that is often associated with the twentieth century was already present in the nineteenth. Scientists made their own bodies objects of experimentation.[22] The psychological laboratories and clinics of figures such as Wilhelm Wundt, William James, and Jean-Martin Charcot devised not only new, controlled

---

[16] Foucault, *Discipline and Punish* (cit. n. 6).

[17] Rose, *The Psychological Complex* (cit. n. 15), 30.

[18] Jan Goldstein, "Saying 'I': Victor Cousin, Caroline Angebert, and the Politics of Selfhood in 19th Century France," in *Rediscovering History: Culture, Politics, and the Psyche,* ed. Michael Roth (Palo Alto, Calif., 1994), 321.

[19] Theodore Porter, *Trust in Numbers: The Pursuit of Objectivity in Science and Public Life* (Princeton, N.J., 1995); Porter, *The Rise of Statistical Thinking, 1820–1900* (Princeton, N.J., 1986); Ian Hacking, "Biopower and the Avalanche of Printed Numbers," *Humanities in Society* 5 (1982): 279–95.

[20] Georges Canguilhem, *The Normal and the Pathological* (New York, 1991).

[21] Olaf Breidbach, *Die Materialisierung des Ichs: Zur Geschichte der Hirnforschung im 19. und 20. Jahrhundert* (Frankfurt, 1997); Michael Hagner and Hans-Jörg Rheinberger, eds., *Die Experimentalisierung des Lebens: Experimentalsysteme in den biologischen Wissenschaften, 1850–1950* (Berlin, 1993); Anne Harrington, *Medicine, Mind, and the Double Brain: A Study in Nineteenth-Century Thought* (Princeton, N.J., 1989).

[22] Simon Schaffer, "Self-Evidence," *Critical Inquiry* 18 (1992): 56–91.

means for analyzing the psychology of individuals but also methods for training students in the laws of subjectivity.[23] This process of acquiring new self-knowledge increasingly became part of the everyday life of the educated classes and went hand in hand with new opportunities for self-fashioning. Indeed, the scientist and the doctor were often seen as the very embodiments of liberal selfhood, the self-made man.[24]

At the same time, this more mutable sense of selfhood was taken up by the cultural avant-garde—by thinkers such as Friedrich Nietzsche and Sigmund Freud—who radicalized the injunction to "know thyself" by mapping the blind spots in the liberal construction of selfhood, the realms of reverie, instinctual life, and the unconscious.[25] They provided the link between the "affective revolution" that was one of the legacies of the Enlightenment and the new reflexive individual who emerged in the nineteenth century. In so doing, figures such as Nietzsche and Freud created a new image of personhood: the atomized self, who attended closely to, yet was often alienated from, his own inner drives, desires, and fears.[26]

For their part, the clinical sciences held out the promise of finding effective responses to the problems thrown up in the industrializing process. Changes in economic life presented an expanded field of opportunities for individuals and exposed them to new kinds of shocks, injuries, and afflictions. Recognition of the extent to which the self was at risk from a host of social ills and from the uneven social distribution of these ills gave further impetus to the emerging human sciences. The latter now involved themselves with a cluster of issues collectively known as the *social question*. The problems of nineteenth-century society—revolution, delinquency, crime, illness, prostitution, alcoholism—seemed ripe for scientific analysis and experimentation.[27] Specialists in fields such as psychiatry, criminology, and sociology took on the task of interpreting and offering solutions to these problems. In most European nations, the state response to these issues was twofold: establish some form of social welfare, and seek the advice of emerging new professions.

A paradigmatic instance of this development was the case of railway medicine. Debates over the psychological consequences of railway accidents and the proper response to them opened up two major lines of inquiry. The first led scientists and clinicians into the enigmas of the psyche, into those problems of the traumatic memory, automatisms, disassociation, hysteria, and malingering that culminated in the work of Charcot, Freud, and Pierre Janet, and like-minded clinicians.[28] The second led to a discourse about the categorical relationship between modernity and risk and to

[23] Kurt Danziger, *Constructing the Subject: Historical Origins of Psychological Research* (Cambridge, 1990).

[24] Roy Porter, introduction to Porter, *Rewriting the Self* (cit. n. 4), 1–14; Lawrence Rothfield, *Vital Signs: Medical Realism in Nineteenth-Century Fiction* (Princeton, N.J., 1992).

[25] J. W. Burrow, *The Crisis of Reason: European Thought, 1848–1914* (New Haven, Conn., 2000).

[26] Lynn Hunt, "The Affective Revolution in 1790s Britain," *Eighteenth-Century Studies* 34 (2001): 491–521; Peter Gay, *The Bourgeois Experience,* vols. 1–5 (New York, 1984–1998); Philippe Ariès and Georges Duby, eds., *A History of Private Life,* vols. 4 and 5 (Cambridge, Mass., 1990–1991).

[27] Paul Rabinow, *French Modern: Norms and Forms of the Social Environment* (Cambridge, Mass., 1989).

[28] Mark S. Micale and Paul Lerner, eds., *Traumatic Pasts: History, Psychiatry, and Trauma in the Modern Age, 1870–1930* (Cambridge, 2001); Marcel Gauchet and Gladys Swain, *Le vrai Charcot: Les chemins imprévus de l'inconscient* (Paris, 1997); Ralph Harrington, "The Neuroses of the Railways," *History Today* 44(7) (1994): 15–21; Christopher G. Goetz, Michel Bonduelle, and Toby Gelfand, *Charcot: Constructing Neurology* (Oxford, 1995).

the passage of social insurance legislation.[29] Out of this discourse on risk was elaborated a complex state welfare apparatus.

Recent scholarship on the history of the welfare state has done much to demonstrate how the everyday practices of social security informed new conceptions of citizenship and political identity. By establishing a very material social contract binding state and citizen, social insurance helped to forge a sense of entitlement among claimants and beneficiaries. In effect, the new welfare state politicized the relationships of individuals to their own bodies and minds.[30] Health, defined as the ability to be productive, became the social norm valued above all others, an integrating national ideology.[31] Achievement and performance became common leitmotifs of selfhood, and means were sought to make workers' motivations, perceptions, and accomplishments more amenable to supervision and improvement. Such ideals were not only lauded by industrialists, however. Lifestyle reformers and social hygienists also sought to help workers internalize new codes of healthy behavior and productive modes of self-formation.[32]

Serving to both inflect and deflect discussions about economic productivity, public health, and national strength, the doctrine of social Darwinism came to play a prominent role in public discourse during the second half of the nineteenth century. The progressivist emphasis on individual betterment through universal education and the inculcation of new standards of hygiene was supported by some proponents of the views of Charles Darwin and Francis Galton. Other specialists, however, challenged the doctrine of liberal universalism, seeing the self more immutably determined and reflected in essential ethnic and racial characteristics.[33] But social Darwinists, too, proposed projects of improvement and intervention, opening new possibilities for public policy: eugenics, sex reform, population policy. This new biopolitical domain gave a new and decidedly ideological cast to existing sexological discourses.[34] At its heart was a concern with maintaining national power, one effect of which was to feed degenerationist fears of declining racial fitness. The specter of degeneration gave scientists a powerful image around which to weave a culturally resonant narrative of decline and to justify the need for action against the threat posed by the "unfit," the "feeble-minded," the "born criminal," and other figures from the degenerationist gallery.[35]

[29] Paul V. Dutton, *Origins of the French Welfare State: The Struggle for Social Reform in France, 1914–1947* (New York, 2002); Anson Rabinbach, *The Human Motor: Energy, Fatigue, and the Origins of Modernity* (New York, 1990); Francois Ewald, *L'Etat Providence* (Paris, 1986).

[30] Greg Eghigian, *Making Security Social: Disability, Insurance, and the Birth of the Social Entitlement State in Germany* (Ann Arbor, Mich., 2000).

[31] Manfred Berg and Geoffrey Cocks, eds., *Medicine and Modernity: Public Health and Medical Care in Nineteenth- and Twentieth-Century Germany* (New York, 1997); Alfons Labisch, *Homo Hygienicus: Gesundheit und Medizin in der Neuzeit* (Frankfurt, 1992).

[32] Michael Hau, *The Cult of Health and Beauty in Germany: A Social History* (Chicago, 2003); Philipp Sarasin, *Reizbare Maschinen: Eine Geschichte des Körpers, 1765–1914* (Frankfurt, 2001).

[33] John P. Jackson Jr. and Nadine M. Weidman, *Race, Racism, and Science: Social Impact and Interaction* (New Brunswick, N.J., 2006); Mike Hawkins, *Social Darwinism in European and American Thought, 1860–1945: Nature as Model and Nature as Threat* (Cambridge, 1997).

[34] Roy Porter and Mikulás Teich, eds., *Sexual Knowledge, Sexual Science: The History of Attitudes to Sexuality* (New York, 1994); Foucault, *History of Sexuality* (cit. n. 11); Thomas Laqueur, *Making Sex: Body and Gender from the Greeks to Freud* (Cambridge, Mass., 1990); George Mosse, *Nationalism and Sexuality: Respectability and Abnormal Sexuality in Modern Europe* (New York, 1985).

[35] Peter Becker and Richard Wetzell, eds., *Criminals and Their Scientists: Essays on the History of Criminology* (New York, 2005); Laurent Mucchielli, ed., *Histoire de la criminologie française* (Paris,

Although eugenic discourse was often accompanied by a critique of welfare's alleged counterselective tendencies, eugenics and welfare were by no means incommensurate with one another.[36] Both were expressions of a transformative and scientific approach to the problems of modern society; they represented complementary forms of social rationalization, an effort to reformulate questions of labor power and reproduction around the problems of efficiency and deficiency.[37] Liberal reformers and scientific professionals wishing to contribute to the nation-building process now began to turn their expertise to more collective solutions to the problems of the modernizing process.

POLITICS AND THE SCIENCES OF THE SELF FROM THE TURN OF THE TWENTIETH CENTURY TO THE SECOND WORLD WAR

In bourgeois circles at least, by the turn of the twentieth century, there was a great deal of talk about there being a crisis of the individual and the self. In eugenic discourse, this was construed in biological terms. In sexual and cultural relations, it was linked with a perceived crisis of masculinity or a crisis of youth. In sociology, the talk was of the risks facing industrial society. In philosophy, questions were being raised about the rational views of selfhood developed by Enlightenment thinkers such as Immanuel Kant.[38]

It is also during that period that American pragmatists such as William James, Charles Horton Cooley, and George Herbert Mead turned away from ideals of the self as a timeless, universal, transcendental, and absolute entity that is above the social whirl in favor of a more dynamic, versatile, and mutable self.[39] Subsequently, cultural anthropology, sociology, psychology, critical theory, cross-cultural studies, gender studies, and literary and cultural studies have pointed to how selves are produced and embedded in culture, language, and society. As Carl Degler points out, the notion of human nature as essentially mutable, changeable, and transitory can turn the self into a political project.[40] American social scientists, at the turn of the twentieth century, favored change, progress, social improvement, and reform. For them, the notion of biological equality, paired with the assumption that human nature is shaped by culture and environment, left room for improvement and reform based on social and

---

1995); Daniel Pick, *Faces of Degeneration: A European Disorder, 1848–1918* (New York, 1989); Robert Nye, *Crime, Madness, and Politics in Modern France: The Medical Concept of National Decline* (Princeton, N.J., 1984).

[36] Mark Adams, *The Wellborn Science: Eugenics in Germany, France, Brazil, and Russia* (New York, 1993); Sheila Faith Weiss, *Race Hygiene and National Efficiency: The Eugenics of Wilhelm Schallmayer* (Berkeley, Calif., 1987); Daniel J. Kevles, *In the Name of Eugenics: Genetics and the Uses of Human Heredity* (New York, 1985).

[37] Paul Julian Weindling, *Health, Race, and German Politics between National Unification and Nazism, 1870–1945* (New York, 1989).

[38] Chandak Sengoopta, *Otto Weininger: Sex, Science, and Self in Imperial Vienna* (Chicago, 2000); Émile Durkheim, *The Division of Labor in Society* (New York, 1997); Jacques Le Rider, *Modernity and Crises of Identity* (New York, 1993); Carl Schorske, *Fin de Siècle Vienna: Politics and Culture* (New York, 1981); Max Weber, *Economy and Society* (San Francisco, 1978).

[39] James Holstein and Jaber Gubrium, *The Self We Live By: Narrative Identity in a Postmodern World* (New York, 2000).

[40] Carl Degler, *In Search of Human Nature: The Decline and Revival of Darwinism in American Social Thought* (Oxford, 1992).

individual intervention. Thus the social self went "hand in glove with ideas about American progress and ingenuity."[41]

Out of these discussions, a new set of discourses and practices surrounding personhood emerged. Increasingly, new narratives of the self abandoned the liberal tradition of seeing individuals as autonomous, rights-bearing, rational actors. Stress was placed on finding programs of social reform that would allow for greater opportunities to monitor, discipline, and integrate individuals into the nation-state.[42] It was during the first three decades of the twentieth century that many of the new sciences of the self first gained public prominence.[43] The ideas, language, and practitioners of psychoanalysis, eugenics, social work, mental hygiene, industrial psychology, sexology, forensic psychiatry, crowd psychology, clinical social work, and criminal anthropology all achieved an unprecedented currency in policy-making circles and the mass media.[44]

If, prior to the war, the role for such specialists in the process of nation-building remained more theoretical than direct, it was perhaps only a matter of time before these sciences were called up by the state. The crucible of the twentieth-century alliance between the human sciences and the state proved to be the Great War. Even with the creation of welfare programs in many Western nations, the degree of state intervention in society had remained relatively circumscribed. But wartime mobilization, coupled with the massive strains that total war produced in the social fabric, irrevocably altered that. In all combatant nations, war led to greater state intervention and a growing concern about productivity, public health, loyalty, and social peace.[45]

Faced with heavy casualties and manpower shortages as well as discipline and morale problems, states had to devise reliable methods to assess the fitness of new recruits, treat war neurotics, and deal with malingerers.[46] It was at this time, for instance, that the intelligence test was adopted for mass use by the U.S. military in an effort to

[41] Holstein and Gubrium, *The Self We Live By* (cit. n. 39), x.

[42] Elizabeth Lunbeck, *The Psychiatric Persuasion: Knowledge, Gender, and Power in Modern America* (Princeton, N.J., 1994).

[43] Mark S. Micale, ed., *The Mind of Modernism: Medicine, Psychology, and the Cultural Arts in Europe and America, 1880–1940* (Palo Alto, Calif., 2004).

[44] The literature on this subject is vast. Among some recent works, see Hans Pols, *Psychiatric Utopias: Masterminding American Mental Hygiene, 1910–1950* (Cambridge, Mass., forthcoming); Andreas Killen, *Berlin Electropolis: Shock, Nerves, and German Modernity* (Berkeley, Calif., 2006); Eric Engstrom, *Clinical Psychiatry in Imperial Germany: A History of Psychiatric Practice* (Ithaca, N.Y., 2004); Andrew Zimmerman, *Anthropology and Antihumanism in Imperial Germany* (Chicago, 2001); Harry Oosterhuis, *Stepchildren of Nature: Krafft-Ebing, Psychiatry, and the Making of Sexual Identity* (Chicago, 2000); Ian Dowbiggin, *Keeping America Sane: Psychiatry and Eugenics in the United States and Canada* (Ithaca, N.Y., 1997); Anne Harrington, *Reenchanted Science: Holism in German Culture from Wilhelm II to Hitler* (Princeton, N.J., 1997); Angus McLaren, *The Trials of Masculinity: Policing Sexual Boundaries, 1870–1930* (Chicago, 1997); Nathan G. Hale, *The Rise and Crisis of Psychoanalysis in the United States: Freud and the Americans, 1917–1985* (New York, 1995); Sander L. Gilman, *The Case of Sigmund Freud: Medicine and Identity at the Fin de Siècle* (Baltimore, 1993), as well as Gilman, *Freud, Race, and Gender* (Princeton, N.J., 1994); Mark S. Micale and Roy Porter, eds., *Discovering the History of Psychiatry* (New York, 1994); Sander L. Gilman and Roy Porter, eds., *Hysteria beyond Freud* (Berkeley, Calif., 1993); Laura Engelstein, *The Keys to Happiness: Sex and the Search for Modernity in Fin-de-Siècle Russia* (Ithaca, N.Y., 1992).

[45] Theodore Kornweibel, *Investigate Everything: Federal Efforts to Compel Black Loyalty during World War I* (Bloomington, Ind., 2002); Roger Cooter, Mark Harrison, and Steve Sturdy, eds., *War, Medicine, and Modernity* (Phoenix Mill, UK, 1998); Joanna Bourke, *Dismembering the Male: Men's Bodies, Britain, and the Great War* (Chicago, 1996).

[46] Paul Lerner, *Hysterical Men: War, Psychiatry, and the Politics of Trauma in Germany, 1890–1930* (Ithaca, N.Y., 2003); Ben Shephard, *A War of Nerves: Soldiers and Psychiatrists in the Twentieth Cen-*

rationalize methods of assessing the war readiness of recruits.[47] At the same time, concerns over the outbreak of venereal diseases prompted the expansion of medical surveillance and control over sexuality.[48]

The wartime relationship forged between state and citizen became the basis for the political and social arrangements of the postwar period. Whether it was Soviet Russia, fascist Italy, or the liberal democracies of Western Europe, the states emerging out of the war were compelled to mobilize every available resource in the project of rebuilding. The new political movements of the day enlisted the participation of human scientists in grand projects of social and personal reconstruction. This was particularly true in liberal democracies, where those who had long sought to reform society along more progressive, scientific lines found that the postwar era opened up a host of possibilities.[49]

The war had upset traditional forms of social order and authority as well as the social attachments and relations that helped people cope with their troubles. Particularly in central Europe, few remained untouched by the traumas of war or its aftermath: revolution, social unrest, inflation, massive unemployment. Newly legitimized by their wartime activities, the human sciences assumed an increasingly prominent role in mediating the dialogue between postwar states and their subjects in the service of projects for social renewal.

War had intensified the consciousness of risk and danger, creating heightened sensitivity to the potential threats to life, health, safety, and future. State welfare programs now extended their reach over the lives of individuals in efforts aimed at social prophylactics. A wide range of programs—vocational counseling, pensions for soldiers, aid for widows and orphans, unemployment benefits, marriage counseling, rehabilitation of juvenile delinquents—were established or reformed as part of the postwar social contract.[50] This served to further politicize the work of the human sciences.

With many political parties and regimes no longer beholden to liberal ideals—indeed, in many cases, committed to the outright destruction of these ideals—policy makers, researchers, and clinicians often reframed the question of the self in collective terms, tying its destiny to that of the nation or the revolution. In this way, the project of personal regeneration became wrapped up in the quest for a thorough social

tury (Cambridge, Mass., 2001); Doris Kauffmann, "Science as Cultural Practice: Psychiatry in the First World War and Weimar Germany," *Journal of Contemporary History* 34 (1999): 125–44; Eric Leed, *No Man's Land: Combat and Identity in World War I* (Cambridge, 1979).

[47] John Carson, "Army Alpha, Army Brass, and the Search for Army Intelligence," *Isis* 84 (1993): 278–309; Michael Sokal, ed., *Psychological Testing and American Society, 1890–1930* (New Brunswick, N.J., 1990).

[48] Lutz Sauerteig, "Sex, Medicine, and Morality during the First World War," in Cooter, Harrison, and Sturdy, *War, Medicine, and Modernity* (cit. n. 45), 167–88.

[49] Peter Fritzsche, "Landscape of Danger, Landscape of Design: Crisis and Modernism in Weimar Germany," in *Dancing on the Volcano: Essays on the Culture of the Weimar Republic,* ed. Thomas W. Kniesche and Stephen Brockmann (Columbia, S.C., 1994), 29–46.

[50] On youth welfare from this period, for instance, see Sarah Fishman, *The Battle for Children: World War II, Youth Crime, and Juvenile Justice in Twentieth-Century France* (Cambridge, Mass., 2002); Corinna Kuhr-Korolev and Stefan Plaggenborg, *Sowjetjugend, 1917–1941: Generation zwischen Revolution und Resignation* (Essen, Germany, 2001); Jon Lawrence and Pat Starkey, eds., *Child Welfare and Social Action in the Nineteenth and Twentieth Centuries: International Perspectives* (Liverpool, 2001); Elizabeth Harvey, *Youth and the Welfare State in Weimar Germany* (Oxford, 1993); Susan Pedersen, *Family, Dependence, and the Origins of the Welfare State in Britain and France, 1914–1945* (Cambridge, 1993); Victor Bailey, *Delinquency and Citizenship: Reclaiming the Young Offender, 1914–1948* (Oxford, 1987).

regeneration. Inspired by World War I, the regenerated self became allied with explicitly ideological concerns and often attached to a rhetoric of heroism.[51]

One prominent expression of this shift crystallized in the figures of the so-called *new man* and *new woman*. Variations on these idealized visions of masculinity and femininity appeared in most industrial nations, but they played particularly crucial roles in the extremist creeds of the left and the right.[52] The new man was a response to battlefield trauma, wartime changes in the gender order, and revolutionary shifts in class structures—a character of strength, courage, and determination who was supposed to emerge naturally from the new social order.[53] The "hardening" of the masculine self that was so central to the rhetoric of the new man strove to eradicate the softness and false humanism associated with liberalism.[54] His counterpart, the new woman, as Andreas Killen's essay in this volume notes, was seen—at least in Germany—in somewhat more ambivalent terms. Yet she, too, was a type amenable to social engineering, whether through conscription into the newly rationalized workplace of the 1920s or, alternatively, into the new ideal of womanly concern, maternal care, and active citizenship.[55] Individuals were enjoined to rethink their lives in accordance with new utopian norms, to inscribe themselves into the narrative of collective renewal, to internalize new blueprints for conduct in the service of a great cause.[56] In all this, physicians, psychologists, pedagogues, and writers played leading roles, interpellating individuals as members of imagined communities to be mobilized for some great common purpose.[57]

At the same time, the new expectations for human being and potential sparked a heightened vigilance about the threats posed by marginal categories of individuals. The other side of the enthusiasm for the new man and the new woman was an obsession with the enemy. An anxiety born during the war now carried over into a preoccupation with domestic adversaries. The monitoring of those deemed unfit or unsuitable for membership in the polity—malingerers, psychopaths, the feebleminded, sex

---

[51] Ruth Ben-Ghiat, *Fascist Modernities: Italy, 1922–1945* (Berkeley, Calif., 2001); Katerina Clark, *The Soviet Novel: History as Ritual* (Bloomington, Ind., 2000); Jay W. Baird, *To Die for Germany: Heroes in the Nazi Pantheon* (Bloomington, Ind., 1990).

[52] Christina Kiaer and Eric Naiman, eds., *Everyday Subjects: Formations of Identity in Early Soviet Culture* (Bloomington, Ind., forthcoming); Torsten Rüting, *Pavlov und der Neue Mensch: Diskurse über Disziplinierung in Sowjetrussland* (Munich, 2002); Mabel Berezin, *Making the Fascist Self: The Political Culture of Interwar Italy* (Ithaca, N.Y., 1997); Jeannine Fiedler, *Social Utopias of the Twenties: Bauhaus, Kibbutz, and the Dream of the New Man* (Wuppertal, Germany, 1995).

[53] Michael S. Kimmel, *Manhood in America: A Cultural History* (New York, 1997); Stephen Kotkin, *Magnetic Mountain: Stalinism as Civilization* (Berkeley, Calif., 1997); George Mosse, *The Image of Man: The Creation of Modern Masculinity* (New York, 1996); Lynne Attwood, *The New Soviet Man and Woman: Sex-Role Socialization in the USSR* (Bloomington, Ind., 1990).

[54] Klaus Theweleit, *Male Fantasies,* vols. 1 and 2 (Minneapolis, 1987, 1989).

[55] Andreas Killen, "Psychiatrists, Telephone Operators, and Traumatic Neurosis in Germany, 1900–1926," *J. Contemp. Hist.* 38 (2003): 201–20; Lynne Attwood, *Creating the New Soviet Woman: Women's Magazines as Engineers of Female Identity, 1922–1953* (New York, 1999); Cornelie Usborne, *Frauenkörper—Volkskörper: Geburtenkontrolle und Bevölkerungspolitik in der Weimarer Republik* (Münster, Germany, 1994).

[56] Jochen Hellbeck and Klaus Heller, eds., *Autobiographical Practices in Russia/Autobiographische Praktiken in Russland* (Göttingen, Germany, 2004); Laura Engelstein and Stephanie Sandler, eds., *Self and Story in Russian History* (Ithaca, N.Y., 2000).

[57] Manfred Kappeler, *Der schreckliche Traum vom vollkommenen Menschen: Rassenhygiene und Eugenik in der sozialen Arbeit* (Marburg, Germany, 2000); Nicola Lepp, Martin Roth, and Klaus Vogel, eds., *Der neue Mensch: Obsessionen des 20. Jahrhunderts* (Ostfildern-Ruit, Germany, 1999); Alex Kozulin, *Psychology in Utopia: Towards a Social History of Soviet Psychology* (Cambridge, Mass., 1984); Raymond Bauer, *The New Man in Soviet Psychology* (Cambridge, Mass., 1952).

offenders, juvenile delinquents—intensified. Nowhere was this more apparent than in Germany, where the institutionalized psychiatric population, for instance, grew from 120,000 in 1924 to some 340,000 by 1939, prompting calls for the forced sterilization and even the "euthanizing" of those considered incurable.[58] Even before Hitler and the Nazis took power in 1933, some social work professionals were demanding sweeping powers that would allow them to identify and detain "wayward" men, women, and youths in special camps.[59]

As Daniel Beer shows in his contribution to this volume, deviance was a particularly prominent concern in the authoritarian regimes of the interwar period. In both Nazi Germany and the Soviet Union, the human sciences were mobilized around the task of uncovering racial and class enemies.[60] Well before the outbreak of war in 1939, both regimes had set up an extensive network of prisons and work and reeducation camps for the purpose of segregating, punishing, and "rehabilitating" those imagined to be threats to society.[61]

Despite the increasingly brutal and lethal policies pursued under Hitler and Stalin, there remained gaps and tensions between and within official rhetoric, scientific practice, and everyday life. One result of recent scholarship has been to offer a more differentiated picture of the fate of individuals in authoritarian systems. Recent work in Soviet studies, for instance, has challenged long-held views according to which the Stalinist totalitarian self represented little more than the negation of liberal individualism. Taking as its central object the "process by which individuals are made and also make themselves into subjects" in Stalinist Russia, this scholarship has sought to restore elements of agency to the Soviet subject.[62] Similarly, recent research on the Third Reich has revealed new elements of complexity and contradiction within the National Socialist politics of the self.[63] Although at first glance it would appear that Nazi policies simply reified the distinction between the normal and the pathological, National Socialist ideology and German social attitudes during the Third Reich remained an amalgam of not just scientific and medical but also political, consumerist, and religious values.[64]

The denouement of many interwar scientific and political projects in state terror has

---

[58] Michael Burleigh, *Death and Deliverance: "Euthanasia" in Germany, c. 1900–1945* (New York, 1994); Hans-Walter Schmuhl, *Rassenhygiene, Nationalsozialismus, Euthanasie: Von der Verhütung zur Vernichtung "lebensunwerten Lebens," 1890–1945* (Göttingen, Germany, 1992).

[59] Matthias Willing, *Das Bewahrungsgesetz (1918–1967): Eine rechtshistorische Studie zur Geschichte der deutschen Fürsorge* (Tübingen, Germany, 2003).

[60] Sheila Fitzpatrick, *Tear off the Masks! Identity and Imposture in Twentieth-Century Russia* (Princeton, N.J., 2005); Robert Gellately and Nathan Stoltzfuss, eds., *Social Outsiders in Nazi Germany* (Princeton, N.J., 2001); Michael Burleigh and Wolfgang Wippermann, *The Racial State: Germany 1933–1945* (New York, 1991).

[61] Anne Applebaum, *Gulag: A History* (New York, 2003); Karin Orth, *Das System der nationalsozialistischen Konzentrationslager: Eine politische Organisationsgeschichte* (Hamburg, Germany, 1999).

[62] Jochen Hellbeck, *Revolution on My Mind: Writing a Diary under Stalin* (Cambridge, Mass., 2006); Anna Krylova, "The Tenacious Liberal Subject in Soviet Studies," *Kritika: Explorations in Russian and Eurasian History* 1 (Winter 2000): 119–46.

[63] See, e.g., Moritz Foellmer, "The Problem of National Solidarity in Interwar Germany," *German History* 23 (2005): 202–31, and his unpublished paper, Foellmer, "Nazism and the Politics of the Self in Berlin"; Dagmar Herzog, "Hubris and Hypocrisy, Incitement and Disavowal: Sexuality and German Fascism," in *Sexuality and German Fascism,* ed. Dagmar Herzog (New York, 2004), 1–20.

[64] Claudia Koonz, *The Nazi Conscience* (Cambridge, Mass., 2003); Richard Steigmann-Gall, *The Holy Reich: Nazi Conceptions of Christianity, 1919–1945* (New York, 2003); Harry Oosterhuis, "Medicine, Male Bonding, and Homosexuality in Nazi Germany," *J. Contemp. Hist.* 32 (1997): 187–205.

led some scholars, such as Detlev Peukert and Jacques Donzelot, to assert a continuity between the progressive, reform-minded spirit of the late nineteenth-century human sciences, on the one hand, and what some would term "totalitarianism," on the other.[65] More recently, however, scholars have questioned this view, noting particularly the relative importance of traditional, religious values and organizations in forging antidemocratic policy.[66] Looking beyond such areas of disagreement, however, the salient point that has emerged from these debates is the need to abandon the notion that the history of the self followed a "special path" in communist Russia and fascist Germany. Indeed, if one result of recent scholarship has been to cast a more skeptical eye on the history of the liberal self, then another has been to "normalize" the individual in ostensibly "totalitarian" society. It is precisely around the topos of science that the possibility of such a comparative "normalization" of the different trajectories of personhood within different polities in the modern era lies.

WARS HOT AND COLD—REVOLUTIONS VIOLENT AND VELVET

As Charles Maier has observed, while most political, social, and cultural trends in recent history have had their beginnings in the nineteenth century, the twentieth century was distinctive in the kinds of moral questions it consistently raised for observers.[67] If the century seems to have effectively called into question the Whig interpretation of history, it is because of the unprecedented forms and scale of coercion and violence that marked it. In the end, World War I and the Russian Revolution proved to be only the beginning of a series of wars, revolutions, coups, mass persecutions, and genocides that challenged how Western publics came to see modern life and themselves.[68]

The human sciences were hardly passive observers in all of this. Social scientists and clinicians were actively involved in shaping, administrating, and assessing the impact of public policies that, in turn, determined life and death questions affecting millions. At no time was this more apparent than during the Second World War.

With the experience of the First World War behind them, states did not hesitate to recruit the assistance of researchers and practitioners once war broke out in 1939. As economies were quickly placed on a war footing, economic planning became ever more entangled in and subordinate to military planning.[69] At the same time, state authorities—both military and civilian—enlisted the services of countless physicians, psychiatrists, psychologists, and pedagogues to assist war efforts in a variety of ways: examining recruits, screening military personnel for special aptitudes or deficiencies,

[65] Detlev Peukert, "The Genesis of the 'Final Solution' from the Spirit of Science," in *Nazism and German Society,* ed. David Crew (New York, 1994), 274–99; Jacques Donzelot, *The Policing of Families* (New York, 1979). See also Bauman, *Modernity and the Holocaust* (cit. n. 3).

[66] Young-Sun Hong, *Welfare, Modernity, and the Weimar State, 1919–1933* (Princeton, N.J., 1998); Edward Ross Dickinson, *The Politics of German Child Welfare from the Empire to the Federal Republic* (Cambridge, Mass., 1996).

[67] Charles S. Maier, "Consigning the Twentieth Century to History: Alternative Narratives for the Modern Era," *Amer. Hist. Rev.* 105 (2000): 807–31.

[68] Beatrice Heuser, "Wars since 1945: An Introduction," *Zeithistorische Forschungen/Studies in Contemporary History,* Online-Ausgabe 2 (2005): H1, http://www.zeithistorische-forschungen.de/16126041-Heuser-1-2005; Eric D. Weitz, *A Century of Genocide: Utopias of Race and Nation* (Princeton, N.J., 2003); Mark Mazower, *Dark Continent: Europe's Twentieth Century* (New York, 2000).

[69] Paul Koistinen, *Arsenal of World War II: The Political Economy of American Warfare, 1940–1945* (Lawrence, Kans., 2004); Mark Harrison, ed., *The Economics of World War II: Six Great Powers in International Comparison* (Cambridge, 1998); John Barber and Mark Harrison, *The Soviet Home Front, 1941–1945: A Social and Economic History of the USSR in World War II* (London, 1991).

providing medical and psychiatric treatment and rehabilitation for wounded soldiers, and conducting research on public attitudes, morale, and the effects of propaganda.[70]

This work had an impact well beyond 1945. As wartime researchers and clinicians brought their experiences back home with them, so too did they help spread new notions about mental health. The contemporary idea of "stress," for instance, comes directly from these war years.[71] At the same time, as the essay by Hans Pols in this volume shows, concerns about the character and adaptive capabilities of servicemen and veterans fed a growing interest in the structure and development of the human personality. By making "personality" as such an object of scientific study and psychotherapeutic intervention, it was researchers during the Second World War who paved the way for the relative ubiquity of the term in the 1950s and the 1960s.[72]

As was the case in the natural sciences, émigré intellectuals, most coming from Germany and Austria in the 1930s, made significant contributions to social scientific research in Allied countries during and immediately after the war.[73] The United States, in particular, was the chief beneficiary of the migration of scores of prominent scholars fleeing Hitler's Germany, including Hannah Arendt, Wolfgang Köhler, Franz Neumann, Theodor Adorno, Herbert Marcuse, and Paul Lazarsfeld. Their perspectives on the relationship of fascism, communism, and capitalism to self-determination and desire influenced a generation of students and scholars in the social sciences and ultimately helped provide a vocabulary for the student protest movements of the 1960s.[74]

Back on the continent of Europe, fascism had effectively eliminated all opposing voices by the time the Nazis invaded Poland. There, the human sciences were easily folded into the war effort. Italian fascism shared with its German counterpart an imperialistic, eugenic outlook that stressed national strength, economic autarky, and racial vitality. Medicine, public hygiene, sociology, and social work were all mobilized not only to justify Italian superiority but also to support the regime's attempts to promote a healthy and growing population.[75]

[70] Klaus Blassneck, *Militärpsychiatrie im Nationalsozialismus: Kriegsneurotiker im Zweiten Weltkrieg* (Würzburg, Germany, 2000); Ben Shephard, " 'Pitiless Psychology': The Role of Prevention in British Military Psychiatry in the Second World War," *History of Psychiatry* 10 (1999): 491–524; Albert R. Gilgen et al., *Soviet and American Psychology during World War II* (Westport, Conn., 1997); Martin L. Levitt, "The Psychology of Children: Twisting the Hull-Birmingham Survey to Influence British Aerial Strategy in World War II," *Psychologie und Geschichte* 7 (1995): 44–59; Mitchell Wilson, "DSM-III and the Transformation of American Psychiatry: A History," *American Journal of Psychiatry* 150 (1993): 399–410; Ulfried Geuter, *The Professionalization of Psychology in Nazi Germany* (New York, 1992); Virgina Yans-McLaughlin, "Science, Democracy, and Ethics: Mobilizing Culture and Personality for World War II," in *Malinowski, Rivers, Benedict, and Others: Essays on Culture and Personality,* ed. George Stocking (Madison, Wis., 1986), 184–217.

[71] Shephard, *A War of Nerves* (cit. n. 46); Hans Binneveld, *From Shell Shock to Combat Stress: A Comparative History of Military Psychiatry* (Amsterdam, 1997); Allan Young, *The Harmony of Illusions: Inventing Post-Traumatic Stress Disorder* (Princeton, N.J., 1995).

[72] On the postwar interest in personality, see David G. Winter and Nicole B. Barenbaum, "History of Modern Personality Theory and Research" in *Handbook of Personality: Theory and Research,* ed. Lawrence A. Pervin and Oliver P. John (New York, 2001), 3–27. See also Warren I. Susman, "Personality and the Meaning of Twentieth Century Culture," in *New Directions in American Intellectual History,* ed. John Higham (Baltimore, 1979), 212–26.

[73] Mitchell Ash and Alfons Söllner, eds., *Forced Migration and Scientific Change: Émigré German-Speaking Scientists and Scholars after 1933* (New York, 1996).

[74] Martin Jay, *The Dialectical Imagination: A History of the Frankfurt School and the Institute of Social Research, 1923–1950* (Berkeley, Calif., 1996); Claus-Dieter Krohn, *Intellectuals in Exile: Refugee Scholars and the New School for Social Research* (Amherst, Mass., 1993).

[75] Silvano Franco, *Legislazione e politica sanitaria del fascismo* (Rome, 2001); Elizabeth Dixon, *Measuring Mamma's Milk: Fascism and the Medicalization of Maternity in Italy* (Ann Arbor, Mich.,

It was National Socialist Germany, as the essay by Geoffrey Cocks highlights, that revealed the lengths to which both the human sciences and political institutions were willing to go in realizing the dream of creating a new man. Not nearly as doctrinally dogmatic as Stalinism, National Socialism as an ideology and set of policies drew its coherence from race hygiene, an explicitly racist form of eugenics.[76] Everything from education to criminal justice to war was understood in terms of positive and negative eugenic functions, all in the service of the *Volksgemeinschaft* (the racially pure national community).

If, as Hitler had noted in *Mein Kampf,* life was nothing but a struggle for survival between the races, then the military campaign was only one part of a much larger war that needed to be fought on a number of fronts. Historians have conclusively shown that professionals from the human sciences often volunteered their knowledge and services to this Nazi struggle for German racial supremacy. The state called upon economists, statisticians, demographers, pedagogues, and social hygienists to promote the "purification" of German scholarship, the ethnic cleansing of eastern Europe, and the fertility of Aryan women.[77] Anthropologists carried out studies of convicts and ethnic "undesirables" such as Sinti and Roma to determine the role of heredity in crime and disease resistance, while medical researchers performed brutal, invasive experiments on detainees, POWs, homosexuals, and children in the name of advancing treatment and surgical techniques.[78] Finally, beginning in 1938 and continuing through the war, the regime actively set out to decisively "cleanse" the European continent of groups defined as posing a threat to Aryan racial health, relying on collaborating psychiatrists, pediatricians, nurses, social workers, and SS personnel to organize the killing of approximately 200,000 disabled, infirm, and mentally ill individuals.[79] The Nazi-orchestrated murder of some 6 million European Jews during World War II itself can be seen as both centerpiece and culmination of the regime's toxic political parasitology.[80]

Even before the Second World War was over, Allied social scientists and health pro-

---

2000); Giorgio Israel and Pietro Nastasi, *Scienza e razza nell'Italia fascista* (Bologna, Italy, 1998); David Horn, *Social Bodies: Science, Reproduction, and Italian Modernity* (Princeton, N.J., 1995). On developments in Spain, see Annette Mülberger and Ana Maria Jacó-Vilela, "Es mejor morir de pie que vivir de rodillas: Emilio Mira y López y la revolución social," *Dynamis* (forthcoming); Rafael Huertas and Carmen Ortiz, eds., *Ciencia y fascismo* (Madrid, 1998).

[76] Götz Aly, Peter Chroust, and Christian Pross, *Cleansing the Fatherland: Nazi Medicine and Racial Hygiene* (Baltimore, 1994); Burleigh and Wippermann, *The Racial State* (cit. n. 60); Robert Proctor, *Racial Hygiene: Medicine under the Nazis* (Cambridge, Mass., 1989).

[77] Rainer Mackensen, *Bevölkerungslehre und Bevölkerungspolitik im "Dritten Reich"* (Opladen, Germany, 2004); Winfried Süss, *Der "Völkskörper" im Krieg: Gesundheitspolitik, Gesundheitsverhältnisse, und Krankenmord im nationalsozialistischen Deutschland, 1939–1945* (Munich, 2003); Michael Kater, *Das "Ahnenerbe" der SS 1935–1945: Ein Beitrag zur Kulturpolitik des Dritten Reiches* (Munich, 2001); Jill Stephenson, *Women in Nazi Germany* (New York, 2001); Michael Burleigh, *Germany Turns Eastward: A Study of Ostforschung in the Third Reich* (Cambridge, 1988).

[78] Weindling, *Nazi Medicine and the Nuremberg Trials* (cit. n. 1); Till Bastian, *Homosexuelle im Dritten Reich: Geschichte einer Verfolgung* (Munich, 2000); Guenter Lewy, *The Nazi Persecution of the Gypsies* (Oxford, 2000); Richard Wetzell, *Inventing the Criminal: A History of German Criminology, 1880–1945* (Chapel Hill, N.C., 2000); Joachim S. Hohmann, *Robert Ritter und die Erben der Kriminalbiologie: "Zigeunerforschung" im Nationalsozialismus und in Westdeutschland im Zeichen des Rassismus* (Frankfurt, 1991).

[79] Henry Friedlander, *The Origins of Nazi Genocide: From Euthanasia to the Final Solution* (Chapel Hill, N.C., 1997); Burleigh, *Death and Deliverance* (cit. n. 58).

[80] Paul Julian Weindling, *Epidemics and Genocide in Eastern Europe, 1890–1945* (Oxford, 2000).

fessionals were readying themselves for the postwar cleanup. In policy-making circles, it was widely believed that Germans were going to require massive reeducation to undo years of National Socialist indoctrination.[81] Efforts in the western zones of occupation, however, showed mixed results, as many Nazi scholars were allowed to remain at their academic posts, while American pollsters showed that large numbers of German citizens retained what were thought to be essentialist, racist, and authoritarian attitudes.[82]

As German historians have noted, the legacy of Nazism and World War II was swiftly transformed into narratives of victimization. Survivors of concentration and death camps drew the attention of psychiatrists and psychologists, who, since 1945, have not only attempted to aid the recovery of those persecuted but also spoken to the traumatic effects of imprisonment and torture on individuals.[83] Perhaps the most surprising development following the war, however, was the ease with which German soldiers and citizens were cast in the roles of victims in the late 1940s and early 1950s. While pediatricians and child psychologists stressed the damaging psychological effects of strategic bombing on German children, psychiatrists helped perpetuate the image of former Wehrmacht soldiers—especially POWs—as mentally damaged men requiring understanding and rehabilitation.[84]

The idea that even German veterans needed to be seen as victims of totalitarianism had its origins not only in military psychiatry but also in attitudes of the cold war, where the lines between friend and enemy were redrawn in short order. Within the span of a few years, the United States and the Soviet Union went from being allies to sworn enemies who questioned one another's legitimacy. In turn, Europe became one of several sites where American and Soviet officials attempted to demonstrate the superiority of their respective worldviews, thereby dividing the continent in two.[85]

With two highly ideological states pitted against one another, the human sciences were easily politicized on both sides of the iron curtain. In the Soviet Union and its satellite states, as Slava Gerovitch's contribution notes, scientific research and education were closely linked to the state and the Communist Party.[86] The official doctrine of Marxism-Leninism became a required orientation point for all work in the social

[81] Uta Gerhardt, "A Hidden Agenda of Recovery: The Psychiatric Conceptualization of Reeducation for Germany in the United States during World War II," *Germ. Hist.* 14 (1996): 297–324.

[82] Steven P. Remy, *The Heidelberg Myth: The Nazification and Denazification of a German University* (Cambridge, 2002); Richard L. Merritt, *Democracy Imposed: U.S. Occupation Policy and the German Public, 1945–1949* (New Haven, Conn., 1995).

[83] Arie Nadler, "The Victim and the Psychologist: Changing Perceptions of Israeli Holocaust Survivors by the Mental Health Community in the Past Fifty Years," *History of Psychology* 12 (2001): 159–81; Robert Krell, Marc Sherman, and Elie Wiesel, *Medical and Psychological Effects of Concentration Camps on Holocaust Survivors* (New Brunswick, N.J., 1997).

[84] Frank Biess, "Survivors of Totalitarianism: Returning POWs and the Reconstruction of Masculine Citizenship in West Germany, 1945–1955," in *The Miracle Years Revisited: A Cultural History of West Germany,* ed. Hanna Schissler (Princeton, N.J., 2001), 57–82; Svenja Goltermann, "Im Wahn der Gewalt: Massentod, Opferdiskurs, und Psychiatrie 1945–1956," in *Nachkrieg in Deutschland,* ed. Klaus Naumann (Hamburg, Germany, 2001), 343–63, as well as Naumann, "Verletzte Körper oder 'Building National Bodies': Kriegsheimkehrer, 'Krankheit,' und Psychiatrie in der westdeutschen Nachkriegsgesellschaft, 1945–1955," *WerkstattGeschichte* 8 (1999): 83–98.

[85] Konrad H. Jarausch and Hannes Siegrist, eds., *Amerikanisierung und Sowjetisierung in Deutschland, 1945–1970* (Frankfurt, 1997).

[86] Janina Lagneau, "Sciences sociales en U.R.S.S.: Institutions, publications, chercheurs," *Revue des Etudes Slaves* 57 (1985): 211–23.

sciences, the chief consequences being an insistence upon consensus and a privileging of natural scientific methods.[87] Yet even while Communist Party leaders in the 1950s decried the "bourgeois" emphasis on individualism in the West, there remained great interest in understanding and social engineering the human being as a subject.[88] After 1961, once the Soviet Academy of Science formally dropped the demand for "party-mindedness" (*partiynost*) in science, fields such as sociology, psychology, and pedagogy experienced a renaissance of research and applied programs focusing on the development of "socialist personalities."[89] Reinforced by détente in the 1970s, the new openness also provided a window for American and Western European scholarship to be reintroduced into the fields of psychiatry and psychotherapy in Eastern Europe.[90] The aim was to keep up with Western professional literature so as to "overtake and surpass" Western scientific advances.[91] This receptiveness, however, came at a price. Growing concerns over international influences and dissidence in the seventies and eighties led Communist Party officials to recruit psychologists and psychiatrists to aid state security agencies in spying on, subverting, interrogating, and incarcerating political opponents.[92]

In the United States as well, the cold war had a direct impact on funding and research in the social sciences.[93] In the decade and a half following the end of the war, an emphasis on applied research promoted the adoption of engineering technologies and reliance on more or less behavioristic and typological approaches to understanding human action.[94] The advent of computers only reinforced this trend.[95] When American observers showed interest in intellectual or clinical developments in the

---

[87] Stefan Busse, *Psychologie in der DDR: Die Verteidigung der Wissenschaft und die Formung der Subjekte* (Weinheim, Germany, 2004); Hubert Laitko, "Wissenschaftspolitik," in *Die SED—Geschichte—Organisation—Politik: Ein Handbuch,* ed. Andreas Herbst, Gerd-Rüdiger Stephan, and Jürgen Winkler (Berlin, 1997), 405–20; David Joravsky, *Russian Psychology: A Critical History* (Oxford, 1989).

[88] Oleg Kharkhordin, *The Collective and the Individual in Russia: A Study of Practices* (Berkeley, Calif., 1999).

[89] Greg Eghigian, "Homo Munitus: The East German Observed," in *Socialist Modern,* ed. Paul Betts and Katherine Pence (Ann Arbor, Mich., in press); Konstantin Ivanov, "Science after Stalin: Forging a New Image of Soviet Science," *Science in Context* 15 (2002): 317–38.

[90] Greg Eghigian, "The Psychologization of the Socialist Self: East German Forensic Psychology and Its Deviants, 1945–1975," *Germ. Hist.* 22 (2004): 181–205; Christine Leuenberger, "Socialist Psychotherapy and Its Dissidents," *Journal of the History of the Behavioral Sciences* 37 (2001): 261–73.

[91] Slava Gerovitch, *From Newspeak to Cyberspeak: A History of Soviet Cybernetics* (Cambridge, Mass., 2002).

[92] Greg Eghigian, "Care and Control in a Communist State: The Place of Politics in East German Psychiatry," in *Psychiatric Cultures Compared: Psychiatry and Mental Health Care in the Twentieth Century,* ed. Marijke Gijswijt-Hofstra et al. (Amsterdam, 2005), 183–99; Sonja Süß, *Politisch mißbraucht? Psychiatrie und Staatssicherheit in der DDR* (Berlin, 1999); Theresa C. Smith, *No Asylum: State Psychiatric Repression in the Former USSR* (New York, 1996); Klaus Behnke and Jürgen Fuchs, eds., *Zersetzung der Seele: Psychologie und Psychiatrie im Dienste der Stasi* (Hamburg, Germany, 1995).

[93] Christopher Simpson, ed., *Universities and Empire: Money and Politics in the Social Sciences during the Cold War* (New York, 1998).

[94] Ron Theodore Robin, *The Making of the Cold War Enemy: Culture and Politics in the Military-Industrial Complex* (Princeton, N.J., 2003); Dorothy Ross, "Changing Contours of the Social Science Disciplines," in *The Cambridge History of Science,* vol. 7, ed. Theodore M. Porter and Dorothy Ross (Cambridge, 2003).

[95] Paul N. Edwards, *The Closed World: Computers and the Politics of Discourse in Cold War America* (Cambridge, Mass., 1996). On similar developments in the USSR, see Gerovitch, *From Newspeak to Cyberspeak* (cit. n. 91).

USSR, it was often to dismiss them as overly ideological or downright unethical.[96] At the same time, in the wake of the Marshall Plan, Western Europe proved to be highly receptive to American influences. Beginning in the mid-1950s, trends in American sociology and social, clinical, and industrial psychology served as the intellectual center of gravity for a generation of investigators and practitioners, particularly in central Europe and the Low Countries.[97]

If the cold war did much to undermine Europe's former preeminence in international politics and science, the effect was only augmented by yet another set of momentous events. Anticolonial movements and decolonization presented a wide range of political, economic, and military challenges for Western countries during the last two-thirds of the century. As indigenous nationalist parties and insurgencies, especially in Southeast Asia, the Middle East, and North Africa successfully defied attempts to preserve the colonial order, they simultaneously called into question the ideas that had justified the imperialist project.

Two fields were especially affected by this critical assessment: psychiatry and anthropology. In clinical psychiatry, it was Frantz Fanon who led the way in the 1950s, noting the essentializing and infantilizing assumptions of Western medicine in its treatment of Africans.[98] Cultural anthropologists at this time came to a similar set of conclusions, sparking a "reflexive turn" in ethnography and a renewed interest in the different ways in which personhood is conceptualized and lived.[99] Since then, ethnopsychiatrists, social scientists, and historians have consistently demonstrated the difficulties in universalizing personality attributes as well as forms of mental illness.[100]

In the wake of the fall of the Berlin wall in 1989 and of the Soviet Union in 1991, the scholarship on decolonization, in tandem with that on the cold war, took on a new significance. The Soviet Union had operated as a vast empire, and the reasons for and consequences of its demise were seen especially by many anthropologists of Eastern Europe as mirroring developments in formerly colonized Africa and Asia.[101] During the 1990s, social scientists in central and eastern Europe took on a leading role in tracking these problems: sociologists studied the effects of economic change on

[96] Gary S. Belkin, "Writing about Their Science: American Interest in Soviet Psychiatry during the Post-Stalin Cold War," *Perspectives in Biology and Medicine* 43 (1999): 31–46.

[97] Pieter J. van Strien, "The American 'Colonization' of Northwest European Social Psychology after World War II," *J. Hist. Behav. Sci.* 33 (1997): 349–63.

[98] David Macey, *Frantz Fanon: A Life* (London, 2000); Hanafy Youssef, "Frantz Fanon and Political Psychiatry," *Hist. Psychiat.* 7 (1996): 525–32.

[99] James D. LeSueur, *Uncivil War: Intellectuals and Identity Politics during the Decolonization of Algeria* (Philadelphia, 2001); James Clifford and George E. Marcus, *Writing Culture: The Poetics and Politics of Ethnography* (Berkeley, Calif., 1986); Michael Carrithers, Steven Colins, and Steven Lukes, eds., *The Category of the Person: Anthropology, Philosophy, History* (New York, 1985).

[100] Steven J. Heine, "Self as a Cultural Product: An Examination of East Asian and North American Selves," *Journal of Personality* 69 (2001): 881–906; Charles Lindholm, *Culture and Identity: The History, Theory, and Practice of Psychological Anthropology* (Boston, 2000), 19–41; Jonathan Hal Sadowsky, *Imperial Bedlam: Institutions of Madness in Colonial Southwest Nigeria* (Berkeley, Calif., 1999); Arthur Kleinman, *Writing at the Margins: Discourse between Anthropology and Medicine* (Berkeley, Calif., 1997); Jock McCullough, *The Empire's New Clothes: Ethnopsychiatry in Colonial Africa* (New York, 1995); Melford E. Spiro, "Is the Western Conception of the Self 'Peculiar' within the Context of the World Cultures?" *Ethos* 21 (1993): 107–53; Kenneth J. Gergen and Keith E. Davis, eds., *The Social Construction of the Person* (New York, 1985).

[101] Daphne Berdahl, Matti Bunzl, and Martha Lampland, eds., *Altering States: Ethnographies of Transition in Eastern Europe and the Former Soviet Union* (Ann Arbor, Mich., 2000); Catherine Wanner, *Burden of Dreams: History and Identity in Post-Soviet Ukraine* (University Park, Pa., 1998).

families and communities, pollsters monitored shifts in public attitudes and values, and psychologists tested and interviewed individuals to assess claims that a psychological Stalinism had fashioned mentally "deformed" citizens.[102] As Christine Leuenberger's essay in this volume demonstrates, the transition from communism to capitalism in East Germany, for instance, has restructured the psychological sciences there without eliminating tensions, gaps, and contradictions between formal stipulations and informal practice. While pre-1989 psychological concepts were harnessed to the project of building scientific socialism and reconstituted post-1989 to fit with the requirements of the West German bureaucratic health care system, in practice formal guidelines and stipulations were often adapted, modified, and frequently superseded by local, interactional, institutional, and cultural practices and contingencies under communism as well as capitalism.[103]

THE WELFARE STATE AND THE AGE OF MASS CONSUMERISM

As World War II showed, the fate of the human sciences in the twentieth century was intimately bound up with that of the state and political economy. Beginning around midcentury, a set of mutually reinforcing changes in European and American statecraft and economic life had a profound impact on professional development, research agendas, clinical practices, and popular attitudes. In particular, the war and postwar reconstruction pressed central states to become far more active in steering economy and society. Whether it was Keynesianism in Great Britain, corporatist planning in France, social market economy in West Germany, or democratic centralism in the Eastern bloc, governments during the years 1945–1975 demonstrated a new willingness to regulate the production and distribution of goods and services.[104] This, in turn, resulted in unprecedented levels of economic growth, an ambitious reform of educational systems, a rise in the average standard of living, and an expansion of the welfare state throughout the United States and Western and Eastern Europe.[105]

Serving as both cause and effect in these developments was the rise of mass consumerism and mass consumer culture.[106] To be sure, consumerism and the concept of human beings as consuming animals were hardly twentieth-century inventions. As

---

[102] Eghigian, "Homo Munitus" (cit. n. 89); Christine Leuenberger, "The Berlin Wall on the Therapists's Couch," *Human Studies* 23 (2000): 99–121. See also *Ten Years after 1989: What Have We Learned?* special issue, *Slavic Review* 58 (Winter 1999); Daphne Berdahl, *Where the World Ended: Re-Unification and Identity in the German Borderland* (Berkeley, Calif., 1999); John Borneman, "Narrative, Genealogy, and the Historical Consciousness: Selfhood in a Disintegrating State," in *Culture/Contexture: Explorations in Anthropology and Literary Studies,* ed. E. Valentine Daniel and Jeffrey M. Peck (Berkeley, Calif., 1996).

[103] Christine Leuenberger, "The End of Socialism and the Reinvention of the Self: A Study of the East German Psychotherapeutic Community in Transition," *Theory and Society* 31 (2002): 257–82.

[104] James C. Van Hook, *Rebuilding Germany: The Creation of the Social Market Economy, 1945–1957* (New York, 2004); Peter C. Caldwell, *Dictatorship, State Planning, and Social Theory in the German Democratic Republic* (New York, 2003); Pekka Sutela, *Economic Thought and Economic Reform in the Soviet Union* (New York, 1991); Peter A. Hall, *Governing the Economy: The Politics of State Intervention in Britain and France* (Oxford, 1986).

[105] Peter Flora and Arnold J. Heidenheimer, *The Development of Welfare States in Europe and America* (New Brunswick, N.J., 1995); Jacques Donzelot, "The Promotion of the Social," *Economy and Society* 17 (1988): 395–427.

[106] Victoria de Grazia, *Irresistible Empire: America's Advance through Twentieth-Century Europe* (Cambridge, Mass., 2005); Lizabeth Cohen, A *Consumer's Republic: The Politics of Mass Consumption in Postwar America* (New York, 2003); Gary Cross, *An All-Consuming Century: Why Commercialism Won in Modern America* (New York, 2002); Martin Daunton and Matthew Hilton, eds., *The*

historians of early modern England have noted, the seventeenth and eighteenth centuries witnessed a "consumer revolution," inspiring an economic paradigm that defined consumption as the sole end of production.[107] Alongside the rise of early forms of consumerism was the ideal of "economic man," a being considered to be fundamentally rational but driven by insatiable needs and a desire to efficiently use scarce resources while maximizing utility and profit.[108] Well before World War II, then, the human sciences had become deeply involved in framing, promoting, and criticizing the desires, tastes, sensibilities, and economic choices of individuals in consumer society.[109]

As the raison d'être of the mid-twentieth-century state moved the public sector into new areas of social engagement, however, there was a palpable growth in the demand for the expert knowledge and services of those working in the human sciences. The rise of the contemporary knowledge-based society, German historian Lutz Raphael has noted, was thus largely brought about by a scientization of the public concern over issues deemed "social."[110] Among the chief beneficiaries of this trend were the psychological sciences. In the United States, for instance, membership in the American Psychological Association grew from 2,739 in 1940 to 30,839 in 1970 (and to some 75,000 members by 1993), and membership in the American Psychiatric Association from 2,423 to 18,407 between 1940 and 1970.[111] Central Europe exhibited a similar trend around this time, with membership in the German Psychological Society going from approximately 2,500 in 1961 to 20,000 in 1984 and more than 40,000 by 1996.[112] At the same time, social service agencies were among the first postwar institutions to adopt psychological testing and personality assessment as routine parts of their administrative work.[113]

One of the chief spurs for the growth of the human sciences, as Harry Oosterhuis's

---

Politics of Consumption: Material Culture and Citizenship in Europe and America (Oxford, 2001); Susan Strasser, Charles McGovern, and Matthias Judt, eds., *Getting and Spending: Consumer Society in the Twentieth Century* (Cambridge, 1998).

[107] Roy Porter, "Consumption: Disease of the Consumer Society?" in *Consumption and the World of Goods*, ed. John Brewer and Roy Porter (London, 1997); Neil McKendrick, John Brewer and J. H. Plumb, *The Birth of a Consumer Society: The Commercialization of Eighteenth Century England* (London, 1982); Joyce Appleby, *Economic Thought in Seventeenth-Century England* (Princeton, N.J., 1978).

[108] Neil J. Smelser and Richard Swedberg, "The Sociological Perspective on the Economy," in *The Handbook of Economic Sociology*, ed. Neil J. Smelser and Richard Swedberg (Princeton, N.J., 1994).

[109] See, e.g., Richard Olson, *The Emergence of the Social Sciences, 1642–1792* (New York, 1993); Dorothy Ross, *The Origins of American Social Science* (Cambridge, 1991); Thomas L. Haskell, *The Emergence of Professional Social Science: The American Social Science Association and the Nineteenth-Century Crisis of Authority* (Chicago, 1977).

[110] Lutz Raphael, "Die Verwissenschaftlichung des Sozialen als methodische und konzeptionelle Herausforderung für eine Sozialgeschichte des 20. Jahrhunderts," *Geschichte und Gesellschaft* 22 (1996): 165–93; Giddens, *Modernity and Self-Identity* (cit. n. 4); Michel Foucault, *The Order of Things* (New York, 1994).

[111] Ellen Herman, *The Romance of American Psychology: Political Culture in the Age of Experts* (Berkeley, Calif., 1995), 2–3. See also James H. Capshew, *Psychologists on the March: Science, Practice, and Professional Identity in America, 1929–1969* (New York, 1999).

[112] Hans Spada, "Lage und Entwicklung der Psychologie in Deutschland, Österreich und der Schweiz," *Psychologische Rundschau* 48 (1997): 1–15.

[113] David Meskill, *Human Economies: Labor Administration, Vocational Training, and Psychological Testing in Germany, 1914–1964* (PhD diss., Harvard Univ., 2003); Roderick D. Buchanan, "On Not 'Giving Psychology Away': The Minnesota Multiphasic Personality Inventory and Public Controversy over Testing in the '60s," *Hist. Psychol.* 5 (2002): 284–309; Trudy Dehue, "Transforming Psychology in the Netherlands I: Why Methodology Changes," *History of the Human Sciences* 4 (1991): 335–49.

article in this volume shows, was a shift in the orientation of mental health work from custodial and palliative treatment to outpatient and preventive care. Over the course of the fifties, sixties, and seventies, this change was most marked in Great Britain, Italy, the Netherlands, the United States, and West Germany, although similar trends were also evident in Eastern Europe.[114] A century-long pattern of increasing numbers of hospitalized psychiatric patients was reversed by a variety of projects designed to integrate and maintain the mentally ill and disabled within the community at large. Supported and promoted by enthusiastic reformers who often saw them as part of a broader progressive reform of social welfare, "community mental health" and "social psychiatry" quickly became rallying points for burgeoning patient and disabled citizens' rights movements.[115]

Much of the new wave of state interventionism was directed at women, children, and families. Ellen Herman's essay in this volume uses the case of child adoption in modern U.S. history to explore how the management of child development became an ambitious governmental project, increasingly subject to scientific inquiry, professional supervision, and legal intervention. At the same time, the human sciences were called upon to reduce risk, manage stigma, and promote permanence and authenticity in adoptive families in which strangers were to be turned into kin. A renewed interest in pursuing such social policies after the war led to greater vigilance about child welfare and the family milieu.[116] Feminists and psychotherapists proved to be instrumental in making the question of sexual abuse a topic for public policy discussions,[117] while social workers became more active in identifying and monitoring "problem families" and "at-risk" young people.[118] In fact, more than any other group of professionals, it was social workers who came to serve as the bridge between social services and criminal law. By the 1990s, their contributions to police and probation departments, juvenile detention, and domestic violence and victim assistance programs had become indispensable.[119]

This postwar growth in social work reveals yet another important site of expansion for the human sciences: criminal justice. Concerns about deviance and antisocial behavior among youths, especially in the 1950s and the 1960s, prompted more delib-

---

[114] Greg Eghigian, "Was There a Communist Psychiatry? Politics and East German Psychiatric Care, 1945–1989," *Harvard Review of Psychiatry* 10 (2002): 364–8.

[115] Franz-Werner Kersting, ed., *Psychiatrie als Gesellschaftsreform: Die Hypothek des Nationalsozialismus und der Aufbruch der sechziger Jahre* (Paderborn, Germany, 2003); Marijke Gijswijt-Hofstra and Roy Porter, eds., *Cultures of Psychiatry and Mental Health Care in Postwar Britain and the Netherlands* (Amsterdam, 1998); Michael Donnelly, *The Politics of Mental Health in Italy* (London, 1992); Gerald N. Grob, *From Asylum to Community: Mental Health Policy in Modern America* (Princeton, N.J., 1991).

[116] On the origins of the maternalist welfare state, see Seth Koven and Sonya Michel, eds., *Mothers of a New World: Maternalist Politics and the Origins of Welfare States* (New York, 1993). See also Viviana A. Zelizer, *Pricing the Priceless Child: The Changing Social Value of Children* (Princeton, N.J., 1994).

[117] Beryl Satter, "The Sexual Abuse Paradigm in Historical Perspective: Passivity and Emotion in Mid-Twentieth-Century America," *Journal of the History of Sexuality* 12 (2003): 424–64.

[118] Alexandra Minna Stern, ed., *Formative Years: Children's Health in the United States, 1880–2000* (Ann Arbor, Mich., 2002); Pat Starkey, *Families and Social Workers: The Work of Family Service Units, 1940–1985* (Liverpool, 2000); John Welshman, "In Search of the 'Problem Family': Public Health and Social Work in England and Wales, 1940–1970," *Social History of Medicine* 9 (1996): 447–65; Agota Horvath, "Routes to Social Work," *East Central Europe* 20–23 (1993–1996): 147–70.

[119] A. R. Roberts and P. Brownell, "A Century of Forensic Social Work: Bridging the Past to the Present," *Social Work* 44 (1999): 359–69.

erate attempts to study, flag, and aggressively combat early signs of delinquency in young people.[120] As Volker Janssen's essay reminds us, however, this crackdown was tempered by a new, "softer" approach toward crime and punishment that involved enlisting the services of social scientists and clinicians. The third-quarter of the twentieth century was the heyday of what David Garland has referred to as "penal welfarism": a clinical approach to criminal deviance, stressing the need to psychologically understand and therapeutically reintegrate offenders.[121] Although American liberals and Western European Social Democrats tended to be the most outspoken proponents of penal welfarism, conservative policy makers, confessional groups, and Eastern European communists also lent their support to the project.[122] Since the late 1970s, however, a radical shift in American and British public opinion and policy toward retributive justice has led institutions there to slowly abandon the rehabilitative ideal of a generation ago.[123]

As mental health care and its patrons moved closer to society, mainstream society was simultaneously drawn to psychotherapeutic knowledge and services. In the years between 1940 and 1980, marriage and family counseling centers, churches, schools and universities, self-help organizations, and private practitioners all brought a wide variety of talk therapies to receptive, largely middle-class clients.[124] At the same time, succeeding generations of psychotropic drugs, particularly antidepressants, were made available to unprecedented numbers of adults and children concerned about keeping up with the demands of contemporary society.[125] The relatively ubiquitous nature of psychological and psychiatric language since the 1950s perhaps reflects what Roger Smith has termed the "internalization of belief in psychological knowledge." This psychologization of individuality, he notes, "altered everyone's subjective world and recreated experience and expectations about what it is to be a person. The

[120] Mimi Ajzenstadt, "Crime, Social Control, and the Process of Social Classification: Juvenile Delinquency/Justice Discourse in Israel, 1948–1970," *Social Problems* 49 (2002): 585–604; Uta G. Poiger, *Jazz, Rock, and Rebels: Cold War Politics and American Culture in a Divided Germany* (Berkeley, Calif., 2000); Simonetta Piccone Stella, "'Rebels without a Cause': Male Youth in Italy Around 1960," *History Workshop Journal* 38 (1994): 157–78.

[121] David Garland, *The Culture of Control: Crime and Social Order in Contemporary Society* (Chicago, 2001).

[122] Verena Zimmermann, *"Den neuen Menschen schaffen": Die Umerziehung von schwererziehbaren und straffälligen Jugendlichen in der DDR, 1945–1990* (Cologne, Germany, 2004); Karl-Michael Walz, *Soziale Strafrechtspflege in Baden: Grundlagen, Entwicklung, und Arbeitsweisen der badischen Straffälligenhilfe in Geschichte und Gegenwart* (Freiburg im Breisgau, Germany, 1999); Philip Jenkins, *Moral Panic: Changing Concepts of the Child Molester in Modern America* (New Haven, Conn., 1998).

[123] James Q. Whitman, *Harsh Justice: Criminal Punishment and the Widening Divide between America and Europe* (Oxford, 2003); Garland, *Culture of Control* (cit. n. 121); Simon A. Cole, "From the Sexual Psychopath Statute to 'Megan's Law': Psychiatric Knowledge in the Diagnosis, Treatment, and Adjudication of Sex Criminals in New Jersey, 1949–1999," *Journal of the History of Medicine and Allied Sciences* 55 (2000): 292–314; Francis A. Allen, *The Decline of the Rehabilitative Ideal: Penal Policy and Social Purpose* (New Haven, Conn., 1981).

[124] Herman, *Romance of American Psychology* (cit. n. 111); Benjamin Ziemann, "The Gospel of Psychology: Therapeutic Concepts and the Scientification of Pastoral Care in the West German Catholic Church, 1950–1980," *Central European History* 39 (2006): 79–106; Rod Janzen, *The Rise and Fall of Synanon: A California Utopia* (Baltimore, 2001); Eva S. Moskowitz, *In Therapy We Trust: America's Obsession with Self-Fulfillment* (Baltimore, 2001); Philip Cushman, *Constructing the Self, Constructing America: A Cultural History of Psychotherapy* (Boston, 1995); Robert S. Wallerstein, *The Talking Cures* (New Haven, Conn., 1995); Donald K. Freedheim, ed., *History of Psychotherapy: A Century of Change* (Washington, D.C., 1992).

[125] David Healy, *The Anti-Depressant Era* (Cambridge, Mass., 1997).

result was an emphasis on 'the personal' in psychological terms, with ramifications in every aspect of life."[126]

If, as Smith and others contend, a psychologization of subjectivity stamps the contemporary history of the individual, recent research indicates that it occurred under the reciprocal influence of the human sciences and civil society.[127] Not only state agencies and therapeutic professionals were invested in probing and managing the inner workings of individuals during the second half of the century. Social movements, companies, educators, and the news media all expressed great curiosity in the attitudes and wishes of the general public. Opinion polling, invented in the interwar years, took off in the 1950s. It rapidly became an essential tool for both corporate marketers wishing to monitor and exploit consumer habits as well as political parties and journalists interested in tracking voter sentiments.[128] Although such scientific surveys are generally associated with the market economies of the United States and Western Europe, studies conducted after the fall of the USSR have shown that communist security agencies and economic planners routinely conducted similar kinds of polling and for similar purposes.[129]

By the late seventies, then, the human sciences had come to provide compelling concepts, terms, explanations, tools, and services for knowing and steering the individual in contemporary society. Particularly in the United States, public intellectuals consistently turned to fields such as psychology, psychiatry, and sociology to inform their analyses of timely social problems, such as prejudice, the generation gap, and race and class relations.[130] In addition, as Sherry Turkle has shown, computing technology has promoted cognitively based theories in fields such as psychology that, in turn, have changed the ways in which researchers and clinicians conceptualize the brain.[131]

That said, the social and medical sciences were not without their critics, who lamented the influence of the former over public debate and policy. Psychotherapy and psychoanalysis were pilloried for promoting moral torpor.[132] Patient groups, lawyers, and even fellow clinicians attacked medicine and psychiatry for undermining personal responsibility and autonomy.[133] Clinical psychologists and psychiatrists were

---

[126] Smith, *Norton History of the Human Sciences* (cit. n. 6), 575.

[127] On changing public values over time, see Ronald Inglehart, *Modernization and Postmodernization: Cultural, Economic, and Political Change in 43 Societies* (Princeton, N.J., 1997).

[128] Sarah Igo, *America Surveyed: The Making of a Mass Public* (Cambridge, Mass., 2006); Christoph Conrad, "Observer les consommateurs: Etudes de marché et histoire de la consommation en Allemagne, des années 1930 aux années 1960," *Mouvement Social* 206 (2004): 17–39; Jon Cowans, "Fear and Loathing in Paris: The Reception of Opinion Polling in France, 1938–1977," *Social Science History* 26 (2002): 71–104; Anja Kruke and Benjamin Ziemann, "Meinungsumfragen in der Konkurrenzdemokratie: Auswirkungen der Demoskopie auf die Volksparteien und den politischen Massenmarkt 1945/49–1990," *Historical Social Research* 26 (2001): 171–9.

[129] See, e.g., Ina Merkel, *Utopie und Bedürfnis: Die Geschichte der Konsumkultur in der DDR* (Cologne, Germany, 1999).

[130] Richard H. King, *Race, Culture, and the Intellectuals, 1940–1970* (Baltimore, 2004); Daryl Michael Scott, *Contempt and Pity: Social Policy and the Image of the Damaged Black Psyche, 1880–1996* (Chapel Hill, N.C., 1997); Christopher Lasch, *The Culture of Narcissism: American Life in an Age of Diminishing Expectations* (New York, 1979); Richard Hofstadter, *The Paranoid Style in American Politics and Other Essays* (New York, 1965); David Riesman, *The Lonely Crowd: A Study of the Changing American Character* (New Haven, Conn., 1950); Theodor Adorno, *The Authoritarian Personality* (New York, 1950).

[131] Sherry Turkle, *Life on the Screen: Identity in the Age of the Internet* (New York, 1995).

[132] Philip Rieff, *The Triumph of the Therapeutic: Uses of Faith after Freud* (New York, 1966).

[133] Robert A. Nye, "The Evolution of the Concept of Medicalization in the Late Twentieth Century," *J. Hist. Behav. Sci.* 39 (2003): 115–29; Duncan B. Double, "The History of Anti-Psychiatry: An Essay

blamed for creating spurious diagnostic categories.[134] Political interest groups and politicians openly questioned the objectivity and legitimacy of a number of prominent studies on sexuality.[135]

Today, new technologies are confronting investigators, health care providers, governments, and individuals with a fresh set of legal and ethical challenges. Aryn Martin's essay in this volume, for instance, traces how the detection of human chimeras (people who contain more than one genetically distinct cell population) have called into question the liberal association of a person with one singular genetic profile. At a time when identification and identity projects that use DNA have become critical to a wide range of public institutions, the possibility that selves can no longer be neatly mapped on to genes and their owners could have ramifications for the general governance and management of citizens and workers.

This being a period of unprecedented developments in science and technology, consumers of medical and scientific knowledge and services now appear, to some ob servers at least, far less willing to simply defer to the authority of experts. Work in science studies has shown that the 1960s constituted a turning point in the interaction between experts and laypeople, as concerned citizens increasingly protested what they took to be the potentially deleterious effects of science and technology.[136] If, however, it seems that the objects of the human sciences have just now begun talking back, we need only look to the essays in this volume to see that neither the human sciences nor political authorities have ever been truly sovereign. Psychiatric patients, war veterans, telephone operators, prisoners, families, confessional groups, entrepreneurs, and experts—in short, a wide range of individuals and groups in civil society— have frequently informed, thwarted, deflected, and outright rejected the ambitions of ideologues and policy makers, researchers, and clinicians. The history of the relationship between politics and the human sciences is clearly ill-suited to being written either as a story of cumulative progress or as one of the disintegrating autonomy of the individual. It is our hope that the essays in this volume can offer and inspire some new narrative possibilities.

Review," *Hist. Psychiat.* 13 (2002): 231–6; Nick Crossley, "R. D. Laing and the British Anti-Psychiatry Movement: A Socio-Historical Analysis," *Social Science and Medicine* 47 (1998): 877–89; Janet Vice, *From Patients to Persons: The Psychiatric Critiques of Thomas Szasz, Peter Sedgwick, and R. D. Laing* (New York, 1992); Norman Dain, "Critics and Dissenters: Relections on 'Anti-Psychiatry' in the United States," *J. Hist. Behav. Sci.* 25 (1989): 3–25.

[134] Herb Kutchins and Stuart A. Kirk, *Making Us Crazy: DSM—The Psychiatric Bible and the Creation of Mental Disorders* (New York, 1997); Hacking, *Rewriting the Soul* (cit. n. 4).

[135] Scott O. Lilienfeld, "When Worlds Collide: Social Science, Politics, and the Rind et al. (1998). Child Sexual Abuse Meta-Analysis," *American Psychologist* 57 (2002): 176–88; Edward O. Laumann, *The Social Organization of Sexuality: Sexual Practices in the United States* (Chicago, 1994).

[136] Trevor Pinch, "Scientific Controversies," in *The International Encyclopedia of the Social and Behavioral Sciences,* ed. Neil J. Smelser and Paul B. Baltes (Oxford, 2001); Steven Epstein, *Impure Science: AIDS, Activism, and the Politics of Knowledge* (Berkeley, Calif., 1998); Brian Wynne, "Public Understanding of Science," in *Handbook of Science and Technology Studies,* ed. Sheila Jasanoff et al. (London, 1995).

Blueprints for Change:

The Human Sciences and the Coercive Transformation of Deviants in Russia, 1890–1930

By Daniel Beer*

ABSTRACT

Drawing on the writings of criminologists and psychiatrists in the late imperial and early Soviet periods, the article argues that Soviet biopsychological constructions of the socially deviant have their origins in the efforts of tsarist liberals to identify and contain the crime and social disorder that accompanied Russia's modernization. While the historiography has traditionally portrayed the Bolshevik Revolution as a tragic overthrow of liberal ideas and values, the article points to important continuities that span the 1917 divide. In the late imperial period, the human sciences began to categorize individuals who posed a biopsychological threat, a "social danger," to the social order. In the wake of the revolution, these ideas became radicalized under the impact of Soviet Marxism to generate indictments of entire social groups and classes.

INTRODUCTION

In 1927, the criminologist G. N. Udal'tsov published a study, "Criminal Offences in the Armed Forces from the Perspective of Pathological Physiology," in the authoritative *Review of Psychiatry, Neurology, and Reflexology,* which related the case of a student at a technical college. Actively religious before the revolution, he had been drafted into the imperial army in 1916 and in 1920 joined the Bolshevik Party. Recently, the young man, who was described as "a passive rank and file worker," had been admitted to the clinic for mental illness at Leningrad's Military-Medical Academy. Udal'tsov's diagnosis of the student's condition is worth citing at length:

> He once again became religious, avidly read the Gospels, and distributed his possessions among his acquaintances. He began to preach against human injustice, in particular emphatically rejecting the Party's program, saying that truth could only be found in the Gospels. He spoke out against war and against all its instruments, in particular the army, declaring military service to be a potential participation in murder. He was arrested as a

* Department of History, Royal Holloway, University of London, Egham, TW20 0EX; daniel.beer @rhul.ac.uk.

Research for this article was generously funded by Downing College, Cambridge. Early versions were presented in 2005 to the Institut für osteuropäische Geschichte at the Humboldt University, Berlin, and to the 37th Annual Convention of the American Association for the Advancement of Slavic Studies in Salt Lake City. I am grateful to Susan Morrissey, Alexandra Oberländer, Steve Smith, Catriona Kelly, David L. Hoffmann, Peter Holquist, and Nóra Milotay for their comments on its various drafts. All translations are my own.

counterrevolutionary for his refusal to serve in the army and for his open dissemination of his views . . . [and] . . . he was sent to the clinic for examination. The clinical diagnosis was "dementia praecox." Here we have an interesting case of the reappearance, as a result of the dissolution of restraints and under the influence of a pathological process, of old reflexes that had been strongly instilled in childhood—religiosity and conservatism.[1]

Leaving to one side the issue of its clinical accuracy, which here can be neither corroborated nor refuted, Udal'tsov's diagnosis is striking for its characterization of the subject's condition in terms of a failure of adaptation to the ideological demands of postrevolutionary society. The anonymous student's exposure to a religious and conservative milieu under the ancien régime had imbibed him with psychological reflexes, or "social instincts" as they were often termed, which resulted in deviant and subversive behavior in the socialist state. Udal'tsov explained this failure of individual adaptation to a changing environment as a consequence of biopsychological deficiencies:

> As a result of phylogenesis, man receives a range of hereditary features, our instincts, unconditional reflexes, on the basis of which he constructs a conditional-reflexive system over the course of his life. A healthy [*polnotsennyi*—literally, "full value"] individual is characterized by a swift and complete adaptation to the surrounding milieu, which is based upon the regulatory work of the higher nervous system.

Individuals with, however, an "easily excitable and functionally unstable cortex" would experience "an acute struggle between excitement and restraint which . . . in individuals with weak nervous systems would result in a disturbance in the cortical balance, an explosion in cortical activity, and a neurotic condition."[2] The language of psychopathology was here integrated into a revolutionary narrative of societal change and individual adaptation. The persistence of a prerevolutionary system of beliefs and values into the Soviet Union of the 1920s resulted in behavior that could be diagnosed as mentally ill, "socially dangerous," and even counterrevolutionary.

This meshing of narratives of societal change and individual deviance was not a new departure in the literature of the human sciences in revolutionary Russia. Udal'tsov's diagnosis of the survivals of the prerevolutionary era in the psychology of some Soviet citizens was, in fact, indebted to theories of social, economic, and political reform and individual adaptation first elaborated in the twilight years of the autocracy by liberal practitioners of the intersecting disciplines of psychiatry, psychology, sociology, jurisprudence, and criminology.

Caught up in the maelstrom of 1914–1921, these Russian liberals and progressives had themselves been seeking to transform the Russian state and society.[3] Their political creed was a mixture of classical liberal adherence to the rule of law, individual autonomy, and respect for private property, infused with a rationalist faith in the power of science to shape the world.[4] Traditionally seen by historians as Macbeth's

---

[1] G. N. Udal'tsov, "Pravonarusheniia v voiskakh s tochki zreniia patologicheskoi fiziologii," *Obozrenie psikhiatrii, nevrologii i refleksologii,* no. 2 (1927): 121.

[2] Ibid., 124.

[3] Peter Holquist, *Making War, Forging Revolution: Russia's Continuum of Crisis, 1914–1921* (Cambridge, Mass., 2002).

[4] I take "liberalism" to mean very broadly the view that the primary aim of the state is to institute a social order that gives the broadest scope to the value of human freedom, where this freedom is understood negatively as the freedom from state interference in the individual's pursuit of her own conception of the good life and is limited only by an equal freedom for all members of that society.

"poor player that frets and struts his hour upon the stage and then is heard no more," the Russian liberal failed on his own terms to achieve the goal of unifying and modernizing Russia in the turbulent years that spanned the turn of the twentieth century. The liberals' political affiliations throughout the years of Russia's constitutional experiment between 1905 and 1917 were primarily to the centrist and left liberal parties such as the Constitutional Democrats and the Octobrists who lost out to their radical rivals in the chaotic year of the February and October revolutions.[5] Yet although the liberals were driven from the political stage in 1917, their intellectual efforts to articulate a vision of a modernized, rationalized society survived the upheavals of 1917 and went on to shape the Soviet regime's program of social transformation.

I argue this case by reading intellectual developments within the Russian human sciences across the period from the close of the nineteenth century through the first three decades of the twentieth century. The extraordinary popularity of science in general in Late Imperial Russia has been the subject of a number of studies.[6] Elizabeth Hachten has observed that "virtually the whole of educated society had suddenly been gripped by an unprecedented enthusiasm for science, an intellectual spasm reflected in the broadened audience for scientific ideas, practices, and potential applications."[7] Although practitioners of the human sciences were frustrated by inadequate state funding, political interference, and the general obduracy of the tsarist regime and its failure to implement their proposed criminal codes and reforms of medical and psychiatric care, on the public stage they were highly influential in articulating the terms in which educated contemporaries came to view the changing world around them.[8]

In the postrevolutionary years, the human sciences found a more generous, if ideologically more inflexible, patron in the Soviet regime. Psychiatrists, criminologists, and psychologists often found themselves working in offices under the auspices of

---

This category is clearly diverse and expansive, and I deal here only with one particular instance of liberalism in vogue with a nebulous social grouping of educated Russians in the fin de siècle. Their rationalist beliefs in the power of science, reason, and order to administer the world were as powerful as their liberalism, but for the sake of brevity, I refer to them as liberals throughout this study.

[5] See Geoffrey Hosking, *The Russian Constitutional Experiment: Government and Duma, 1907–1914* (Cambridge, 1973); William Rosenberg, *Liberals in the Russian Revolution: The Constitutional-Democratic Party, 1917–1921* (Princeton, N.J., 1974).

[6] Alexander Vucinich, *Science in Russian Culture, 1861–1917* (Palo Alto, Calif., 1970); Vucinich, *Social Thought in Tsarist Russia: The Quest for a General Science of Society, 1861–1917* (Chicago, 1976); Vucinich, *Darwin in Russian Thought* (Berkeley, Calif., 1988); Daniel Todes, *Darwin without Malthus: The Struggle for Existence in Russian Evolutionary Thought* (Oxford, 1989); Torsten Rüting, *Pavlov und der Neue Mensch: Diskurse über Disziplinierung in Sowjetrussland* (Munich, 2002).

[7] Elizabeth A. Hachten, "In Service to Science and Society: Scientists and the Public in Late Nineteenth-Century Russia," *Science and Civil Society, Osiris,* 17 (2002): 182.

[8] Julie V. Brown, "The Professionalism of Russian Psychiatry, 1857–1911" (PhD diss., Univ. of Pennsylvania, 1981); Brown, "Psychiatrists and the State in Tsarist Russia," in *Social Control and the State,* ed. A. Scull and S. Cohen (Oxford, 1985), 267–87; Brown, "Social Influences on Psychiatric Theory and Practice in Late Imperial Russia," in *Health and Society in Revolutionary Russia,* ed. Susan Gross Solomon and John F. Hutchinson (Bloomington, Ind., 1990), 27–44; Laura Engelstein, *The Keys to Happiness: Sex and the Search for Modernity in Fin-de-Siècle Russia* (Ithaca, N.Y., 1992); Dan Healey, *Homosexual Desire in Revolutionary Russia: The Regulation of Sexual and Gender Dissent* (Chicago, 2001); Irina Sirotkina, *Diagnosing Literary Genius: A Cultural History of Psychiatry in Russia, 1880–1930* (Baltimore, 2002); Daniel Beer, "'Microbes of the Mind': Moral Contagion in Late Imperial Russia," *Journal of Modern History* (forthcoming, 2007). For examples of this influence, see Aleksandr S. Izgoev, "Ob intelligentnoi molodezhi (Zametki ob ee byte i nastroeniiakh)," in *Vekhi. Iz glubiny* (Moscow, 1991), 100; Daniel Beer, "The Medicalization of Deviance in the Russian Orthodox Church, 1880–1905," *Kritika* 5 (Summer 2004): 451–83.

repressive judicial organs of the new state.[9] In the 1920s, they even published their psychiatric studies of criminals in volumes edited by the people's commissar of internal affairs, bearing the imprimatur of state approval and endorsement.[10] In the early Soviet period, these studies offered influential interventions into party debates about the morality and the behavior of Soviet youth, and their ideas were reflected in the party press and in literature intended to shape the formation of Soviet selfhood.[11]

From the 1880s to the 1930s, then, the intersecting disciplines of anthropology, psychiatry, psychology, criminology, jurisprudence, and sociology offered contemporaries a lens through which to examine the threats posed by rapid social change, the perceived spread of crime and subversion in Russia, and the challenges of building a unified society.[12] In their exploration of deviance and its causes, the practitioners of these disciplines constructed a conceptual framework within which social and individual disorders appeared both mutually constitutive and mutually sustaining. If individual deviants were powerfully shaped by their environments, they in turn wielded powers of destabilization and destruction that were awesome in their reach and intensity. The existence of unenlightened, irrational, defective, recalcitrant, and disruptive constituencies in late imperial society also represented a significant obstacle in the path of liberal reform. Accordingly, the human sciences came to articulate strategies for the attainment of national emancipation and harmony, predicated on the gradual transformation not simply of the empire's social, economic, and political structures but, ultimately, of its human population as well. This progressive vision of modernization and improvement provided the Bolsheviks with an arsenal of ideas that they then extended and radicalized into a program of massive and limitless social reconstruction.

For the most basic continuity between Late Imperial Russian liberalism and early Soviet radicalism lay in the assumption that human material could and should be remolded. Peter Fritzsche has remarked that "renovation and experimentation are distinctive modernist practices but because they presume both the extreme malleability and impermanence of the material world and are also often undertaken in conditions of apprehension, they can serve dangerously adventurous ends."[13] "Renovation," or

[9] Iu. Kasatkin, "Ocherk istorii izucheniia prestupnosti v SSSR," in *Problemy iskoreniia prestupnosti,* ed. Vladimir N. Kudriavtsev (Moscow, 1965), 187–225.

[10] Vladimir N. Tolmachev, ed., *Khuliganstvo i khuligany. Sbornik* (Moscow, 1929).

[11] See, e.g., Martyn N. Liadov, *Voprosy byta* (Moscow, 1925), 22, 28–9.

[12] Nancy Mandelker Frieden, *Russian Physicians in an Era of Reform and Revolution, 1856–1905* (Princeton, N.J., 1981); Mark B. Adams, "Eugenics as Social Medicine in Revolutionary Russia: Prophets, Patrons, and the Dialectics of Discipline-Building," in Solomon and Hutchinson, *Health and Society* (cit. n. 8), 200–23; Harley Balzer, "The Problem of Professions in Imperial Russia," in *Educated Society and the Quest for Public Identity in Late Imperial Russia,* ed. Edith W. Clowes, Samuel D. Kassow, and James L. West (Princeton, N.J., 1991), 183–98; John F. Hutchinson, *Politics and Public Health in Revolutionary Russia, 1890–1918* (Baltimore, 1990); Hutchinson, "Politics and Medical Professionalization after 1905," in *Russia's Missing Middle Class: The Professions in Russian History,* ed. Harley D. Balzer (Armonk, N.Y., 1996), 89–116; Hutchinson, "Tsarist Russia and the Bacteriological Revolution," *Journal of the History of Medicine and Allied Sciences* 40 (1985): 420–39; Hutchinson, "Who Killed Cock Robin? An Inquiry into the Death of Zemstvo Medicine," in Solomon and Hutchinson, *Health and Society* (cit. n. 8), 3–26; Susan Gross Solomon, "David and Goliath in Soviet Public Health: The Rivalry of Social Hygienists and Psychiatrists for Authority over the *Bytovoi* Alcoholic," *Soviet Studies* 41 (April 1989): 254–75; Solomon, "The Expert and the State in Russian Public Health: Continuities and Changes across the Revolutionary Divide," in *The History of Public Health and the Modern State,* ed. Dorothy Porter (Amsterdam, 1994), 183–223; Hachten, "In Service to Science and Society" (cit. n. 7).

[13] Peter Fritzsche, "Did Weimar Fail?" *J. Mod. Hist* 68 (Sept. 1996): 649.

ozdorovlenie (literally, healthification), was a mobilizing slogan of both liberal reformers and Soviet radicals during the revolutionary era.[14] Before and after 1917, modernizing elites struggled to articulate a vision of a harmonious and stable society, in part by identifying the forces impeding its emergence. These forces, it was argued, were embodied in the person of the deviant. Accordingly, the project of investigating, categorizing, and ultimately, transforming the deviant is to be understood as a central feature of ongoing attempts to renovate the social order.

DEGENERATIONAL ACCOUNTS OF CRIME

Scrutiny of the deviant in Russia began in earnest under the impact of a study published in Italy in 1876, *L'uomo delinquente,* by Cesare Lombroso (1835–1909). Although Lombroso's status as a pioneer in the field of criminology can be overstated, the publication of his study had such an impact on European criminology that no subsequent study over the following decades could ignore its precepts, even if only to condemn them.[15] During the 1870s, Lombroso had revived the study of phrenology, initially pioneered by the German physicians Franz Joseph Gall (1758–1828) and Johann G. Spurzheim (1776–1832), extending it into an evolutionary theory of racial development.[16] Lombroso applied this phrenology to a specific theory of evolutionary development that classified different social groups in a hierarchy of social "savagery." The phrenology of the criminal, which Lombroso identified in 1870, illuminated an evolutionary past quite different from those of his or her nondeviant contemporaries. Criminals languished in a primitive state of evolutionary development and were throwbacks to a long lost age of brutality, amorality, and anarchy, a state Lombroso termed "atavism."[17]

The local and immediate concerns of educated contemporaries inflected Lombroso's reception in Russia.[18] Russian criminologists and psychiatrists responded positively to Lombroso's insistence that the criminal, rather than simply the crime, should be the object of legal and medical attention. Many of them also accepted that there was a link between physical and moral deformity and that the born criminal,

---

[14] John Hutchinson has shown the currency of the slogan and its use by progressive zemstvo physicians keen to press for greater autonomy. It gradually came to serve as a synonym for democratic reforms. *Politics and Public Health* (cit. n. 12), xix–xx.

[15] Cesare Lombroso, *L'uomo delinquente studiato in rapporta alla anthropologia, alla medicina legale ed alle discipline carcerarie* (Milan, 1876). On Lombroso, see Daniel Pick, *Faces of Degeneration: A European Disorder, c.1848–c.1918* (Cambridge, 1989), 113–39; David Horn, *The Criminal Body: Lombroso and the Anatomy of Deviance* (London, 2003); Mary Gibson, *Cesare Lombroso and the Origins of Biological Criminology* (London, 2002). On Lombroso's reception in Germany, see Mariacarla Gadebusch Bondio, *Die Rezeption der kriminalanthropologischen Theorien von Cesare Lombroso in Deutschland in 1880–1914* (Husum, Germany, 1995); in France, see Laurent Mucchielli, ed., *L'histoire de la criminologie française* (Paris, 1994).

[16] F. J. Gall and J. C. Spurzheim, *Observations sur la phrénologie, ou la connaissance de l'homme moral et intellectuel, fondée sur les fonctions du système nerveux* (Strasbourg and Paris, 1810).

[17] Richard Wetzell, *Inventing the Criminal: A History of German Criminology, 1880–1945* (Chapel Hill, N.C., 2000), 30. For a scathing refutation of Lombroso's theories, albeit one that pays scant regard to the wider culture in which they were developed, see Stephen Jay Gould, *The Mismeasure of Man,* rev. ed. (New York, 1996), 151–73.

[18] Ironically, the works of Lombroso were not translated into Russian until after his influence had begun to wane. However, the vast majority of Russian psychiatrists and jurists had already read his works in the Italian original or in French translation. See Chezare Lombrozo, *Prestuplenie,* trans. G. I. Gordon (St. Petersburg, 1900).

posited by Lombroso, did exist. Yet this broad acceptance of certain features of Lombroso's theory coexisted with an uncompromising rejection of others. With the exception of a few dissenting voices, Russian experts were extremely skeptical about the existence of a separate criminal class, characterized by atavistic biological and psychological features. They were, from the outset, also reluctant to accept that criminality was biologically encoded in the individual and would manifest itself in destructive and illegal acts, independently of social context. In terms of its historical narrative, the notion of atavism abstracted the individual from the social context within which the crimes were committed, thereby removing any responsibility from the shoulders of society. It cast criminals as evolutionary freight, surplus to society's requirements and beyond the reach of societal influence. Such a theory sat uncomfortably with the liberal *mission civilatrice* of Russian educated society, centered as it was on a belief in the power of education and material improvement to transform society and its members.

In their rejection of atavism, Russian psychiatrists and criminologists turned to degeneration theory, which argued that individual biopsychological disorders could be acquired over a lifetime, then transmitted from one generation to the next in increasingly acute manifestations. Degeneration theory offered the psychiatrists and criminologists a structure within which environmental forces could be accommodated and, indeed, accorded an etiological primacy of sorts, while enabling the Russians to retain their emphasis on the biopsychological constitution of the criminal as a central causal factor in the etiology of deviance. In a major study published in 1890, *Psychopathological Types and Their Relationship with Crime and Its Different Forms,* Russia's leading criminal anthropologist, Dmitrii Andreevich Dril' (1846–1910), explained that degeneration could account for all the atavistic features of Lombroso's born criminal: "All the organs of the body are afflicted with degeneration, including the cranial and cerebral brain. Hypertrophy, atrophy, and the arrested development of individual organs and their complete disappearance [producing] deformities; those anatomical-physiological stigmata appear, on the basis of which Lombroso discerns "the born criminal."[19] A year later, Isaak Grigor'evich Orshanskii (1851–?) argued that the demonstration of diseased changes from a normal type over the course of a number of generations amounted to a stark refutation of Lombroso's theory of atavism: "there is no reason to see in that unfortunate group of people, standing at the frontier of health and illness, a mystical return to the past, so-called atavism. Rather we should see in them a socially diseased group . . . in essence, the debris of our culture, the victims of the harmful sides of civilization and the offspring of these victims."[20] Criminal anthropology in Russia retained this strong emphasis on the normative influence of environmental forces throughout the remaining years of the autocracy.

Degenerative pathologies of the deviant might have been firmly rooted in the contemporary social order, but they still set him or her apart from the normal, healthy majority of human beings. This difference came to be defined in terms of a criminal predisposition, often characterized by pathologies of the will. Pavel Isaevich Liublinskii (1882–1938), a law professor at St. Petersburg University, argued in 1912 that

[19] Dmtrii A. Dril', *Psikhofizicheskie tipy i ikh sootnoshenii s prestupnost'iu i ee raznovidnostiami* (Moscow, 1890), 15.

[20] Isaak Grigor'evich Orshanskii, *Nashie prestupniki i uchenie Lombrozo* (St. Petersburg, 1891), 17.

degeneration led to a weakening of willpower and an increased susceptibility to vice: "the majority of the representatives of this criminal class are not susceptible to moral or intellectual pressures, are deprived of willpower and the capacity to restrain themselves, such that even knowing of the criminal nature of their actions, they are unable to resist."[21] Three years later, the criminologist Susanna Al'fonsovna Ukshe (1885–?) echoed these comments, affirming the existence of

> a group of degenerates, who are unbalanced in the area of willpower, and who frequently enter into conflict with the law. There can be no doubt that the power of an individual to resist negative external conditions is directly proportionate to the degree of perfection of his psychophysical organization [*psikhofizicheskaia organizatsiia*], and that a person who is sick and unbalanced will find it extraordinarily difficult to orient his activities in accordance with the law.

For Ukshe, the quality of an individual's "psychophysical organization" was the most eloquent indicator of his or her prospects for leading a life of normality and legality.[22]

The notion of a criminal predisposition allowed Russian criminologists and psychiatrists to soft-pedal the notion of radical biopsychological otherness that underpinned Lombroso's understanding of crime. Criminals were indeed distinct from the mass of healthy individuals, but the differences might often remain latent unless catalyzed into action by social forces. When attempting to explain the conversion of a predisposition to criminality into its active pursuit, theorists often invoked Darwin's metaphor of the "struggle for existence." Pavel Ivanovich Kovalevskii (1849–1923), professor of psychiatry at the University of Kharkov, observed in 1903 that the children of those afflicted with the degenerative condition were born with "weak, defective, debilitated psychophysical constitutions. When they . . . take part in the struggle for existence . . . [t]hey are unable to find sufficient strength within themselves for a determined struggle for the right to live and so they swiftly embark on the slippery path of vice and crime."[23] Lev Grigor'evich Orshanskii explained that "criminals suffer from all the diseases that erode the health of the mass of the people, but they suffer to such an extent that they necessarily become unable to adapt to the given order of things."[24] Society bore a historical responsibility both for the original genesis of criminal predispositions and for allowing conditions to obtain that then converted this predisposition into criminal activity.

If the defective biopsychological constitution of the criminal was ultimately traceable to defects in the organization of the social order, crime itself became a symptom of broader social pathologies, as Dril' evocatively wrote:

> Young and adult criminals are born and live in the midst of society; they are its reflection; they are "the flesh of its flesh, and the bones of its bones." Under the influence of peculi-

[21] Pavel I. Liublinskii, "Novaia mera bor'by s vyrozhdeniem i prestupnost'iu," *Russkaia mysl,'* no. 3 (1912): 53.

[22] Susanna A. Ukshe, "Vyrozhdenie, ego rol' v prestupnosti i mery bor'by s nim," *Vestnik obshchestvennoi gigieny, sudebnoi i prakticheskoi meditsiny,* no. 42 (June 1915): 808, 800.

[23] Pavel I. Kovalevskii, *Vyrozhdenie i vozrozhdenie: Prestupnik i bor'ba s prestupnostiiu (Sotsial'no-psikhologicheskie eskizy),* 2nd ed. (St. Petersburg, 1903), 287.

[24] Lev G. Orshanskii, *Kriminal'naia antropologiia i sudebnaia meditsina* (St. Petersburg, 1903), 25.

arities of the living conditions created by society, their peculiar natures are prepared and develop, pushing them, under certain circumstances, to crime.[25]

Indeed, implicit in this entire body of criminological writings was the notion that social forces were to blame for the genesis of criminal impulses. Yet the buffer of heredity enabled many theorists to sidestep the direct question of contemporary society's complicity in the existence of criminality. References to social hardship or the challenges of industrial modernity were discussed but without the identification of a particular agency that could mitigate them. By the turn of the twentieth century, however, clinicians were ever more explicit in their articulation of a language not simply of social causation but of social responsibility.

Kovalevskii insisted, for example, upon the essential indivisibility of the criminal and the social order in language that raised the issue of responsibility rather than simply causation:

> We are the flesh of our parents' flesh, we are the blood of the society's blood in which we lived and were raised. . . . It is true that criminals are guilty of violating society's interests. But are not the parents, who gave birth to them with a nervous system that expressed a malevolent will, guilty? Is not the society, which permitted their education in an environment of corruption, drunkenness, poverty, prostitution, thievery, and so on, guilty?[26]

Several factors explain this concentration on social influences at the root of biopsychological anomalies in the criminal. First, there was a discernible theoretical impulse within degenerational accounts of deviance to investigate the relationship between individual disorders and the wider society in which they had emerged. Second, widespread ambivalence about the empire's experience of modernization tended to generate arguments that expressed dissatisfaction or disdain for the social conditions of contemporary Russia. These grew more voluble over the course of the late imperial period as the forces of capitalism and urbanization began to change the face of Russian society and generate concentrations of poverty and squalor in Russia's cities.[27]

Perhaps most important, however, society's responsibility for crime also signaled its power ultimately to eliminate it. Russian clinicians declared the contemporary social order to be the source of the disease of crime; social reform the cure. In his attack on the Lombrosian school in 1890, Petr Narkizovich Obninskii (1837–1904) condemned the "hopeless theory of born criminality," contending that the new emphasis on the criminogenic properties of the environment was

> a bright and cheerful [theory], which once again opens up to the confused practitioners [of the legal system] who had been condemned to inaction a broad range of possibilities of fruitfully taking measures against those factors and stimulants of criminality. The latter have now been born out of the narrow immobile sphere of the physical imperfections

---

[25] Dmitrii A. Dril', *Prestupnost' i prestupniki (ugolovno-psikhologicheskie etiudy)* (St. Petersburg, 1895), 247.

[26] Kovalevskii, *Vyrozhdenie i vozrozhdenie* (cit. n. 23), 238.

[27] See, e.g., Evstafii M. Dement'ev, *Fabrika: chto ona daet naseleniiu i chto ona u nego beret* (Moscow, 1893), 225–40; Tat'iana Vladimirovna Boiko, *Rabochie Rossii i kul'tura: Polemika na stranitsakh konservativnoi i liberal'noi periodiki nachala XX veka* (Moscow, 1997).

and deviations of a given criminal type into a realm of extra-personal and social phenomena, with which it is possible to conduct a strategic and deliberate struggle and which
it is possible to eliminate.[28]

Writing a decade later, Kovalevskii articulated the nature of this struggle. Parental
and social responsibility for the actions of criminals conferred not merely "a duty on
society to ensure that these people do not harm it, that they are removed from its midst;
society's obligation is also to correct those unfortunate individuals and to turn them
into law-abiding and useful citizens. But a greater, a higher, and an even more binding obligation is to renovate [ozdorovlenie] society itself, so that it does not give birth
to and raise such deformed and unfortunate members."[29]

The application of degeneration theory to the genesis of crime thus held out the possibility of its eventual eradication. If all criminal deviance was understood ultimately
to be the result of an unhealthy social environment, then the theory contained an important utopian urge for prophylactic improvement. That the theory should enjoy such
a purchase on the minds of contemporary Russian scientists, legal theorists, and clinicians is not surprising given the ethos of the professional disciplines in the post-
emancipation era (post-1861). Support for strategic reform of the social order in the
interests of a biopsychological improvement of its constituent members was central
to the essential Enlightenment belief in human perfectibility that characterized so
much of Russian progressive thought in the second half of the nineteenth and early
twentieth centuries. To condemn an entire criminal class of Russians to moral and
social oblivion would have been an admission, at least in part, of the limited effects of
social reform.

THE "SOCIALLY DANGEROUS"

Yet the theoretical reinforcement of the ethical impulse to reform still left an important issue unaddressed. The human sciences necessarily had to confront the problem
of what to do with those individuals whose moral and mental faculties were so de-
graded that they posed a threat to the health of society. Not only did these deviants
threaten to inseminate broader society with their diseased heredity, but they also posed
an immediate danger to the well-being of their contemporaries. Across the continent,
the classical deterrence theory, which posited an essentially rational criminal, making
calculations based upon self-interest before engaging in criminal activity, gradually
gave way to a theory of punishment based on the concept of social defense, pioneered
in Germany in the 1890s.

Richard Wetzell has traced the increasing influence of a utilitarian idea of protecting society championed by a reformist movement within the German legal establishment. Spearheaded by the law professor Franz von Liszt (1851–1919), the movement charged that the existing German penal system, with its fixed prison sentences
prescribed by the penal code, was ineffective in protecting society from crime. Liszt
and his fellow reformers argued that this goal could best be achieved by a criminal
justice system in which the punishment fit the criminal rather than the crime. Accordingly, they proposed a system of individualized punishments, in which first-time

[28] Petr N. Obninskii, "Illiuzii pozitivizma," *Zhurnal grazhdanskogo i ugolovnogo prava*, March
1890, 2.
[29] Kovalevskii, *Vyrozhdenie i vozrozhdenie* (cit. n. 23), 239.

offenders would receive suspended sentences for the purpose of specific deterrence, repeat offenders who seemed both in need of, and capable of, rehabilitation would be subject to a prison sentence for the purpose of rehabilitation; and repeat offenders who seemed incorrigible would be subject to indefinite detention for the sake of incapacitation.[30]

Liszt's influence in Russia was substantial; his works were translated and eagerly discussed by Russian criminologists and psychiatrists.[31] In 1901, for example, the criminologist Bronislav I. Vorotynskii summarized Liszt's findings that existing penal policy made no impact on the rate of recidivism and even encouraged it: "such negative results from the application of the existing system of penal repression flow from the fact that we currently judge and punish the crime, rather than the criminal according to his crime, as the new positive school demands."[32]

It should not be surprising that Lombroso's disciples were at the forefront of calls for the introduction of social defense.[33] More surprisingly, however, as Robert A. Nye and Ruth Harris have each noted, the idea of social defense became very popular and influential in France, where the medicolegal establishment had spearheaded the onslaught against Lombroso's ideas of fixed hereditary criminality in the 1880s and the 1890s.[34] In Russia, a similar paradox emerged in which the very psychiatrists and jurists who had embraced degenerational accounts of the etiology of crime endorsed social defense not merely as an unsavory social necessity but as something positively in the interests of the deviant.

The Pan-European context that saw the rise to prominence of the idea of social defense coincided with the explosion of social disorder and political violence that accompanied Russia's entry into the twentieth century. Joan Neuberger has examined the "moral panic" surrounding the issue of hooliganism in the period from 1900 to the outbreak of the First World War. Educated Russian society was shocked by the combination of easy violence and lack of social deference that they encountered on the streets of their towns and cities at the hands of lower-class men.[35] There was also an enormous increase in politically motivated crime, beginning with the revival of the terrorist campaign, initially directed against servitors of the state by a collection of left-wing extremists and culminating in the inchoate violence of the 1905 Revolution

[30] Wetzell, *Inventing the Criminal* (cit. n. 17), 75.

[31] Frants von List [Franz von Liszt], *Nakazanie i ego tseli* (St. Petersburg, 1895); Liszt, *Prestuplenie, kak sotsial'no-patologicheskoe iavlenie* (St. Petersburg, 1903); Liszt, *Uchebnik ugolovnogo prava: Obshchaia chast'*, trans. from 12th German ed. by R. El'iashchevich (Moscow, 1903). See also the enthusiastic review of Liszt's ideas of social defense in Vladimir V. Przheval'skii, *Prof. Frants List i ego vozzreniia na prestuplenie i nakazanie* (St. Petersburg, 1895).

[32] Bronislav I. Vorotynskii, "Biologicheskie i sotsial'nye faktory prestupnosti," *Zhurnal Ministerstva Iustitsii*, no. 7 (Sept. 1901): 21–2.

[33] In his influential *Criminal Sociology,* which was first published in 1897, Enrico Ferri (1856–1929) declared that "punishment ought not to be the visitation of a crime by retribution, but rather a defence of society adapted to the danger personified by the criminal." Enrico Ferri, *Criminal Sociology* (New York, 1897), 208. For translations of Ferri's work, see, e.g., Ferri, *Prestuplenie i prestupniki v nauke i v zhizhni* (Odessa, 1890); Ferri, *Ugolovnaia sotsiologiia,* with a foreword by E. Ferri and Dmitrii A. Dril', trans. from 4th Italian ed. (St. Petersburg, 1910–12).

[34] Ruth Harris, *Murders and Madness: Medicine, Law, and Society in the Fin de Siècle* (Oxford, 1989), 3. For Harris's account of "social defence" in fin de siècle France, see 105–24. Robert A. Nye, *Crime, Madness, and Politics in Modern France: The Medical Concept of National Decline* (Princeton, N.J., 1984), 191.

[35] Joan Neuberger, *Hooliganism: Crime, Culture, and Power in St. Petersburg, 1900–1914* (Berkeley, Calif., 1993).

and the two bloody years of virtual civil war that raged across the empire until 1907.[36] Anna Geifman has calculated that during a one-year period beginning in October 1905, a total of 3,611 government officials of all ranks were killed and wounded throughout the empire.[37]

Such an explosion of criminality concentrated minds in Russia, as elsewhere in Europe, on the question of whether the existing legal system was adequate to the task of containing the disorder.[38] Criminal anthropology set great store by its transcendence of the notions of retribution and punishment and its concern with what Dril' termed "the necessity of protecting society from the evil of crime."[39] Vorotynskii declared that the great achievement of the criminal anthropologists was that they "positively demonstrated that we should not see crime as an abstract juridical concept but as a pathological feature of social life." A medicalized understanding of criminal deviance logically invoked a language of prophylactics and therapy when identifying the penal measures society might take to protect itself from a particularly inveterate criminal class:

> No one disputes any longer that there do exist criminals with morally corrupted natures, individuals who are mentally deformed from birth. Their criminal activities are determined by their unfortunate heredity, their deformed psychophysical constitution. These individuals should not be punished but rather corrected or cured; they should not be terrorized with severe penalties but simply isolated, as it were, removed from everyday life with a view to protecting society from their harmful influence.[40]

As a consequence, Vorotynskii argued a few years later, "the struggle with crime should express itself primarily in terms of measures of a preventive character."[41] Kovalevskii agreed that the fight against crime should not be directed only toward the removal of those social causes of degeneration, such as poverty and alcoholism. "Born criminals with inherited diseased natures" should, "in view of the extreme danger they pose to the wider society, be isolated from the entire world in special institutions, in special prisons."[42]

Isolation was generally framed in paternalistic terms, as a removal of the deviant from the unhealthy or demanding environment that fostered the sustenance—and the genesis—of the diseased criminal condition. In an article published in 1912, Gustav Vladimirovich Reitts (1876–?), a doctor at the hospital of St. Nikolai the Miracle Worker in St. Petersburg, argued that the existing prison system was in no position to deal with pathological criminals because "the source of their conflict with society lies

---

[36] For descriptions of the massive levels of violence and social disorder, see Abraham Ascher, *The Revolution of 1905: Russia in Disarray* (Palo Alto, Calif., 1988).

[37] Anna Geifman, *Thou Shalt Kill: Revolutionary Terrorism in Russia, 1894–1917* (Princeton, N.J., 1993), 21. See also Jonathan Daly, "Political Crime in Late Imperial Russia," *J. Mod. Hist.* 74 (March 2002): 62–100.

[38] For an institutional history of the Russian penal system, see Peter H. Solomon, ed., *Reforming Justice in Russia, 1864–1994: Power, Culture, and the Limits of Legal Order* (New York, 1996); Bruce F. Adams, *The Politics of Punishment: Prison Reform in Russia, 1863–1917* (DeKalb, Ill., 1996).

[39] Dmitrii A. Dril', "Chto govorilos na mezhdunarodnom ugolovno-antropologicheskom kongresse v Briussele," *Russkaia mysl,* Feb. 1893, 60.

[40] Bronislav I. Vorotynskii, "Psikho-fizicheskie osobennosti prestupnika-degenerata," *Uchennye zapiski Kazanskogo Universiteta,* 1900, book 3, 101.

[41] Vorotynskii, "Biologicheskie i sotsial'nye" (cit. n. 32), 24.

[42] Kovalevskii, *Vyrozhdenie i vozrozhdenie* (cit. n. 23), 327, 356.

within them and is hereditary, unresponsive to external influences; the current penal system is designed with normal, psychologically healthy individuals in mind, not the ill and half-deranged." The answer, for Reitts, was the construction "of a specific and artificial environment, adapted to the individuality of the criminal, in which all the harmful influences of contemporary life would be removed and in which the subject could exist without harming either himself or others." Yet, paradoxically, it was precisely such a benevolent approach to the treatment of the individual deviant that justified "not a brief but a prolonged period of residence in this environment."[43]

Indeed, in many quarters, part of the ethical justification for indefinite preventive detention was paradoxically a continuing belief in the essential reformability of deviants. The criminologist Aleksandr Aleksandrovich Zhizhilenko (1873–?) proposed the construction of labor colonies "in which dangerous recidivists could be accommodated. In as much as these individuals show themselves to be accessible to corresponding forms of influence, we should not lose hope that they might be restored to an honest way of life and therefore the regime in these institutions should be organized such that it allows for their possible correction."[44]

Even the eminent psychiatrist Vladimir Mikhailovich Bekhterev (1857–1927), one of the most eloquent critics of the iniquities of the tsarist social order, affirmed in 1912 that although society's culpability for the genesis of deviance deprived it of the right to punish criminals, it was entitled to "isolate" those who posed a major threat to its well-being: "in some cases in the interests of education and in others in the interests of treatment but under no circumstances in order to punish. . . . If there is nothing from which the criminals can be cured or nothing about which they can be educated, as in the case of political crimes, then isolate them if it is helpful, but do not hang them and execute them: you do not have the right!"[45]

Some criminologists and psychiatrists were more alive to the risks of social defense. In his paper to the Eighth General Meeting of the Russian Group of the International Union of Criminologists in 1910, the eminent jurist Vladimir Dmitrievich Nabokov (1869–1922) cautioned against the indiscriminate use of the term "socially dangerous" in the absence of a strengthened *Rechtsstaat* in Russia. He reminded his audience that Russians were all too familiar with the state's penchant for labels such as "unreliable" and "individuals harmful for public peace" and the "harmful and corrupt members of village societies." "We know," he declared, "what preventive measures can turn into when applied to individuals deemed by one or other institution to be 'dangerous.'" As a consequence, the notion of "socially dangerous" could not be allowed to shape penal policy in the absence of a consolidation of individual rights within society, otherwise the result would be "a deformed deviation from the function of a *Rechtsstaat*."[46] Yet even Nabokov remained convinced that the strategic, precise, and measured use of the term "socially dangerous" would minimize the opportunity

[43] Gustav V. Reitts, "Patologicheskaia prestupnost i umen'shennaia vmeniaemost'," *Sovremennaia psikhiatriia,* June 1912, 439.

[44] Aleksandr A. Zhizhilenko, "Mery sotsial'noi zashchity v otnoshenii opasnykh prestupnikov," *Pravo,* pt. 1, no. 35 (1910): 2078–91; pt. 2, no. 36 (1910): 2136–43; pt. 3, no. 37 (1910): 2167–77, on 2171.

[45] Vladimir M. Bekhterev, *Ob "ektivno-psikhologicheskii" metod v primenenii k izucheniiu prestupnosti* (St. Petersburg, 1912), 64.

[46] Vladimir D. Nabokov, *Ob "opasnom sostoianii" prestupnika kak kriterii mer sotsial'noi zashchity* (St. Petersburg, 1910), 29–30.

for its abuse and would represent a significant advance on the current clumsy and indiscriminate penal policy that fostered recidivism in the empire's prisons on a massive scale.[47]

Two apparently contradictory theoretical impulses emerged from the application of
degeneration theory to criminality in late imperial Russia. The first was an attempt to
identify threats to social stability and progress represented by those whose biopsychological predisposition impelled them to engage in criminal acts detrimental to the
health of the body social. The second was an acknowledgement that society was ultimately the source of these disorders; the body social that Russian psychiatrists and
criminologists sought to defend from deviance had itself spawned the criminal delinquents that threatened it. These arguments resulted in calls for two related measures,
each indebted to an abiding belief in the capacity of rationality and science to reorder
not only the structures of society but also the biopsychological material that constituted its population. The first measure was reform of the socioeconomic conditions
understood to unleash the degenerative syndrome and then to drive biopsychologically weakened individuals to acts of crime. The second measure was the isolation of
those with a criminal predisposition in the interests of collective health and the introduction of preventive detention justified in a language of medical prophylactics and
therapy. Both theoretical urges, the reform of society and the isolation of the deviant,
would find their apotheosis in the revolutionary project of the Bolsheviks.

REVOLUTIONARY ADAPTATION

After the Russian Revolution, the human sciences continued to regard deviance as the
consequence of failure to adapt to the surrounding environment. This assessment was,
of course, familiar from prerevolutionary studies of criminality, which had stressed
that individuals whose biopsychology was impaired were unable to compete legally
in the struggle for existence. The criminologist Evgenii Konstantinovich Krasnushkin
(1885–1951) noted that "every living creature needs to adapt its constitution and its
reactions to the surrounding environment since this adaptation allows it to preserve
itself in this environment. When this capacity to adapt is disturbed or weakened, a diseased condition comes into being."[48] One of Krasnushkin's colleagues, Aleksandr
Karlovich Lents (1882–1952), similarly maintained that "in order to adapt to the surrounding environment an individual needs the constant formation in his cerebral cortex of those conditional reflexes and conditional inhibitors, which are the organism's
strategic responses to changes in the external circumstances." He defined psychopaths
as "individuals with an unstable nervous system who have difficulty in adapting to
their social environment and easily come into all manner of conflicts with it . . . Their
chief characteristic is *social inadaptability.*"[49]

In the wake of 1917, however, the concept of adaptation began to acquire an addi-
tional and powerful resonance. The revolution itself had radically redefined the terms
on which society functioned and the nature of the struggle for existence within it. This
narrative of revolutionary adaptation addressed prominent concerns within the new
regime about the need to overcome the legacy of the past. The issue of adaptability to

---

[47] Ibid., 36–7.
[48] Evgenii K. Kranushkin, *Sudebno-psikhiatricheskie ocherki* (Moscow, 1925), 103.
[49] Aleksandr K. Lents, *Kriminal'nye psikhopaty (sotsiopaty),* with a foreword by V. Osipov
(Leningrad, 1927), 52, 24 (emphasis in original).

the postrevolutionary conjuncture thereby became a gauge by which to measure the danger posed by various social constituencies in Soviet society.[50] In 1926, a psychiatrist of prerevolutionary standing, Viktor Petrovich Osipov (1871–1947), declared: "The October Revolution was an enormous leap forward, which immediately and with dizzying speed propelled the country through a series of evolutionary forms. New demands on the behavior of citizens emerged together with the new order; the great majority of citizens accepted these demands, accepted them seriously, and began to adapt to them, elaborating new [reflexive] combinations that were to force out and replace the old ones."[51]

In a similar argument, Udal'tsov believed the entire revolutionary project rested upon its success or failure in facilitating the adaptation of individuals to the new social order. The construction of the army was analogous to this broader process, as he explained in a passage worth citing at length:

> The army draws recruits from a huge mass of people with the most diverse types of reaction. The same set of demands confronts them all. What is required is a swift and complete adaptation to new environmental conditions, which represent a series of powerful stimulants. To a degree, the individuality of the person is suppressed, and there are great demands for the development of a social reflex; the individual is sacrificed to the interests of the collective. As a result of both this break in the old, habitual foundations of human behavior, acquired over the course of a lifetime under the system of reaction, and the attempt to generate a new knowledge, a new chain of conditional reflexes, . . . we encounter an entire series of emerging conflicts.[52]

Biopsychological explanations of mental illness and deviance came to provide a model for discussions of conventional criminality, revealing an explanatory scope that acknowledged no boundaries. In a study of swindlers published in 1927, Mikhail Aleksandrovich Chalisov argued that their crimes were biologically embedded in their constitutions: "a significant percentage are characterized by alcoholic heredity, instability, suggestibility, which means a certain insufficiency [*nepolnotsennost'*]." This biological deficiency was only compounded by the experience of the early years of Soviet power:

> After the civil war with its devastation, . . . people had to adapt to the new circumstances. . . . In addition, after a long period in which the restraining influences dissipated, new

---

[50] It also enabled the human sciences to respond to the frustratingly elusive nature of class designations in the postrevolutionary era. Sheila Fitzpatrick has noted how the party struggled to "reclassify" society in the wake of the upheavals of the revolution and civil war, which had led so many individuals, willingly or unwillingly, to abandon their prerevolutionary stations and professions, only to reappear under an entirely different guise in the early years of the Soviet regime. The issue of revolutionary adaptation constituted a theoretical means of scrutinizing each Soviet citizen, of peering behind his or her occupation to discern "true" class affiliation, evidenced by the individual's psychological reflexes and social instincts. Sheila Fitzpatrick, "The Problem of Class Identity in NEP Society," in *Russia in the Era of NEP: Explorations in Soviet Society and Culture,* ed. Sheila Fitzpatrick, Alexander Rabinowitch, and Richard Stites (Bloomington, Ind., 1991), 12–33; Fitzpatrick, "Ascribing Class: The Construction of Social Identity in Soviet Russia," *J. Mod. Hist.* 65 (Dec. 1993): 745–70.

[51] Vladimir Osipov, "O kontrrevoliutsionnom komplekse u dushevno-bol'nykh," *Obozrenie psikhiatrii, nevrologii i refleksologii,* no. 2 (1926): 91. At the time of writing, Osipov was professor of psychiatry at Kazan University and the Military Medical Academy in Leningrad. One of the most influential psychiatrists in the Soviet Union, he went on to become a member of the Soviet Academy of Sciences in 1939.

[52] Udal'tsov, "Pravonarusheniia v voiskakh" (cit. n. 1), 119–20.

ones had not yet managed sufficiently to establish themselves in order to direct tendencies and urges along a particular channel. Only individuals with sufficient biological resources were able to develop new restraining mechanisms relatively quickly; biologically deficient individuals, such as our swindlers, were unable . . . to develop these useful mechanisms quickly enough.[53]

Individual adaptation was one issue; class adaptation, another. Lents pointed out that the sustained influence of a given milieu, especially in early childhood, served to unify the biopsychology of members of a particular class: "from the first year of life, every one of us is subject to the formative influence of the people around us, especially parents and teachers. In games with other children, the growing child gains another wave of social influences, every one of which leaves a certain mark in the nervous system." As a consequence, "in each individual there is a certain fairly extensive group of features which bind him together with people of the same sex, class, nationality, profession, and epoch."[54] So far, so good. But such a view could easily shade into a blanket denunciation of the failure of certain social constituencies to adapt to the revolution, as Krasnushkin declared in the same year: "the offspring of the bourgeoisie of the tsarist era, déclassé by the revolution, are frequently biologically defective/of poor quality [*nepolnotsennyi*]."[55]

Such a model of revolutionary adaptation lent itself to an analysis not simply of conventional criminality but also of politically motivated opposition to the regime. A rejection of the values and practices of the new Soviet order could be interpreted as a biopsychologically determined failure to adapt to them, not unlike the adaptive failures of common criminals. Such was the view Osipov put forward in an article titled "The Counterrevolutionary Complex in the Mentally Ill," published in 1926 in the authoritative *Review of Psychiatry*. Like Orshanskii and Udal'tsov, Osipov observed that the experience of revolution had unleashed atavistic forces in those who had experienced its tumult and destruction:

> The turbulent early years of Soviet power represent a good touchstone against which to measure changes in the stability of the social cortical superstructures [*nadstroika*]. . . . Under the influence of the different conditions of that time, a great many people's psychological combinations, linked with their hereditary reflexes, began to dominate over the later cortical architecture, and they were reduced to their primitive state, revealing behavior of which even people who had known them would have not thought them capable. Those difficult times have changed, and together with them, these people, although by no means all, once again acquired their cortical image; yet many had sunk to such depths that they were no longer able to raise themselves to the necessary cortical level.[56]

The distinction between those individuals able to recover psychologically and those who were not lay in their biological constitution: "only physiologically, biologically sound people succeeded in preserving themselves in this purifying fire of revolution and to walk out of the revolutionary flames even more stable than they were before of it." Osipov invoked the authority of physiologist Ivan Petrovich Pavlov

[53] Mikhail A. Chalisov, "Opyt bio-sotsial'nogo obsledovaniia rastratchikov v Rostov n-D," in *Voprosy izucheniia prestupnosti na Severnom Kavkaze* (1927), 2:74.

[54] Lents, *Kriminal'nye psikhopaty* (cit. n. 49), 9.

[55] Evgenii K. Kranushkin, "Chto takoe prestupnik," in *Prestupnik i prestupnosti: Sbornik I* (Moscow, 1926), 26.

[56] Osipov, "O kontrrevoliutsionnom komplekse" (cit. n. 51), 86.

(1849–1936) to explain the psychophysiological responses of some individuals to the revolutionary changes: "When changes to the surrounding environment are too abrupt, and the nervous system is deluged by an entire flood of new, unexpected, unfamiliar stimuli, which destroy the old orientation but which still require time in order to consolidate themselves, I. P. Pavlov has argued that confusion, distraction, [and] agitation ensue." Osipov explained the opposition of certain social groups to the new order with reference to their biopsychological condition on the eve of revolution:

> Those who were ideologically well prepared for the Revolution were not so adversely affected by the changes. Yet other people, whose old reflexive combinations had established themselves so firmly that the new reflexes differed from them too markedly. . . required too long a period for their education; examples of these people are to be found in extreme conservatives . . . Their inability to change their old combinations of psychological reflexes often expressed itself in counterrevolutionary behavior.[57]

Osipov explained divergent political responses to the October Revolution with reference to the biopsychological condition of different social groups: "in the revolutionary period, a series of political configurations emerge—each social group assimilates the new stimuli in as far as its preparation, its [reflexive] combinations that were established earlier, allow."[58] Political opposition to the October Revolution was thus a reflection of the biopsychological divergence of groups of Russian citizens from the values and practices in the prerevolutionary era. Osipov's thesis might have been controversial, but in an important sense, it simply stretched to their logical conclusion the arguments put forward by the majority of his contemporaries: an individual's or group's experiences over a sustained period of time combined with hereditary material to produce biopsychological changes in the constitution. These changes then determined the person's responses to environmental changes and facilitated or restricted his or her adaptive capacity.

COERCIVE HEALING: THE CIVILIZING PROCESS

Theories of psychopathology thus generated a thoroughgoing medicalization of criminality in the early years of the Soviet regime. The criminologist Timofei Efimovich Segalov argued that "the deprivation of freedom is understood in terms of isolation as a measure of social defense, and that naturally in such an understanding, isolation in a corrective-labor institution and isolation in an asylum are of equal value and meaning."[59]

Prior to 1917, deep disquiet about the biopsychological damage inflicted by late tsarist capitalism tended to mitigate the force of arguments in favor of individual biopsychological factors. Not only did sustained exposure to the depredations of the factory and the slum weaken individuals but the brutal nature of the struggle for existence then denied them any opportunity to compete legally as well. In the arguments

[57] Ibid., 86–7.
[58] Ibid., 87.
[59] Timofei E. Segalov, "Mery sotsial'noi zashchity meditsinskogo kharaktera po UK 1926 g.," in *Problemy prestupnosti,* no. 3, ed. Evsei G. Shirvindt, F. Traskovich, and M. Gernet (Moscow, 1927), 174. For a discussion of the notion of "social danger" and its place in Soviet jurisprudence, see Iu. I. Liapunov, *Obshchestvennaia opasnost' deianiia kak universal'naia kategoriia sovetskogo ugolovnogo prava* (Moscow, 1989); Paul M. Hagenloh, " 'Socially Harmful Elements' and the Great Terror," in *Stalinism: New Directions,* ed. Sheila Fitzpatrick (London, 2000), 288–90.

that flowed from this analysis, social defense was understood in terms of a holding operation intended to contain criminality while the essential project of transforming the social order and its capitalist economic foundations was undertaken. In the wake of 1917, the second component of this general response to crime was deemed essentially accomplished; social transformation was already under way. More important, by the 1920s, liberal sympathy for the plight of the economic and social losers under prerevolutionary capitalism had mutated into entrenched hostility to those who appeared marked by its influence. Deviants would no longer charitably be deemed unable to compete within a bruising and unequal struggle for survival. Failures to adjust to the new order now were evidence of an inability to adapt to the revolution itself, of an antipathy toward its values. Under these conditions, what had before 1917 amounted to a liberal plea for reform and tolerance based upon a twinned strategy of social defense and social change was reconfigured within the deeply discriminating discursive framework of Soviet Marxism.

As the old order was institutionally dismantled, the civil war won, and the former class disenfranchised, exiled, or slain, statements to the effect that criminality flowed from both endogenous and exogenous factors harbored an increasing tendency to refer to the former at the expense of the latter. In a study of youth crime published in 1923, Liublinskii, still a professor of law at Petrograd State University, argued that individuals who had failed to respond to the curative powers of the revolution or had become corrupted in its wake could no longer expect the indulgence of the authorities: "Already developed defectiveness and antisocial behavior demand a more complex and specialized intervention, which cannot be achieved with the usual medico-pedagogical methods, but by a long process of reeducation, correction, and social isolation."[60] Osipov similarly argued that not all hooligans were to be seen as merely the unfortunate victims of their corrupting environments and employed categories familiar from degeneration theory to assert the need for the "coercive healing" of the hooligan: "The hooligan is created by the conditions of the surrounding environment, by those circumstances that evolved unfavorably. Yet among the hooligans there are also clearly pathological types—morally defective, unbalanced, degenerate individuals with various manifestations of diseased deviations from the norm."[61]

This insistence on importance of the biopsychological constitution of the individual as a primary causal factor in the etiology of crime served additionally to blur the distinction between actual and potential perpetrators. In 1929, for example, Ivan Nikolaevich Vvedenskii (1875–1960) declared that "the criminal act itself is irrelevant—what matters are attitudes and tendencies that pose a threat to the health of society."[62] The criminologist Boris Samoilovich Utevskii (1887–?) came in 1927 to the view that "there is no basis upon which to distinguish between punishment for a crime that has just been committed by a dangerous criminal and preventive detention, served after the punishment."[63] This conflation of actual and potential criminal was central to

[60] Pavel I. Liublinskii, *Bor'ba s prestupnost'iu v detskom i iunosheskom vozraste (sotsial'no-pravovye ocherki)* (Moscow, 1923), 255.

[61] Vladimir Osipov, "K voprosu o khuliganstve," in *Khuliganstvo i prestuplenie: Sbornik statei,* ed. Lev G. Orshanskii, Aleksandr A. Zhizhilenko, and I. Ia. Derzibashev (Moscow, 1927), 88.

[62] Ivan N. Vvedenskii, "Prinuditel'noe lechenie dushevno-bol'nykh i psikhopatov," in *Dushevno-bol'nye pravonarushiteli i prinuditel'noe lechenie,* ed. V. Gannushkin (Moscow, 1929), 13, 19.

[63] Boris S. Utevskii, "Retsidiv i professional'naia prestupnost'," in Shirvindt, Traskovich, and Gernet, *Problemy prestupnosti* (cit. n. 59), 107.

the development of the legal concept of "social danger." Mark Zakharovich Kaplinskii observed that "in order to decide the question of social danger and take measures of social defense of a medical character, we must on every occasion take an individually detailed approach to each case, considering not merely the manifestations of social danger but also the potential urges and capacities fixed within the individual."[64]

The concept of "social danger" and suggestions for a corresponding raft of legal policies known as "social defense" had come into vogue in Europe in the first decade of the twentieth century in legal and psychiatric circles. These precepts were implemented in the first two legal codes of the new Soviet state. Dmitrii Aleksandrovich Amenitskii (1875–?) applauded the 1922 legal code for its recommendation of the incarceration of "mentally and morally defective" criminals in special institutions as a form of social defense.[65] Like their prerevolutionary predecessors, many Soviet psychiatrists and criminologists hailed the replacement of the concept of criminal responsibility with that of social danger as a progressive development in the interests of both the criminal and wider society. Udal'tsov declared that "we have to welcome in every way the concept of social danger, which has been able to replace the concept of criminal responsibility. People who manifest antisocial reactions, and are therefore the enemies of society, should be isolated with the aim of coercively healing or re-educating them."[66] Osipov boasted:

> Our legal system has assimilated this biosocial perspective [*biosotsial'naia tochka zreniia*], the logical consequence of which is a root and branch change in the approach to the question of *punishment* for crime. The concept of punishment, as with retribution for the violation of social and personal interests, has been replaced by the assertion of measures of *social defense* from harmful and dangerous antisocial elements.[67]

Accordingly, criminology and psychiatry came to regard violations of the law as the manifestation of biosocial illnesses to be treated by the coercive power of the state. Indeed, the arguments in favor of social defense could, on some accounts, supersede the claims of the legal process. Kaplinskii declared that "treatment" could not be subordinated to the law: "If a person is sick, if he or she is in need of treatment, if someone's behavior, determined by [that person's] illness, brings him or her into conflict with [others], brings him or her into conflict with the law, then treatment must begin even before the start of the legal process."[68]

Amenitskii was more alive to the pernicious potential of the indeterminacy of the term "social danger," claiming that it needed to be treated "with great care." In what surely stood as an acutely prophetic commentary on the state's policies of gathering

---

[64] Mark Z. Kaplinskii, "K voprosu o prinuditel'nom lechenii: K redaktsii 24 i 26 st. U.K.," in Gannushkin, *Dushevno-bol'nye pravonarushiteli* (cit. n. 62), 61.

[65] Dmitrii A. Amenitskii, "K voprosu o prinuditel'nom lechenii i o sotsial'no-opasnykh dushevno-bol'nykh i psikhopatakh," in Gannushkin, *Dushevno-bol'nye pravonarushiteli* (cit. n. 62), 24. The rise of social defense in European "positivist" penal codes of the 1920s was especially marked in Italy under the influence of Enrico Ferri. Pick, *Faces of Degeneration* (cit. n. 15), 145–52. Ferri's proposed codex was much discussed by criminologists in the Soviet Union. See, e.g., Aleksandr K. Lents, "Zadachi i plan sovremennoi kriminal'no-psikhiatricheskoi ekspertizy," in *Sudebno-meditsinskaia ekspertiza. Trudy II Vserossiiskogo s"ezda sudebno-meditsinskikh ekspertov, Moskva, 25 fevralia—3 marta 1926,* ed. Ia. L. Leibovich (Ulyanovsk, 1926), 112. On the German case, see Wetzell, *Inventing the Criminal* (cit. n. 17), 125–78.

[66] Udal'tsov, "Pravonarusheniia v voiskakh" (cit. n. 1), 132.

[67] Osipov, "K voprosu o khuliganstve" (cit. n. 61), 85 (emphasis in original).

[68] Kaplinskii, "K voprosu" (cit. n. 64), 58.

and retaining discriminatory information about individuals and groups in Soviet society, Amenitskii noted that

> it was enough to attach the label 'socially dangerous' to a particular offender, whether he had only committed one crime or was a recidivist, or even for reasons of a previous judgment, for his position outside the walls of a hospital or a prison to be completely intolerable. . . . Declaring a particular subject to be socially dangerous on the basis of an old judgment, without taking account of the possibility of his having changed and having regenerated the individual over the preceding lengthy period is for many a more severe punishment than a long period of incarceration in prison.[69]

Amenitskii perceived that the limitless nature of social defense might merge fatefully with a definitional vagueness leading to "all manner of possible errors . . . an arbitrary and excessively subjective interpretation of the concept by individuals in the localities."[70] Such cautionary voices were, however, few and far between.

Most criminologists and psychiatrists hailed the sanction of the unlimited use of repressive force generated by the biomedical discourse of curing the deviant as a productive and useful engine of individual and social transformation. Osipov gave an example of this precept in practice: "When psychiatric expertise establishes that a hooligan has a diseased condition, he should be subject to coercive treatment [*prinuditel'noe lechenie*], best of all, in the framework of curative-labor colonies [*lechebno-trudovykh kolonii*]."[71] Coercion was understood, therefore, to wield curative, reconstitutive powers; the more intense its visitation, the more rehabilitative benefits the deviant might accrue. Indeed, Chalisov saw the entire repressive apparatus of the state as an inalienable constituent feature of the broader concern with forging a new citizenry, what Vadim Volkov and David L. Hoffmann have termed, following Norbert Elias, "the Stalinist civilizing process."[72] "Measures of social defense" elaborated precisely "the forms of external restraint" that the prerevolutionary era had failed to instill in the population. Chalisov argued that "in those cases in which, as a consequence of biological peculiarities, the internal restraints have developed badly, then we have to turn to the establishment of external ones, which the state does and with real success." Addressing the particular issue of swindlers, he endorsed "an intensification of the repression . . . a measure of general coercion that acts as a considerable restraint on them."[73] The head of the Soviet penal administration, Evsei Gustafovich Shirvindt (1891–?), also saw the prison system as a workshop in which to forge a new citizen. He explained that social defense served not merely the "general prevention [of crime] but also "the adaptation of the criminal to communal living conditions."[74] Osipov saw the function of social defense in analogous terms: "The struggle with crime should be, and is, carried out along two lines—the strategy of social defense and the correction of the criminal in the sense of inoculating him with new social habits, solid foundations that will turn him, if not into a useful, then into a harmless member of so-

---

[69] Amenitskii, "K voprosu o prinuditel'nom lechenii" (cit. n. 65), 31.

[70] Ibid., 40.

[71] Osipov, "K voprosu o khuliganstve" (cit. n. 61), 88.

[72] Vadim Volkov, "The Concept of *Kul'turnost*": Notes on the Stalinist Civilising Process," in Fitzpatrick, *Stalinism* (cit. n. 59), 210–30; David L. Hoffmann, *Stalinist Values: The Cultural Norms of Soviet Modernity, 1917–1941* (Ithaca, N.Y., 2002).

[73] Chalisov, "Opyt bio-sotsial'nogo obsledovaniia" (cit. n. 53), 74.

[74] Evsei G. Shirvindt, "O problemakh prestupnosti," *Problemy prestupnosti,* no.1 (1926): 10.

ciety."[75] In his foreword to Lents's *Criminal Psychopaths,* Osipov argued for the "organization of curative labor, in which individuals would be able to acquire useful habits [*navyky*], in as much as they are capable of doing so, adapted to labor and gradually become educated."[76]

BIOSOCIAL HYGIENE IN THE MODERN AGE

Such precepts nurtured an impulse to universal applicability in the late 1920s, generating arguments that understood biosocial hygiene in terms of the very essence of societal progress. In a volume edited and with a foreword by the people's commissar of internal affairs, Vladimir Nikolaevich Tolmachev (1886–1937), Segalov argued that the pressures of modern, urban, industrial existence meant that the state had to discharge certain prophylactic functions that had earlier been a natural constituent feature of less developed societies:

> It would be appropriate here to turn our attention to a certain natural prophylaxis [*estestvennaia profilaktika*], by which, during the precapitalist era, unbalanced and psychopathic-unfortunate elements [*psikopaticheski-neudachlivye elementy*] were swept out of the general life of citizens . . . [T]he particular prophylactic expediency [*profilakticheskaia tselesoobraznost'*] for the healing [*ozdorovlenie*] of the contemporaries of that age of medieval institutions was in the form of a huge army of brothels and monasteries. All the unbalanced, mischievous, unfortunate, and criminally minded humanity had the opportunity to leave the world and pass behind the walls of the monasteries or the dens of iniquity or move to the fringes of the state.[77]

Segalov noted ruefully that the capitalist era, with its system of conscription and a unitary tax, had deprived these unfortunates of the opportunity of "stealing away from life's labors and taxes. . . . [T]he more developed a state is, the more effort each individual has to make to keep in step with his cocitizens, at an ever increasing pace. . . . Those left without any particular solicitous supervision, the unbalanced, traumatized, psychopathic personalities, represent an ever greater social danger [*sotsial'naia opasnost'*]."[78] The new demands of production, compounded by the dire shortage of housing in the Soviet Union, left the authorities with no option except to isolate those disruptive and corrupting elements of the population whose conduct and ideas might prove harmful to the socialist project: "Securing the life and work of the healthy—by means of isolating the sick [*vydelenie bol'nykh*], by means of filtering out those who have not adapted to particular labor processes, particular living conditions, a prophylactic returning to health [*profilakticheskoe ozdorovlenie*] of the population—is the essential task."[79] Segalov's call for the reestablishment of a neofeudal system of social marginalization and exclusion lent, of course, a theoretical sanction to the

---

[75] Osipov, "K voprosu o khuliganstve" (cit. n. 61), 88.

[76] Vladimir Osipov, "Predislovie," in Lents, *Kriminal'nye psikhopaty* (cit. n. 49), 7.

[77] Timofei Segalov, "Prestupnoe khuliganstvo i khuliganskie prestupleniia," in Tolmachev, *Khuliganstvo i khuligany* (cit. n. 10), 73–4. On Tolmachev's career and fate, see Lynne Viola, "A Tale of Two Men: Bergavinov, Tolmachev, and the Bergavinov Commission," *Europe-Asia Studies* 52 (Dec. 2000): 1449–66.

[78] Ibid. For a recent article stressing the expansion of taxation as a hallmark of Russian modernity, see Yanni Kotsonis, " 'No Place to Go': Taxation and State Transformation in Late Imperial and Early Soviet Russia," *J. Mod. Hist.* 76 (Sept. 2004): 531–77.

[79] Segalov, "Prestupnoe khuliganstvo" (cit. n. 77), 73–4.

burgeoning Soviet gulag. Yet in Segalov's eyes and in the eyes of his colleagues, the elaboration of a system of curative-labor colonies, administered by the state, was a way of usurping the natural course of societal evolution, of accelerating and directing society toward rational and harmonious ends. The greater the degree of uniformity and stability to which the state aspired, the more intense and sustained the mechanisms of repression it endorsed.

By the eve of the regime's frontal assault on the vestiges of the old order in the Soviet Union—the campaigns against kulaks, *nepmen* in the collectivization and industrialization drives, and against priests, bourgeois specialists, and "former people" in the Cultural Revolution—the human sciences had elaborated a vision of modernity defined in a language of social excision and coercive rehabilitation. Indeed, they came to conceive the entire revolutionary project in terms of social renovation [*ozdorovlenie*], moral and social purification, and forced adaptation to the new socialist order.

BRIDGING THE 1917 DIVIDE

Such a view leads us to a reconceptualization of the thorny relationship between the late imperial and early Soviet regimes. Laura Engelstein has articulated what is probably a majority opinion among historians of the revolution:

> whether one believes that Russian liberalism ought to have triumphed, had any reasonable chance of succeeding, or was inevitably doomed, the liberal project has an unavoidable pathos in the social and political environment of late imperial Russia. . . . [T]he ultimate drama of their situation rests perhaps in the nation's failure to enter the difficult and flawed terrain of postabsolutist public life, its failure to create a polity in which citizens might have struggled with the imperfections of the civic condition in terms supplied by the arsenal of liberal thought.[80]

Yet as the Russian empire disintegrated and became reconstituted under the Soviet regime, its citizens did continue to struggle "with the imperfections of the civic condition in terms supplied by the arsenal of liberal thought." Liberalism in the late imperial period did not "fail"; rather, it contained—precisely in its modernist preoccupations with science, crisis, and solutions—a dangerous potential that could be radicalized and implemented in unforeseen ways.[81]

Unforeseen, yet not entirely unexpected. Radicalization was, so to speak, hardwired into the Bolshevik mainframe. Peter Holquist has noted that "Bolshevism was distinct not so much because it was ideological, or even utopian, but on account of its specifically Manichean and adversarial nature."[82] The late imperial human sciences certainly did elaborate theories of social categorization and strategies for the normalization of society but, while they did undoubtedly contain a utopian impulse, their practitioners remained loyal to the idea of reform as a gradual way of managing social problems. Indeed, the critique of the status quo firmly embedded in most bio-

---

[80] Engelstein, *Keys to Happiness* (cit. n. 8), 8.

[81] I distance myself from the influential thesis that the totalitarianism of both Nazism and Stalinism was preordained in the very project of modernity. Zygmunt Bauman, *Modernity and Ambivalence* (Cambridge, 1991); Detlev J. K. Peukert, "The Genesis of the 'Final Solution' from the Spirit of Science," in *Reevaluating the Third Reich,* ed. Thomas Childers and Jane Caplan (New York, 1993), 238.

[82] Peter Holquist, "Violent Russia, Deadly Marxism? Russia in the Epoch of Violence, 1905–21," *Kritika* 4 (Summer 2003): 652.

medical theories of deviance meant that all coercive attempts to rehabilitate socially marginal or dangerous elements were advocated in a mood of paternal indulgence.

The Bolsheviks infused this ambiguous legacy, however, with the concrete and discrete facts of the Marxist imagination so that there emerged, notionally at least, far more explicit borders of civil society and political community than had existed before 1917.[83] Moreover, whereas the liberals' sense of social responsibility for the iniquities of the late imperial social order had impeded them from reacting with unadulterated hostility to the deviants they confronted, the Bolsheviks had no such qualms. Their Manichean worldview, predicated on the Marxist division of society into classes, allowed no quarter in the struggle to cleanse Soviet society of the pathologies of the old order.

This transformative program was increasingly understood in terms of a coercive rehabilitation of those amenable to treatment and the isolation (and even elimination) of those who were not. The Russian Revolution itself witnessed an intensive campaign of reconfiguring the values and meanings attached to the possession of different socioeconomic and political identities, which led to the generation of "deviant classes" and "socially dangerous" individuals on a massive scale. The human sciences continued in the postrevolutionary period to forge a discursive axis around which threats and solutions could be identified and acted upon. They offered a particular articulation of Marxism's division of society into classes in conflict with each other, furnishing the abstract language of collectivities with a concrete, tangible quality that reverberated with urgency and the immediacy of class struggle.

Rather than viewing the late imperial period as a liberal experiment that failed, opening up the path to the Bolshevik dictatorship, it is more instructively understood in terms of a laboratory of modernity that bequeathed an ambiguous legacy to its successor regime. The Bolsheviks' ideology of class struggle and social transformation did not enslave or subjugate the human sciences to its own oppressive ends; rather, it meshed with them, generating theories of change and sanctions for repression.

Liberalism may have proved a spectacular political failure in the Russian Revolution. Yet it was the unwitting architect of significant features of the project that triumphed over it. The early years of the Soviet regime serve not so much as the verification of the failure of Russian liberalism as the validation of its dangerous potential.

---

[83] Michael Halberstam has noted that "the concrete and comprehensive ideologies of totalitarian movements appear less ephemeral, less ideal, and more 'real' than the abstract political principles of liberalism that emphasise the pragmatics of the political process and the indeterminacy of human nature." Halberstam, "Totalitarianism as a Problem for the Modern Conception of Politics," *Political Theory* 26 (Aug. 1998): 472.

Weimar Psychotechnics between Americanism and Fascism

By Andreas Killen*

ABSTRACT

In the aftermath of the Great War, the new science of psychotechnics was enlisted in the construction of the democratic social order that emerged from the ruins of German authoritarianism, a key component of the fragile social compromise between capital and labor that was a foundation of the Weimar Republic. During the 1920s, representatives of this branch of social engineering promised to use science to conjure away the workplace conflicts that had wracked Wilhelmine Germany and to usher in a new era of social harmony and productivity. Advertised as modern, rational, and humane, psychotechnics became a cornerstone of the rationalization movement that, originating in America, swept Germany in the 1920s. This article, focusing on interactions between psychotechnicians and female switchboard operators, places the objectives and contradictions of this new science of the working self within the context of wider debates about Germany's postwar economic and political restructuring as well as the process of personal restructuring associated with the so-called new woman. Ultimately, the article shows that the psychotechnicians' failure to realize their aims within the Weimar system led them to reposition their science as handmaiden to the new National Socialist order, which they embraced as the best means of bringing about this process of restructuring.

INTRODUCTION

There is a scene in Fritz Lang's *Metropolis* in which an exhausted worker, pushed to the limit of his energies, collapses at the controls of a pressure gauge, an event that triggers a colossal power outage, eventually shutting down the metropolis and threatening to plunge its inhabitants into class warfare. The scene illustrates a variant of what Anson Rabinbach has described as the thermodynamic calculus of modernity, according to which the body itself, and not the social relations of the workplace, became the arena of labor power. By extension, the body's breakdown constituted an event that threatened industrial civilization with its demise.[1]

It is this potential for what Louis Mumford termed the "gigantic breakdowns and

* City College of New York, CUNY, North Academic Center, Room 5/144, Convent Ave. at 138th Street, New York, NY 10031; akillen@ccny.cuny.edu.

The author would like to thank Christine Leuenberger and Greg Eghigian, as well as the two anonymous readers, for their help with and comments on earlier versions of this article.

[1] Anson Rabinbach, *The Human Motor: Energy, Fatigue, and the Origins of Modernity* (New York, 1990), 11.

stoppages" of modern life that forms the subject of this paper.[2] Specifically, I want to look at a particular response to this potential for breakdown, one embraced by the fragile democracy that emerged in Germany after World War I. This response went by the name *psychotechnics*—a new science of the workplace whose emergence was prompted by a desire to rationalize and optimize the mental components of labor power. During the Weimar period, representatives of this branch of social engineering promised to use science to conjure away the workplace conflicts that had wracked Wilhelmine Germany and to usher in a streamlined new order of social harmony and productivity. They achieved significant backing from a state anxious to rebuild the war-torn nation on a new basis of class collaboration. Advertised as modern, rational, and humane, a means of winning workers' consent, psychotechnics was made a cornerstone of the rationalization movement that swept Germany in the 1920s.

German psychotechnicians made up a broad coalition of experts, who used the authority granted them by the state, industry, and labor to analyze and propose scientific solutions to a multitude of social problems. A key figure in this movement was Fritz Giese, a prominent psychologist whose experimental studies formed part of a broader search for a more stable, rational social order to replace the one that had collapsed in Germany at the end of the war. In this article, I examine Giese's psychotechnical researches and relate this work to his intellectual migration from enthusiast of American-style rationalization to supporter of National Socialism. In particular, I focus on his studies of a special class of modern worker, the female switchboard operator. From its inception, psychotechnics had been closely linked to the discovery of "mental labor" as a new category of modern work that imposed unique demands on the senses, nerves, and mind. It was probably for this reason that switchboard operators, who were seen as exemplifying this category, served as a special object of study for the new science. However, these women were also perceived as uniquely representative of the new social types thrown up by the modernization process, exemplars of a new form of personhood, whose attitudes, abilities, and behavior constituted, as one rationalization expert put it in the early 1920s, "a psychosociological problem" of the first order.[3]

In what follows, I also examine the reception given by switchboard operators themselves to the new science. At least initially, these women seem to have taken its claims at face value and to have welcomed its promise of a more rational society. Closely identified with the state that employed them, they readily embraced the project of social reconstruction it had undertaken. They were also amenable to calls for personal restructuring, seeming to serve, in the words of Weimar sociologist Siegfried Kracauer, virtually as "natural objects" of psychotechnics.[4] Simultaneously emancipated and regimented, switchboard operators embodied a new image of female personhood

---

[2] Louis Mumford, *The City in History* (New York 1961), 544. For a discussion of the place of the "breakdown"—nervous, technological, and social—within the social imaginary of early twentieth-century Germany, see Andreas Killen, "From Shock to *Schreck*: Psychiatrists, Telephone Operators, and Traumatic Neurosis, 1900–1926," *Journal of Contemporary History* 38 (2003): 201–20.

[3] Niederschrift über die Verhandlung, betreffend Förderung der Arbeitswissenschaft in den Verkehrsverwaltungen, 26 April 1920, im RAM, Dr. Tiburtius, Bl. 9, Schule für Fernsprechgehilfinnen, Reichspostministerium (hereafter cited as RPM) 5110, Bundesarchiv Potsdam, Germany (hereafter cited as BAP).

[4] Siegfried Kracauer, *The Salaried Masses: Duty and Distraction in Weimar Germany* (New York, 1998), 37.

that Kracauer associated with the Tiller Girls, the American-style precision dance team that was a sensation in Weimar Germany. Kracauer was not alone in making such a connection. Fritz Giese also devoted a widely read work of cultural criticism, *Girlkultur,* to the Tiller Girls, whose serialized forms and rhythmic movements reproduced on stage those of the operators ranged at the switchboard.[5] For both Giese and Kracauer, the Tiller Girls were emblematic of a distinctively American technique of organizing the masses. They represented in the world of entertainment what the operators represented in the world of the modern workplace: variations on the "mass ornament," the wish image of a perfectly rationalized society and self.[6] Yet despite their initial receptivity to this wish image, these women ultimately withdrew their consent from the project of rationalization that animated Weimar social policy, a project whose collapse helped precipitate the demise of the Weimar Republic itself. In the wake of this collapse, German psychotechnicians would embrace National Socialism and its politics of the self.

PSYCHOTECHNICS AND OPERATORS IN THE PREWAR PERIOD

Psychotechnics emerged in the early twentieth century as a more or less natural extension of the scientific management movement launched by Frederick W. Taylor. The term itself was first introduced into psychology by William Stern in 1900. Stern conceived of psychotechnics as a tool for studying individual differences and as an integral part of the larger field of "human management." The psychiatrist Emil Kraepelin had already made his own contribution to this field with his invention of the so-called work curve and fatigue curve, which measured individual variations in performance over time.[7] It was not, however, until Hugo Münsterberg published his *Psychology and Industrial Efficiency* in 1913 that the science of psychotechnics received a more systematic, comprehensive treatment.[8] In the belief that Taylorism tended to neglect considerations such as workers' aptitude and satisfaction, the German-born, Harvard-based Münsterberg saw industrial psychology as a means of integrating the diverse aspects of the human factor into the equations of modern labor power. He thus positioned psychotechnics midway between industrial efficiency and social reform.[9] By claiming this political middle ground, he believed the new science could avoid the pitfalls of either "reckless capitalism on the one side" or "feeble sentimentality on the other."[10] Psychotechnics, in his mind, offered remedies to the modern pathologies of work, whether those associated with unbridled Taylorism or those stemming from worker radicalism. Münsterberg was commissioned by several companies, including Boston's streetcar company and Bell Telephone, to study their personnel and make recommendations for selecting candidates and improving service. His book included

---

[5] Fritz Giese, *Girlkultur: Vergleiche zwischen amerikanischem und europäischem Rhythmus und Lebensgefühl* (Munich, 1925).

[6] Siegfried Kracauer, "The Mass Ornament," in *The Mass Ornament* (Cambridge, Mass., 1995), 75–86.

[7] Robert Chestnut, "Psychotechnik: Industrial Psychology in the Weimar Republic," *Proceedings of the Annual Convention of the American Psychological Association* 7 (1972): 781–2; Rabinbach, *Human Motor* (cit. n. 1).

[8] Hugo Münsterberg, *Psychology and Industrial Efficiency* (Boston, 1913). This was originally published in German as Hugo Münsterberg, *Psychologie und Wirtschaftsleben* (Leipzig, 1912).

[9] Chestnut, "Psychotechnik" (cit. n. 7), 781.

[10] Rabinbach, *Human Motor* (cit. n. 1), 165.

a chapter on telephone operators, providing data on their reaction times, attention, and aptitude.[11] While accepting the basic premises of Taylorism, he also took seriously the discontent it occasioned. One of the documents he examined was a report on a strike by Bell operators in Toronto. Münsterberg's study of the circumstances surrounding this strike convinced him that scientific means were required to address the grievances of these workers. His work breathed the reformist spirit of the Progressive Era, positioning human science as a crucial link in the relationship between democracy and efficiency.[12]

In Germany, where the implementation of Taylorism generated intense controversy, Münsterberg's proposals, with their promise of a more humane form of scientific management, were taken up by industrial experts in many fields.[13] They found resonance across the political spectrum, from liberal reform-minded engineers and socialists to those Jeffrey Herf has called "reactionary modernists." The proposals also met with considerable interest from trade unions. Representative of the latter was the reception accorded the new field by telephone operators. As in the United States, these women had, by the early 1900s, become objects of scientific and sociological scrutiny. Young and usually single, the operator was seen as the bearer of a new sensibility, and her attitudes toward work, leisure, and marriage were scrutinized for clues to the mass psychology of the modern era. Social conservatives, worried about the consequences of women's forsaking their familial duties, seized upon irregularities in service as evidence that women were not suited for work that demanded such concentration and responsibility. Operators were repeatedly forced to defend themselves against charges of "confusion" in the switchboard exchanges.[14]

Despite such concerns, the Imperial Postal Ministry (Reichspostministerium, RPM), which oversaw Germany's telephone system, continued to hire female operators. (By 1924, with 65,000 female employees, the RPM would outstrip the Reichsbahn as the state's largest employer of women.)[15] In recognition of the importance of this new branch of work, it granted them status as a special category of salaried worker, with legal privileges in the areas of insurance and labor rights. Nonetheless, RPM authorities shared the concern about this new class of employee. Worried that unsupervised exchanges with male callers could endanger the morals of these young women, the RPM imposed linguistic discipline on them by means of a lexicon of strictly routinized responses such as *"Hier Amt!"* This was just one of a series of

[11] For more on Münsterberg, see Rabinbach, *Human Motor* (cit. n. 1), 189–95; Matthew Hale, *Human Science and Social Order: Hugo Münsterberg and the Origins of Applied Psychology* (Philadelphia, 1980).

[12] On this relation, see Charles Maier, "Between Taylorism and Technocracy," in *In Search of Stability: Explorations in Historical Political Economy* (Cambridge, 1987), 27.

[13] On the controversy surrounding Taylorism, see Hans Ebert and Karin Hausen, "Georg Schlesinger und die Rationalisierungsbewegung in Deutschland," in *Wissenschaft und Gesellschaft: Beiträge zur Geschichte der Technischen Universität Berlin 1879–1979,* ed. R. Rürup (Berlin, 1979), 1:319.

[14] See "Presse contra Telephon," *Unter den Reichsadler* 3 (1911): 87; "Berliner Telephonverhältnisse," 5 (1913): 233–5. Well into the 1920s, operators continued to be identified with communicative disorder: "crosstalk" and faulty connections as well as unsupervised communication between men and women.

[15] The literature on operators includes Ursula Nienhaus, *Vater Staat und seine Gehilfinnen: Die Politik mit der Frauenarbeit bei der deutschen Post (1864–1945)* (Frankfurt am Main, 1995); Nienhaus, "Unter dem Reichsadler: Postbeamtinnen und ihre Organisation, 1908–1933," *1999: Zeitschrift für Sozialgeschichte des 20. und 21. Jahrhunderts* 3 (1990): 56–79; Helmut Gold and Annette Koch, eds., *Das Fräulein vom Amt* (Munich, 1993).

measures designed to help coordinate the rapid, precise actions of hundreds of women to minimize disruption of service. By the 1910s, telephone exchanges had become minutely regulated, highly rationalized spaces.[16] In the interests both of efficiency and of "moral hygiene," operators' speed in answering calls was timed and their interactions with callers monitored. The following account, from a Berlin paper in 1911, describes a form of surveillance recently introduced into the telephone system:

> Everything, even the slightest offense, every impoliteness on the part of the operator, bad connections, careless work, and so on, will be noticed by the monitor, daily comments will be made on the records and sent to the relevant telephone offices . . . on the basis of which steps will be taken against those responsible.[17]

The operators' representatives in the Reichstag protested this heightening of workplace discipline but to little avail.[18] As more than one doctor pointed out, it was hardly surprising that working under such conditions wore down the women and left them chronically exhausted, virtually programmed for nervous breakdown.[19]

Operators proved more amenable to psychotechnical methods of workplace management. In 1914, the pages of their publication *Unter den Reichsadler* included a favorable review of Hugo Münsterberg's *Psychology and Industrial Efficiency*. It took an open mind toward his claim to offer a solution to the problems of the modern workplace and went so far as to defend psychotechnics against accusations that it threatened conceptions of the self or turned work into a soulless occupation that would "drain human labor power dry."[20] Referring to charges routinely leveled against operators in the press, the review cited Münsterberg's work to suggest that such charges remained mired in old-fashioned attitudes. By turning switchboard work into an object of scientific analysis, Münsterberg demonstrated that the work itself, rather than the women who performed it, caused most of the problems in service. Whereas Münsterberg had established 150 calls per hour as an upper limit for most operators, the article pointed out that in Berlin periods of 600 calls per hour were not uncommon. The consequences of such overload were spelled out in a passage from Münsterberg's text, highlighted in the review:

> It is only natural that such rapid, yet subtle, activity under such high tension . . . can be carried out only by a relatively small number of human nervous systems. The inability to maintain one's attention for a long time under such tense conditions . . . or to correctly remember the desired numbers, leads . . . to exhaustion and finally to nervous breakdown of the operator and confusion in the exchange.[21]

---

[16] Nienhaus, *Vater Staat und seine Gehilfinnen* (cit. n. 15), 110–5.

[17] "Schule fur Fernsprechgehilfinnen," RPM 47.01, 5109, BAP. The new system employed a modified galvanoscope that enabled the supervisor to monitor all the processes involved in connecting two parties. Notes on similar devices can be found elsewhere in the archives. The director of one exchange claimed to have designed a contrivance, called a "neuroscope," that took electrical readings of the nerve endings on the scalp to determine the operator's abilities. See RPM 47.01, 14509, letter, April 1929, BAP.

[18] Reichstag Sitzung 36, 27 March 1912, Abg. Bruhn. RPM 47.01, 5109, BAP.

[19] The RPM archives contain numerous psychiatric case histories of this period. See "Schreckwirkung als Ursache nervöser Störung," RPM 47.01, 14575/1, BAP. See also Killen, "From Shock to *Schreck*" (cit. n. 2).

[20] "Professor Hugo Münsterberg in 'Psychologie und Wirtschaftsleben,'" *Unter den Reichsadler* 6 (1914): 101–4.

[21] Ibid., 104. See also Münsterberg, *Psychology and Industrial Efficiency* (cit. n. 8), 98.

If "confusion" existed, the problem lay not with the women themselves but with the effects of overwork. The ramifications of overwork, moreover, were serious, leading not just to confusion but, in extreme cases, to generalized breakdown with potentially grave consequences. While conservatives tended to trace the causes of such breakdown to a collective weakness or hysteria that was explicitly gendered, Münsterberg addressed the topic without reference to the female constitution. Operators thus perceived in psychotechnics a nonpartisan scientific spirit that they welcomed as an ally in their disputes with conservatives. In the conflict between tradition and modernity, they believed psychotechnics to be on their side.

RATIONALIZATION, AMERICANISM, AND PSYCHOTECHNICS
IN THE WEIMAR REPUBLIC

Münsterberg's book defined an experimental agenda and laid the prewar foundations for industrial psychology. It was the Great War itself that transformed his work into a field with wide social application in Germany. With its enormous labor demands, the war gave a tremendous boost to advocates of Taylorism, which found champions in the directors of the War Raw Materials Office, Walter Rathenau and Wichard von Moellendorff. The war also acted as a stimulus on the field of psychotechnics. Starting in 1915, methods adapted from those pioneered by Münsterberg were used to test German military personnel, especially pilots and transport personnel. This undertaking was spearheaded by scientists working at the recently founded Kaiser Wilhelm Institute (KWI) for Labor Physiology in Berlin, under the leadership of Max Rubner. Rubner placed fatigue research at the center of what he envisioned as a more humane version of scientific management than that practiced by the Americans.[22] Under the auspices of the KWI, it was during the war that most of the individuals who became key figures in the psychotechnics movement of the 1920s first gained practical experience with the new science.

It has long been known how profoundly the war left its imprint on the modern psyche.[23] Less widely appreciated has been its role in engendering a new awareness of the mental aspects of modern labor power. In the course of the war, there emerged in all nations a heightened need for measures of mental labor and techniques for mobilizing it as part of a more efficient allocation of social resources.[24] This was spelled out by one of the KWI's researchers, the physiologist Heinrich Boruttau, in whose estimation the war had inaugurated a hitherto undreamed of rationalization of human resources. At the heart of this development lay a consequential shift from work understood as a mechanical activity to a conception of work that stressed its psychological dimensions.[25] Citing Rubner, Boruttau observed that cultural development made hu-

[22] On the wartime uses of psychotechnics, see Rabinbach, *Human Motor* (cit. n. 1), chap. 10.

[23] See, e.g., Paul Fussel, *The Great War and Modern Memory* (New York, 1975); Paul Lerner, *Hysterical Men: War, Psychiatry, and the Politics of Trauma in Germany 1890–1930* (Ithaca, N.Y., 2003); Joanna Bourke, *Dismembering the Male: Men's Bodies, Britain, and the Great War* (Chicago, 1996).

[24] For developments elsewhere, see John Carson, "Army Alpha, Army Brass, and the Search for Army Intelligence," *Isis* 84 (1993): 278–309; Nikolas Rose, *Governing the Soul: The Shaping of the Private Self* (London, 1990).

[25] For a programmatic statement, by one of the KWI's researchers, of the need to apply scientific methods to the study of work processes, particularly its mental aspects, see Adolf von Harnack, "Abschrift," 5 Nov. 1913, Merseburg, Rep. 76 VL. Sekt. 2 Tit. 23 LITT A., No. 115, Kaiser-Wilhelm Institut für Arbeitsphysiologie, Zentrales Staatsarchiv, Potsdam, Germany (hereafter cited as KWI Arbeitsphysiologie, ZsTA).

man performance as pure mechanical labor increasingly less important than mental labor.[26] More and more, individuals simply monitored the mechanical work performed by machines, under conditions that nevertheless imposed great strain on their minds, nerves, and senses.[27] By meticulously analyzing such work, Boruttau hoped to mobilize the energies of the working individual to a quite astonishing degree:

> By avoiding . . . unnecessary processes of mental association, by switching off recollections that represent unnecessary ballast for the memory, in short by avoiding unnecessary detours of the thought processes, efficiency will be achieved and this will result in . . . an intensification of mental performance according to the principles of an improved economy of the underlying material and energetic processes.[28]

His book included a ringing endorsement of psychotechnics, which he regarded as the most promising means of harnessing this new form of social capital.[29]

Spurred on by wartime developments, the field of psychotechnics came into its own in the postwar period. Wartime destruction and postwar turmoil cleared away resistance to organizational changes in the workplace and opened a wide field to the prophets of the rationalization movement. Despite the failure of the factory councils that had emerged in the war's latter stages to revolutionize German society, workers and their representatives wielded considerable influence over Weimar social policy. They were determined to move beyond the autocratic, paternalist styles of industrial management and the authoritarian state that dominated Wilhelmine Germany. For the time being at least, German industrialists were in no position to argue. One of the founding compromises of the Weimar Republic, the so-called Central Working Association, institutionalized a new spirit of cooperation between labor and business on which the rationalization experts were able to capitalize.

As Detlev Peukert has written, the new Weimar order assigned science the role of midwife to a wholesale restructuring of society. Modern science was to act as mediator between the state, capital, and labor and thus solve, in a manner based on American models, the social problems of industrial civilization.[30] "Americanism"—the enthusiasm for all things American that overtook postwar Germany—encompassed mass culture, domestic appliances, and architecture, but above all it meant rationalization in the workplace: Taylorism, Fordism, and psychotechnics.[31] The Social Democrat Gustav Bauer, who headed the Ministry of Labor from 1920 to 1928, predicted that objections to the Taylor system would end, "now that the democratization of Germany has secured a wide economic influence for workers":

[26] For Rubner's views on mental labor and his rejection of the "vulgar machine model," see "Das arbeitsphysiologische und arbeitshygienische Institut," 16 May 1912, KWI Arbeitsphysiologie, ZsTA.

[27] Heinrich Boruttau, *Die Arbeitsleistungen des Menschen* (Leipzig, 1916), 67. The modernity of mental work was suggested to him by the analogy between nervous and telephone systems.

[28] Ibid., 71.

[29] Psychotechnician Walter Moede echoed Boruttau in arguing that machines had taken a physical load off workers but placed greater demands on the senses and attention. Moede, cited in Ulfried Geuter, *The Professionalization of German Psychology in Nazi Germany,* (Cambridge, 1992), 86.

[30] Detlev Peukert, *The Weimar Republic* (New York, 1989).

[31] See Alf Lütke, Inge Marssolek, and Adelheid von Saldern, eds., *Amerikanisierung: Traum und Alptraum in Deutschland des 20. Jahrhunderts* (Stuttgart, 1996); Detlev Peukert, "'Amerika,' oder der Traum der Rationalisierung," in *Max Webers Diagnose der Moderne* (Göttingen, Germany, 1989); Mary Nolan, *Visions of Modernity: American Business and the Modernization of Germany* (New York, 1994).

From this vantage point the Taylor-system will achieve a significance hitherto under-appreciated, namely as an instrument of peaceful national liberation in the hands of a democratic and, of course, socialized state.[32]

Although advocates of the American model rarely spelled out its political implications quite so directly, their acceptance of the postwar democratic state and of the basic principles of worker participation and of equality of rights had an implicit political dimension. Americanism, in various modified forms, became part of the postwar search undertaken in many European nations for a politics of the third way that steered a middle ground between capitalism and socialism. Yet however powerful Americanism became as a rhetorical construct, it nevertheless concealed significant ambiguities. This was particularly true in Germany, where the belief that a commonality of economic interests could be established under the auspices of a democratic state concerned with the task of reconstruction ignored both the shakiness of the truce between labor and capital and the underlying weakness of the new state.[33]

Some of these ambiguities are discernible in the rhetoric surrounding one of the key symbols of this new social order, the *new man,* and his counterpart, the *new woman.*[34] As part of the project of reimagining German society, politicians, pedagogues, journalists, and scientists conjured an idealized new type who would reflect the values of the postwar experiment in social democracy. Economic restructuring, as Mary Nolan has written, would be accompanied by personal restructuring; emancipated from the authoritarian prewar order, the new worker would embrace the dynamic, productivist Weimar system and reorganize his private life in accordance with the principles of discipline and efficiency.[35] He would thereby become transformed from worker into citizen. At the same time, however, this idealized new form of selfhood was an object of considerable misgiving, insofar as it was seen as the product of impersonal forces and processes that remained only partly understood. Whether it was the promise of revolutionary freedom that emerged briefly at war's end, or the process of rationalization that the war had accelerated, these forces were regarded with a complicated mixture of hope and anxiety.[36]

It was the figure of the new woman, in particular, who, in giving a face to these processes, embodied the deepest anxieties and hopes of the postwar era. Wartime changes in the sphere of industrial relations had mirrored changes in the sphere of gender relations. In Germany, as Atina Grossmann notes, much of the postwar sense of rupture crystallized around the figure of the young, emancipated, self-aware new woman. The opportunities for work that wartime afforded women, together with the new Weimar constitution, which gave them the vote, seemed to many to turn the old German gender order on its head. The result was an intense fascination with the new woman.[37]

[32] Peter Hinrichs and Lothar Peter, *Industrieller Friede? Arbeitswissenschaft, Rationalisierung, und Arbeiterbewegung in der Weimarer Republik* (Cologne, Germany, 1976), 54.

[33] See Richard Bessel, *Germany after the First World War* (Oxford, 1993).

[34] On postwar rhetoric of the new man, see George Mosse, *The Image of Man: The Creation of Modern Masculinity* (Oxford, 1996).

[35] Nolan, *Visions of Modernity* (cit. n. 31), 6.

[36] For a general discussion, see Peter Fritzsche, "Landscape of Danger, Landscape of Design: Crisis and Modernism in Weimar Germany," in *Dancing on the Volcano: Essays on the Culture of the Weimar Republic,* ed. Thomas W. Kniesche and Stephen Brockmann (Columbia, S.C., 1994), 29–46.

[37] On the new woman, see Atina Grossmann, "*Girlkultur* or Thoroughly Rationalized Female," in *Women in Culture and Politics,* ed. Judith Friedlander (Bloomington, Ind., 1986), 62–80; Grossmann,

Americanism and the new woman were brought together in a key work of Weimar cultural analysis, psychotechnician Fritz Giese's *Girlkultur*. This text, published in 1925, offered an analysis of the precision dance team the Tiller Girls. Originating in England, this team had been imitated by numerous American groups, and for Giese it was these Girls—rather than the endlessly glorified Fordist methods of production— that exemplified the American phenomenon. At a time when Germany was making a sharp break with the past, the Girls seemed emblematic of the future and the new human types that would people it. In their tightly choreographed movements and se-rialized forms, Giese discerned something like a pure distillation of the modern era. The Girls beckoned Germans on to a new era of sobriety, functionalism, and eugenic consciousness. Giese concluded that any hope of national regeneration after the post-war breakdown rested on close study of the American model, its techniques of organ-izing the masses, and its promise of a new type of personhood.[38]

If Giese's book coded mass society as female, it was nevertheless a new, positive image of femininity that he identified with the Girls, one lacking any trace of senti-mentality or weakness.[39] He welcomed the absence of ornamentation in their attire and movements, a trait he saw as evidence of their harmony with the austere, func-tionalist spirit of the times. Their patterned movements had nothing to do with the wildly expressionist forms of dance and body culture that had flourished in Germany during the abortive revolution; rather, as the natural outgrowth of a democratic nation, they expressed the elemental rhythms of the human organism captured by the instru-ments of modern science, the pneumograph and the kymograph.[40] In place of willful individualism, they celebrated the virtues and pleasures of a self-ideal molded by the rhythms of modern machinery and by the industrial principles of serialization and efficiency.[41] Like few other texts of this period, *Girlkultur* reveals how the American model tantalized industrial experts with visions of a new social order, even while they passed over the question of whether rationalization was a process that could be con-trolled for socially desirable ends.

Such visions, and the ambiguities inherent in them, permeated the German psy-chotechnical movement. What united the members of this movement was a belief that their field contained a blueprint for Germany's economic regeneration. They also believed their approach to the labor question to be superior to the purely economic ra-tionality of Taylorism. Beyond that, their approaches diverged, often in significant ways. Roughly speaking, the movement was divided into two main camps, the en-gineers and the academics.[42] The former included Georg Schlesinger, head of the

---

"The New Woman and the Rationalization of Sexuality in Weimar Germany," in *Powers of Desire: The Politics of Sexuality*, ed. A. Smitow, C. Stansell, and S. Thompson (New York, 1983), 153–71; Renate Bridenthal, Atina Grossmann, and Marion Kaplan, eds., *When Biology Became Destiny: Women in Weimar and Nazi Germany* (New York, 1984); Ute Frevert, *Women in German History* (New York, 1989), 176–85; Nolan, *Visions of Modernity* (cit. n. 31), chap. 10.

[38] Giese, *Girlkultur* (cit. n. 5), 142.

[39] For another discussion of the coding of the "masses" as female, see Klaus Theweleit, *Male Fan-tasies*, vol. 1, *Women, Floods, Bodies, History* (Minneapolis, Minn., 1987).

[40] Giese, *Girlkultur* (cit. n. 5), 21.

[41] See Frank Trommler, "The Creation of a Culture of Sachlichkeit," in *Society, Culture, and the State in Germany 1870–1930*, ed. G. Eley (Ann Arbor, Mich., 1997), 465–86; Peter Jelavich, "'Girls and Crisis': The Political Aesthetics of the Kick-line in Weimar Berlin," in *Rediscovering History: Culture, Politics, and the Psyche*, ed. M. Roth (Palo Alto, Calif., 1994), 224–42.

[42] Geuter, *Professionalization of German Psychology* (cit. n. 29), 165. See also Anson Rabinbach, "Betriebspsychologie zwischen Psychotechnik und Politik während der Weimarer Republik: Der Fall

Institute for Industrial Psychotechnics at the Technical University in Berlin-Charlottenburg. In his approach, Schlesinger, known as the German Taylor for his enthusiasm for scientific management, closely resembled Münsterberg, who envisioned workers as "working machines, perpetually exhausting and regenerating themselves."[43] The leadership of his institute would later be assumed by the like-minded Walter Moede. Others, such as Otto Lipmann, an academic psychologist whose political leanings were socialist, took more seriously the field's claims to mediate between labor and capital, and tried to downplay the "Americanist" aspects of psychotechnics—those stressing efficiency at all costs—and to privilege instead a more "German" regard for the worker's well-being.[44]

Perhaps more than any other single figure, Fritz Giese strove to synthesize the divergent tendencies within the German psychotechnical movement.[45] Born in 1890, Giese studied under psychologist Wilhelm Wundt at Leipzig. After serving during the war in a hospital for soldiers with brain injuries, Giese created an institute for applied psychology in Halle, before becoming a professor in Stuttgart. In addition to teaching and lecturing on psychotechnics and work science, he published on numerous subjects, ranging, in addition to the cultural analysis of *Girlkultur*, from the psychology of education and gender differences to a late study of Nietzsche. Politically, he fits the "reactionary modernist" mold: starting out in the 1920s as a technocrat and a modernist, he would, in the 1930s, embrace the Nazi movement.[46]

Like most of his peers, Giese conceived of psychotechnics as a select branch of social engineering that would help compensate for some of the "irrationalities" of capitalist methods of production.[47] In war-battered Germany, maximizing labor power, Giese felt, held the key to national regeneration, and psychotechnics the key to maximizing labor power. In 1919, Giese was presented with an opportunity to test his claims when he was commissioned by the RPM to carry out a study of its personnel in Berlin. This commission reflected mounting concern on the part of the authorities about both growing discord and health care costs.[48] The trade union unrest of the war's latter stages had, in the case of the operators, culminated in a strike at the Berlin exchanges that briefly cut central state authorities off from the rest of the nation, a strike that was only ended with the help of a loyal military telegraph unit.[49] The realization

Otto Lipmann," in *Betriebsärzte und produktionsbezogene Gesundheitspolitik*, ed. Dietrich Milles (Bremerhaven, Germany, 1992), 41–64. On the politics of Weimar engineers, see Jeffrey Herf, *Reactionary Modernism: Technology, Culture, and Politics in Weimar and the Third Reich* (Cambridge, 1984).

[43] Geuter, *Professionalization of German Psychology* (cit. n. 29), 86. See also S. Jaeger and I. Stäuble, "Die Psychotechnik und ihre gesellschaftlichen Entwicklungsbedingungen," in *Die Psychologie des 20. Jahrhunderts*, ed. F. Stoll (Zurich, 1981), 53–94. Schlesinger's proposed solution to the labor problems in the telephone system was to automate service as far as possible. See "Schule für Fernsprechgehilfinnen," meeting 20 April 1920, RPM 5110, BAP. For more on Schlesinger, see Ebert and Hausen, "Georg Schlesinger" (cit. n. 13).

[44] Geuter, *Professionalization of German Psychology* (cit. n. 29), 86. For an analysis of Lipmann's work, see Rabinbach, "Betriebspsychologie" (cit. n. 42).

[45] For biographical details on Giese, see Geuter, *Professionalization of German Psychology* (cit. n. 29), 215–36; F. Schulz, "Giese," in *Neue Deutsche Biographie* (Berlin, 1964), 6:378–9.

[46] See Herf, *Reactionary Modernism* (cit. n. 42).

[47] Fritz Giese, *Theorie der Psychotechnik* (Braunschweig, Germany, 1925), 23; cited in Rabinbach, "Betriebspsychologie" (cit. n. 42), 47.

[48] Killen, "From Shock to *Schreck*" (cit. n. 2).

[49] Frank Thomas, "The German Telephone System," in *The Development of Large Technological Systems*, ed. R. Mayntz and T. Hughes (Frankfurt am Main, 1988), 192. Harry Kessler's memoirs contain an account of a strike that shut down Berlin's power stations in January 1919, knocking out

of the social paralysis such work stoppages could cause led, in the postwar era, to a search for a new formula for labor relations. This was the backdrop against which Giese conducted his psychotechnical investigations into improving service at telephone exchanges.[50] The results were published in 1923 in *Observations on Occupational Psychology in the Imperial Telegraph Service.*

Early in this book, an image appears that serves as a leitmotif for Giese's project. The image unites two kinds of information in the graphical form of curves: the first, the energy curve illustrating peak hours of electricity usage throughout Berlin; the second, the fatigue curve measuring operators' performance throughout the day. By superimposing these curves on one another, Giese showed that demand and performance closely matched during most of the morning shift, but in the afternoon, as fatigue set in, they increasingly diverged.[51] In this highly compressed form lay the "image of objectivity" to which psychotechnics laid claim, this image being for its practitioners the safeguard of its political neutrality and of the privileged status of the human sciences within the new social order.[52] This image is rendered in the distinctive scientific shorthand of the curve: a pure line distilling both operator's capabilities and the metropolis' energy consumption into graphical form.[53] The simplicity of this image, and the scientific authority it lays claim to, may be seen, in part, as a response to the postwar period of social turmoil and breakdown.[54]

The central problem Giese sought to address was that of fatigue.[55] Far more than simply a physiological condition, fatigue was a "psychosociological" phenomenon influenced by a complex set of variables, including work hours, wages, monotony, attitude, satisfaction, and nervous disposition. Each of these variables required comprehensive study. For instance, Giese wrote, it was important to "be clear about what it means . . . to call 'Amt' sometimes 770 times per hour, and one must multiply this number daily by 7, in order to convey through this trifling point a sense of the monotony of this work."[56] By sapping the worker's energy, monotony made her less resistant to the effects of fatigue.

---

phone service and plunging the city into darkness for several hours: "Fourteen hundred workers at the electricity plants brought the machine of national administration to a halt to a degree never achieved by the Spartacists' armored cars, machine-guns, and marauding methods." See *Berlin in Lights: The Diaries of Count Harry Kessler,* trans. and ed. Charles Kessler (New York, 1999), 61.

[50] He later continued his research at Halle and Leipzig. See Giese to RPM, 21 Nov. 1919, 8 June 1920, letters, Schule für Fernsprechgehilfinnen, RPM 5110, BAP.

[51] Fritz Giese, *Berufspsychologische Beobachtungen im Reichstelegraphendienst* (Leipzig, 1923), 22–3. For general discussion of the significance of inscription devices in the "psy" sciences, see Rose, *Governing the Soul* (cit. n. 24); Bruno Latour, "Visualization and Cognition: Thinking with Hands and Eyes," in *Knowledge and Society,* ed. H. Kushlick, vol. 6 (Greenwich, 1987).

[52] See Rose, *Governing the Soul* (cit. n. 24), 132 ff.

[53] For more on the "curve," see "Über die Kurve der möglichen Rangverschiebungen," report, 30 June 1920, RPM 5110, BAP.

[54] Many of the rationalization projects of this era circle obsessively around scenarios, both real and imagined, of stoppages and breakdowns. At a meeting at the Ministry of Labor (Reicharbeitsministerium, RAM), Schlesinger reported that in Munich, telephone service in entire neighborhoods was occasionally shut down due to overloading of the personnel. A note scribbled in the margin of the minutes of this meeting in the RPM archives contains the following question: "Where did Schlesinger get this unbelievable story?" See RPM 5110, 26 April 1920, BAP. The possibility that such incidents might represent a form of work stoppage worried the authorities; some operators certainly represented their nervous breakdowns in such terms. See, e.g., "Als meine Nerven streikten," *Unter den Reichsadler* 23 (1931): 186.

[55] Giese, *Berufspsychologische Beobachtungen* (cit. n. 51), 30–1.

[56] Ibid., 24.

Using Münsterberg's research as his point of departure, Giese followed the basic Taylorist strategy of disassembling work into parts and then reassembling them for optimal efficiency. He established that there were close to twenty different elements of switchboard work, such as visual recognition of an incoming call, inserting the jack into the correct place on the switchboard, and answering the caller. Each engaged a different combination of the operator's mental and sensory faculties: hearing, vision, attention, and memory. Giese calculated the time necessary to perform these tasks and the psychophysical profile associated with each. Out of these he created norms for selecting candidates and for improving the efficiency of those already employed.

Giese's research, however, went considerably beyond what he perceived as the limitations of the "human-motor" model favored by his American counterpart. In a context in which the trade unions had to be treated as legitimate partners in the quest for national efficiency, Giese argued that the science of work could no longer ignore the questions the unions raised. Labor power could no longer be considered in purely mechanical terms as reducible to calculations of energy expenditure; properly understood, labor power was a political and psychological problem that engaged the worker's entire being, including her satisfaction, health, mental well-being, and general moral sensibility. Transforming the skills and attitudes of the worker meant transforming her psyche, in a fashion that could only be done with her consent. By situating the problem of labor within a more organically conceived notion of the self and of the larger project of national regeneration, Giese's investigations opened up a wide discursive field.[57]

On the objective side, Giese showed that the "work curve" was determined by variables ranging from demand to the equipment itself, work hours, and more sociological factors. Variations in phone usage between one district and another, for instance, reflected the complex social topography of the capital city. The district of Wilmersdorf, with its demimonde and its bars and clubs, produced a work curve sharply at variance with that of Berlin's central business district. A further condition that distinguished Berlin from other cities was the freedom enjoyed by adolescents in the capital, where even the children "ring each other up" so as "to chatter about some triviality or other"—a "burdening of the exchange" that reflected conditions that were "purely sociological (and in part even racial-biological)."[58] Psychotechnics could not afford to ignore such factors; for this reason, he concluded, the science of work "must always proceed along cultural-psychological lines."[59]

In keeping with this conclusion, Giese was prepared to go very far in investigating the subjective aspects of telephony. In addition to charting the operator's reaction times, attention, and attitude, he also studied the effects of daydreaming, menstruation, and "moral character" on performance. It was not just conduct in the workplace that was at issue; the free time available to these women since the passage of the eight-hour day at the outset of the Weimar period made their leisure activities—whether film-going, shopping, attending lectures, or sporting activities—a vitally important area of research. Each of these variables had its own regularity, its distinctive curve, to be charted and added to the overall profile.

---

[57] For more on how this reflected a broader shift in the psy sciences, see Rose, *Governing the Soul* (cit. n. 24), 65–70.

[58] Giese, *Berufspsychologische Beobachtungen* (cit. n. 51), 23.

[59] Ibid.

This may be illustrated by considering one of the tests he administered to the Reichs-post personnel, which involved measuring the operator's response to sexual stimuli. The operator was hooked up to a pneumograph and given an album of postcards depicting erotic scenes. Through a peephole, the investigator watched while the subject leafed through the album. The pneumograph matched the process of turning from one picture to the next to her respiration curve. From this curve, Giese extracted what he called "ethical-erotic diagnoses" of the personnel.[60] Such diagnoses reflected the administration's desire for information about the moral character of the personnel, including their romantic and private lives. Although the results indicated to Giese that many of the women suffered from ethical defects, he remained certain that specialized training and close supervision could turn them into dependable workers capable of the most exacting concentration and responsibility.

A principal theme of Giese's writings was the subject of rhythm and the importance of alternating cycles of work and rest to maintaining good work hygiene.[61] Giese wrote that these cycles belonged with other regularities such as breath, pulse, and menstruation, all of which were part of a primordial cycle (*Urrhythmus*) of tension and relaxation, excitement and calm, whose regular alternation corresponded to the curve captured by a pneumograph.[62] The task of psychotechnics was to bring about the closest possible meshing of the human with the technical agents of productivity, thus to better integrate workers into the social-technical circuitry of modern industrial civilization. One way that Giese addressed this was by stressing the value of physical culture to workplace hygiene. In the early 1920s, exercise and fitness programs had been introduced into the workplace in many sectors of German industry, as a way to alleviate stress and monotony and to counteract the problem of *entgeistete* or *entseelte Arbeit*—work emptied of psychological meaning or satisfaction.[63] Giese was a strong advocate of gymnastics and devoted a virtually endless stream of writings to the subject.[64] Taylorism and gymnastics, he wrote in one article, approached the same task from different sides: training the body and its movements to promote purposeful activity. By strengthening the muscles, increasing the body's capacity for performance through training, and raising tolerance for fatigue, gymnastics could, if properly conceived, promote a kind of "Taylorization of the body."[65] Heightening the worker's awareness of physical activity as a time-and-motion–based process, physical culture could teach the worker to internalize the new rhythms of the modern workplace. Although he felt that most present-day gymnastics systems failed to accomplish this goal—largely because they did not address the mental processes that accompanied all physical activity—he set the task of future gymnastics as instilling a "typology of movement," modeled upon the time-and-motion studies of scientific management, taking account of the many mental conditions of work and minimizing the effects of fatigue. The German science of work would replace the "human motor" model with a

[60] Ibid., 57–60.

[61] This emphasis on rhythm reflects the influence of Karl Bücher's book *Arbeit und Rhythmus* (Leipzig, Germany, 1909).

[62] Giese, *Girlkultur* (cit. n. 5), 21.

[63] Joan Campbell, *Joy in Work, German Work: The National Debate, 1800–1945* (Princeton, N.J., 1989).

[64] See, e.g., Fritz Giese, *Körperseele; Gedanken über persönliche Gestaltung* (Munich, 1924); Giese, "Wiedergeburt des Körpers," in *Das Reich des Kindes* (Berlin, 1930).

[65] Fritz Giese, "Gymnastik und Taylorsystem," in *Weibliche Körperbildung und Bewegungskunst auf Grundlage des Systems Mensendieck* (Munich, 1924), 156.

concern for the "whole person," body and mind. Properly understood, gymnastics would help offset the specialization that was a necessary result of the modern division of labor, by strengthening—against the "one-sidedness" of modern work—the body-mind connection, in a fashion that was rational, modern, and humane. In this manner Giese forged a connection between the rehabilitation of body and mind and the subject's ability to internalize and augment her own welfare and bodily efficiency.

A rather similar motive may be detected in his approach to the issue of discipline and how best to achieve it without sacrificing the worker's well-being or consent. Whereas in the case of his "ethical-erotic diagnoses" Giese positioned the operator as the object of an intrusive gaze, here he sought to train the eye of the operator herself on her own performance. This meant, in the first place, establishing the legitimacy of psychotechnics vis-à-vis the RPM's existing methods of workplace surveillance. Using questionnaires, Giese polled supervisors concerning the performance of operators. He then compared these judgments with the results of his own tests. Giese found the supervisors' judgments unreliable and, in more than 50 percent of the cases, contradicted by his own observations. Moreover, given the surreptitious nature of the existing monitoring system, the supervisors' judgments created discontent among the personnel.[66] If psychotechnics was to have "social-political value," he argued, it would have to overcome this discontent and take its stand on "the middle ground between dogmatic *Bureaucratismus* and revolutionary freedom."[67]

Giese's own methods drew upon the full repertoire of aptitude testing techniques and made extensive use of graphical devices like kymographs, tachistoscopes, and pneumographs. Although we cannot here consider all the methodological implications bound up in the use of such devices, one at least shall be highlighted. It concerns his recasting of the surveillance issue and may be characterized as follows: while the Reichspost's methods remained external to the subject, Giese sought to fashion a form of internalized discipline.[68]

This may be best illustrated by considering one test Giese used that involved a device called a "self-registering attention-measuring apparatus."[69] This consisted of a series of kymographs attached to the operator's fingers by electrical wires. The results of the operator's work were displayed on a meter that was visible to her, thus enabling her both to observe the measurable progress of her own performance and, in principle, to improve on it.[70] In essence, it allowed the operator to read her own personal curve, an innovation that, coupled with wage incentives, would promote efficiency.[71] By simplifying the test so radically that it could be virtually self-administered, Giese

---

[66] Giese, *Berufspsychologische Beobachtungen* (cit. n. 51), 15. For an expression of this discontent, see "Kontrollbeobachtung im Fernsprechbetrieb," *Unter den Reichsadler* 14 (1922): 86–8. Such forms of surveillance, it was argued, turned the employees into "soulless machines."

[67] Giese, *Berufspsychologische Beobachtungen* (cit. n. 51), 16, 32.

[68] Ibid., 32. See also "Psychotechnische Eignungsprüfung," 1924–1926, RPM 47.01, 13524, BAP.

[69] Giese, *Berufspsychologische Beobachtungen* (cit. n. 51), 47.

[70] For a fuller description, see Fritz Giese, *Handbuch der psychotechnischer Eignungsprüfungen* (Jena, Germany, 1927), 206. Giese extracted a general norm from his research on operators, obtained from the so-called telegram-test, which measured ability to condense messages into telegram form as an indicator of the subject's internalization of the principles of work economy. This test became incorporated into the examination used for selecting candidates for employment. The tests, according to one advocate, "can also be used for purposes of training." See report, 30 June 1920, IV-75, RPM 47.01, 5110, BAP.

[71] For account of similar methods used at DINTA, see Geuter, *Professionalization of German Psychology* (cit. n. 29), 88.

enlisted the subject as consenting partner in her own rationalization. In the discourse of psychotechnics, workers were no longer simply amenable to having things done to them; they could do things to themselves.[72] This way of conceptualizing the "worker question" (*Arbeiterfrage*) reflected both the new level of sophistication attained by the science of work and the political sea change in Germany from an authoritarian to a democratic order. The question remained of what role the state would play as arbitrator in the sphere of industrial relations to ensure that self-rationalization did not degenerate into a new form of workplace exploitation.

THE RISE AND FALL OF PSYCHOTECHNICS AT THE SWITCHBOARD

In the early 1920s psychotechnicians spearheaded an ambitious program of aptitude-testing and occupational counseling throughout German industry. State, industry, and labor embraced this program as integral to the project of national reconstruction. The new field received the imprimatur of the Weimar state in 1920, when the Ministry of Labor (Reichsarbeitsministerium, RAM) created a special agency to deal with psychotechnics. A meeting at the Prussian Ministry for Science, Arts, and Education in late May 1920, with representatives from the RPM, the Reichsbahn, and several institutes for applied psychology in attendance, evinced support for this undertaking. Occupational psychology, the discussants concluded, was to play a key role in overcoming "the authoritarian state" in the present "democratic and socialist age." It would do so by creating a society in which social position was determined exclusively by aptitude and training.[73] A further key meeting at the RAM gave the final go-ahead to this undertaking.[74] By 1922, 170 psychological testing stations had been created across Germany, under Ministry of Labor auspices. A veritable psychotechnical craze was born, attended by great fanfare and a flurry of publications in leading journals. German trade unions endorsed the rationalization movement in the belief that science guaranteed "the best relation between labor power and the labor process." The support of labor affirmed its vital role in the creation of a scientifically regulated society from which conflict would be banished.[75]

The Reichspost stood in the forefront of these developments. A fully operational psychological testing station was set up at RPM headquarters in Berlin in 1922, and over the next two years, six more were created in Hamburg, Leipzig, Frankfurt, Breslau, Cologne, and Stuttgart.[76] Yet despite the seemingly propitious circumstances and high expectations, the experiment with psychotechnics in the Reichspost was destined to be a relatively short-lived affair. By 1926, all but the original laboratory in Berlin had been discontinued.[77] At the end of the 1920s, this, too, was closed. Although they continued to remain active, German psychotechnicians increasingly reoriented themselves toward characterological studies, which moved the focus away from performance and psychological "functions" and toward leadership qualities,

---

[72] On the significance of this shift, see Rose, *Governing the Soul* (cit. n. 24), 7.

[73] Geuter, *Professionalization of German Psychology* (cit. n. 29), 47.

[74] See Schule für Fernsprechgehilfinnen, Niederschrift über die Verhandlung, betreffend Förderung der Arbeitswissenschaft in den Verkehrsverwaltungen, am 26 April 1920, im Reichsarbeitsministerium, RPM 5110, BAP.

[75] Rabinbach, *Human Motor* (cit. n. 1), 279; Nolan, *Visions of Modernity* (cit. n. 31), 6.

[76] Schule für Fernsprechgehilfinnen, letter, 31 Aug. 1920, RPM 5110, BAP.

[77] For a chronology of these developments, see Psychotechnische Eignungsprüfungen 1925–1929, letter and report, dated 15 Oct. 1927, RPM 14509, BAP.

strength of character, motivation, and will—a significant shift in the nature of the self under investigation.[78]

Several factors account for these developments. In addition to the state's increasingly dire financial circumstances, as well as the general disillusionment with Americanism that set in after the world economic crisis in 1929, these factors included divisions within the field itself. In the late 1920s, a controversy erupted between the two main factions—the engineers and the university psychologists—over charges that the former had compromised the field's scientific legitimacy through excessive partisanship on behalf of employers, by using psychotechnical methods to ferret out malingerers and politically motivated troublemakers.[79] Yet even before the crisis, indeed as early as the mid-1920s, psychotechnics seems to have lost the legitimacy it enjoyed up to then among German workers. As Giese was to note in 1925, "Today it is exceptionally hard to carry out time-and-motion studies in factories. One is constantly confronted with the word 'Taylor' and the workers become distrusting and even rebellious. . . . *Psychotechnik* has become in their eyes a mechanization of spirit, an oppression of will and so forth."[80] The trade unions' withdrawal of support by 1925 became an important factor in the breakdown of the social consensus around psychotechnics.

This withdrawal of support reflected the emergence of a new political conjuncture in which the unions felt they had lost the protection of the state.[81] The real high-water period of the rationalization movement in Germany coincided with the creation of a new coalition, lasting from 1924 to 1928, that was dominated by bourgeois and conservative parties in which the Social Democratic Party did not participate.[82] The initiative was now seized by organizations such as the German Institute for Technical Labor Training (Deutsches Institut für Technische Arbeitsschulung, DINTA), which was deeply hostile to the trade unions and which adapted psychotechnics to the task of creating a reorganized workplace under the absolute control of engineers. DINTA's approach represented a classic form of reactionary modernism: coupling technological modernity with authoritarian politics, the institute sought to fashion a new worker who was both willing to participate in his or her own rationalization and politically reliable.[83] Although the organization's activities found a ready reception among industrialists who were eager to restore their authority in the workplace, it met with resistance among those who saw DINTA-style rationalization as a thinly veiled form of exploitation.

The withdrawal of support by trade unions may be illustrated by turning once again to the reception given to the new science by the switchboard operators. As we saw, operators were, in the prewar era, ready—perhaps too ready—to take industrial psychology's claims at face value. This attitude continued into the postwar era, with operators remaining generally receptive to the arguments of the rationalization

[78] See Geuter, *Professionalization of German Psychology* (cit. n. 29), 86; Rabinbach, *Human Motor* (cit. n. 1), 278–88.

[79] Rabinbach, "Betriebspsychologie" (cit. n. 42), 59–60. Moreover, the results were simply disappointing. In a letter written to the minister of the Interior in response to an inquiry regarding the advisability of establishing a central Imperial Office for Psychotechnics, a representative of the RPM wrote that such a venture seemed premature based on the results to date. Psychotechnische Eignungsprüfungen 1925–1929, letter, 12 March 1928, RPM 14509, BAP.

[80] Giese, cited in Chestnut, "Psychotechnik" (cit. n. 7), 782.

[81] Ibid.

[82] Maier, "Between Taylorism and Technocracy" (cit. n. 12), 48.

[83] Nolan, *Visions of Modernity* (cit. n. 31), 196.

experts. They acknowledged the principle of "the right man for the right job" as the most effective way of organizing modern labor, despite the hardship it meant in individual cases.[84] As Ursula Nienhaus has written, these women saw themselves as an industrial vanguard and, as such, accepted the rationality of factory mechanisms and time-and-motion-studied work.[85] Moreover, at least during the early 1920s, with the nation still feeling the aftershocks of the war, switchboard operators were prepared to accept the necessity of personal sacrifice, greater efficiency, and streamlining of the labor process. Calls for sacrifice, as they understood it, were but a first step within a larger process of social reconstruction that contained an implicit promise of emancipation. The project of personal restructuring was thus closely entwined with the grand Weimar project of social restructuring.

Interpellated by both state and science as members of a national community faced with a project defined in productivist terms, operators responded in the same terms. Thus one article in their publication, titled "The Conservation of the Body's Energies," praised gymnastics for strengthening the nerves and increasing efficiency.[86] Such measures, it was argued, were particularly important for mental laborers such as the operators, whose jobs made such high demands on their nervous systems. The principle invoked here was based not on the nineteenth-century German tradition of gymnastics as political training but on more modern, hygienic precepts concerning the value of physical culture as an antidote to the wear and tear of work and to the problem of *entgeistete* work. Subsequent articles suggested that it was not so much the demands of the work, however great, as its "one-sidedness" that led to the deterioration of performance. "Overspecialization" led to a disastrous split between mind and body, disturbing the normal rhythm of exertion and relaxation. "Modern man," wrote one author, "has degenerated thanks to a one-sided civilization and must return to a more balanced training of both body and mind."[87] Physical culture, like other aspects of contemporary work science that conceived of the person as a "whole" whose parts, body and mind, were in constant interaction, helped offset the "one-sidedness" of industrial civilization.

Operators recognized themselves as beneficiaries of the larger cultural project of rationalization, even if they sometimes questioned the manner and speed of its implementation. It was only when this project became linked with an attack on their jobs that they began to turn a skeptical eye on the claims of the psychotechnicians. Following the fiscal emergency of 1923, the German state stepped up its rationalization campaign and, at the same time, began cutting back public sector employment. The contracting labor market hit female employees hard. The operators responded by writing to the RPM in November 1923, in the midst of that year's catastrophic hyperinflation, to voice their objection to cutbacks and the intensification of the work process.[88] But to little avail. In 1924, 13,000 female employees (18 percent of the total) were let go by the RPM.[89]

---

[84] See "Die Psychotechnische Auslese," *Unter den Reichsadler* 13 (1921): 373.

[85] Nienhaus, *Vater Staat und seine Gehilfinnen* (cit. n. 15), 171.

[86] "Die Erhaltung der Körperkräfte," *Unter den Reichsadler* 11 (1919): 268.

[87] "Die erste Tagung für die körperliche Erziehung der Frau," *Unter den Reichsadler* 17 (1925): 134–7.

[88] See Bl. 125, RPM 47.01, 14575/1, BAP. See also Nienhaus, *Vater Staat und seine Gehilfinnen* (cit. n. 15), 142–51.

[89] See Nienhaus, *Vater Staat und seine Gehilfinnen* (cit. n. 15), 74.

This move was interpreted as a signal that the administration was restructuring the workforce in the interests of making it younger and easier to discipline. Faced with a loss of job security, the operators began to realize that they were also faced with the erosion of their status as a special class of worker, with privileges that set them apart from other workers. In an article published in 1924, their leader, Else Kohlshorn, suggested "over-rationalization" was being used in the service of engineering a more politically pliable workforce. The "systematic repression," as she called it, of the female personnel—particularly, the uncompensated release of "troublemakers" or newly married and ailing operators whose health was impaired by their "nerve-shattering" jobs—was being facilitated by the scientifically intensified work of the remaining personnel.[90]

By the mid-1920s, operators had begun to chafe openly at the regime of aptitude testing. RPM minister Ernst Feyerabend attempted to reassure them that the psychological testing stations would not serve merely research purposes.[91] But these reassurances did little to allay their concerns. In 1926, one internal memo summed up the situation in the following terms:

> The workforce is adamantly opposed to the psychotechnical aptitude tests. [T]hey question the need for administering the tests to those already employed. They feel that such employees should not be psychotechnically tested at all, if no consequences for the employment of the personnel are to ensue, and the tests simply serve the purpose of research.[92]

Within the Reichspost, opposition was chalked up to "radical elements" in the workforce. But opposition had by this time become widespread. For the employees, it was not only the neutrality of the tests that was in question; the entire rationale for them had become suspect.

From the mid-1920s onward the operators' attitudes toward rationalization, its ends, and its means took an increasingly critical tone. After 1926, the rewriting of the nation's social insurance laws, which had hitherto given operators suffering from job-related stress and nervous disorders recourse to Germany's insurance courts, deprived them of this instrument in their conflicts with the administration.[93] The dismantling of the psychological testing stations was accompanied by stepped-up efforts to introduce greater economies: automation of service where possible, and further rationalization of the workforce.[94]

---

[90] Else Kohlshorn, "Die Lage der weibliche Post- und Telegraphenbeamtinnen," *Archiv für Frauenarbeit* 1 (1924): 80–1. Another objection raised by the operators was that the investment in psychotechnics competed with funds for medical benefits. See *Verband* to RPM, letter, 10 Nov. 1925, RPM 14509, BAP.

[91] Bl. 106, RPM 14509, BAP.

[92] Report, with letter of 3 June 1926, RPM 14509, BAP.

[93] See Killen, "From Shock to *Schreck*" (cit. n. 2); Greg Eghigian, "Die Bürokratie und das Enstehen von Krankheit: Die Politik und die 'Rentenneurosen' 1890–1926," in *Stadt und Gesundheit: Zum Wandel von 'Volksgesundheit' und kommunaler Gesundheitspolitik im 19. und frühen 20. Jahrhundert*, ed. J. Reulecke and A. Gräfin zu Castell Rüdenhausen (Stuttgart, 1991), 203–23. Tröger argues that prudent concern for workers during the Weimar period was not the result of humanitarian concern but stemmed from the expense of prematurely worn-out workers caused by the generous social insurance system. Anne-Marie Tröger, "The Creation of a Female Assembly-Line Proletariat," in Bridenthal, Grossmann, and Kaplan, *When Biology Became Destiny* (cit. n. 37), 247.

[94] Schlesinger argued most strenuously for automation as the solution to the RP's labor problems. See note 43.

The catastrophic economic situation that developed in Germany after 1929 further deepened the sense of disillusionment, especially as it became clear that America, the promised land of rationalization, had not been spared a similar fate.[95] While the administration steadfastly denied that "overburdening" constituted a serious problem in the telephone system, and the Reichspost's doctors mounted a campaign designed to stigmatize operators claiming illness as malingerers, editorials in the operators' publication painted a picture of a deeply demoralized workforce afflicted by a virtual epidemic of shattered nerves.[96] One article after another described conditions as intolerable and suggested that the "human economy" of the scientific managers had degenerated into an inhuman form of workplace discipline that, despite paying lip service to the notion of "German rationalization," merely reproduced all the worst features of Taylorism.[97]

It was no wonder, ran one editorial, that callers complained constantly of poor service. Rationalization did not alleviate the problem of fatigue; rather, it intensified it, producing pathological exhaustion and predictable consequences: bad connections, crossed lines, confusion, and worse.[98] The article proceeded to discuss an expert opinion that had been commissioned by the Reichsbahn from Fritz Giese following a recent train wreck near Cologne. It is a striking measure of the severity of the crisis within his field that Giese, once among its most energetic advocates, had by this time come to doubt whether its objectives could be achieved within the present system. By 1928, he had become a member of DINTA, and his approach to the field of industrial relations was becoming increasingly reactionary. In the expert opinion cited in the editorial, this shift manifested itself in the form of a criticism of over-rationalization. Whereas in his earlier writings he had celebrated statistics, abstraction, and mathematics as essential tools of the rationalization project, Giese now expressed reservations concerning what he called the "mathematization of the self": the remorseless attempt to force the "personal curve" to match the "work curve."[99] The operators, fully appreciative of the human costs of this attempt, welcomed Giese's philosophical defense of the self.

Nowhere is the disillusionment of the switchboard personnel better illustrated than in a satiric article that appeared in 1930, titled "*Die Kurve.*" In it, the anonymous author describes a dream in which, worn out by her job and too tired to go to the lecture she had planned to attend, she unexpectedly received a visit from the moon. What ensues is a fantastical conversation in which the moon reminds her of "the law of the curve." "In heaven's name!" protests the woman. "Leave me alone! Productivity curve, sickness curve, adjustment of work hours to the traffic curve—those are watchwords that I, as a working woman, have heard all too often." And yet, the moon responds, "you seem to know nothing of the big curve that keeps the cosmos in motion." What

[95] On the growing criticism of Americanism in Germany after 1929, see Nolan, *Visions of Modernity* (cit. n. 31), 227–35.

[96] See "Überlastung liegt nicht vor," *Unter den Reichsadler* 20 (1928): 179–80. Many of these articles depict the switchboard itself as a kind of persecutory agent.

[97] "It was already clear in the 1920s," writes Anne-Marie Tröger, "that no human would be able to endure Taylorised work for a full normal work life from age 15 to 65 without major damage to the nervous system." See Tröger, "Creation of a Female Assembly-Line Proletariat" (cit. n. 93), 257.

[98] See "Die Rationalisierung bei der deutschen Reichspost in der Tätigkeitsgebieten des weiblichen Personals," *Unter den Reichsadler* 20 (1928): 345–7.

[99] See Giese, *Girlkultur* (cit. n. 5), 87; "Das Ermüdungsproblem und die Rationalisierung," *Unter den Reichsadler* 20 (1928): 289–90.

the moon evidently implies is that the author has neglected her well-being by failing to bring her cycle of exertion and rest into harmony with the "energy vibrations" of the cosmos. "And why should I not enjoy some peaceful twilight hours during the full moon?" she asks. "Because," the moon informs her, "at this time my cosmic positive electricity tugs most strongly at the earth's negative energy. Every living thing exists in a strong force field and oscillates therein. Think of all the things you could achieve at the zenith of my influence!"[100]

The moon, gendered male in German, assumes the position of scientist in this fantasy, albeit in a guise that mocks the language of the rationalization experts. The recitation of terms such as productivity curve and sickness curve suggests that the rhetoric itself has begun to induce a kind of brain fatigue. But the moon, as symbol of the monthly cycle, also reminds the woman of a higher duty, beyond that of the workplace, recalling her to her maternal responsibilities and to the intricate balance between production and reproduction. This reminder was one that the operators were becoming increasingly sensitive to, as the burdens of the new woman role and the human costs of the rationalization project began to arouse nostalgia for more traditional forms of female personhood.[101] The tone of resignation in this article would only deepen over the next three years, which were difficult ones for the Reichspost personnel. Those who were married became the object of a fierce campaign against "double earners," while the rest saw their benefits slashed or were laid off. With the coming of the new regime, their leaders were fired and their trade union shut down. By 1934, female employment in the Reichspost had sunk to an all-time low.[102]

FROM AMERICANISM TO NATIONAL SOCIALISM

The collapse of the Weimar rationalization project became a significant factor within the constellation of forces precipitating the demise of a state that had been closely identified with rationalization and its promise of an efficient and democratic new order. Yet certain features of this project were to survive and flourish under the Nazi regime that came to power in 1933. If the psychotechnicians' promise of mediating between labor and capital to create a new social order ultimately proved hard to deliver on, their quest for such an order did not end. With fervor perhaps unmatched in any other branch of psychology, they embraced the new regime.[103] The shattering of what Peukert calls the "grand ambitions of Weimar social engineering" was accompanied by a paradigm shift within the human sciences: the replacement of the "rationalization discourse" by the "selection discourse."[104] Industrial psychology was among those fields leading the way, its methods appropriated for purposes of indoctrination

---

[100] "Die Kurve," *Unter den Reichsadler* 22 (1930): 130.

[101] Nolan, *Visions of Modernity* (cit. n. 31), chap. 10; Bridenthal, Grossmann, and Kaplan, *When Biology Became Destiny* (cit. n. 37).

[102] Nienhaus, "Unter dem Reichsadler" (cit. n. 15) ; Nienhaus, *Vater Staat und seine Gehilfinnen* (cit. n. 15), chap. 8. For a discussion of Nazi policy toward female employment, see Tröger, "Creation of a Female Assembly-Line Proletariat" (cit. n. 93). As she points out, unions accepted rationalization almost euphorically during the 1920s, but by 1933 one union study stated, "The machine has become a curse." "It must be assumed," Tröger concludes, "that the continuing process of rationalization would have encountered great difficulties without the Nazi regime's violent suppression of the labor movement." Ibid., 246.

[103] Jaeger and Stäuble, "Die Psychotechnik" (cit. n. 43), 91; Rabinbach, "Betriebspsychologie" (cit. n. 42), 62.

[104] Peukert, "Amerika" (cit. n. 31), 81.

by organizations such as DINTA, now subsumed into the German Labor Front (Deutsche Arbeitsfront, DAF) created by the Nazis.[105] Under the DAF, psychotechnics finally received its own institute, something that had eluded the field throughout the 1920s. As the Society for Psychotechnics put it in a statement issued in 1933: "All practitioners and scientists in the fields of applied psychology and psychotechnics who are for the new state must come together. Innumerable new tasks are awaiting realization."[106] Indeed, the new political dispensation presented the still-nascent field with tremendous opportunities to prove its worth to the national community. As Ulfried Geuter has shown, it was under the Nazis that psychology finally became a fully professionalized discipline. Frustrated in its development during the 1920s, psychology enjoyed its greatest success under the Nazis.

Particularly fruitful was psychology's alliance with the Wehrmacht, under whose auspices it played a key role in the militarization of German society undertaken by the new regime. A centralized institute was created in the army for conducting mass tests of troops and selecting officers, as a result of which, between 1933 and 1941, the number of psychologists in Germany grew from 33 to around 500, most employed by the army.[107] Psychotechnics now became absorbed into the larger field of characterology, which stressed military criteria such as leadership, motivation, and subordination to officers. This was in keeping with an overall shift in the nature of the self under investigation, from an emphasis on performance to a "diagnostics of the total personality." One aspect of this entailed an effort to redesign Kraepelin's "work curve" to measure the effect of will on performance.[108]

Giese, for one, remained skeptical that will could be measured using psychotechnical methods. This did not reflect any lack of commitment on his part to the new regime; on the contrary, he found it quite congenial to the increasingly metaphysical conception of labor power he had been moving toward in his later writings. Indeed, Giese welcomed National Socialism with particular enthusiasm, going so far as to proclaim that 1933 represented the Year 1 of history.[109] In 1934 he devoted a monograph on Nietzsche to the führer, noting that he regularly taught a seminar on the political psychology of *Mein Kampf* and that Hitler had donated signed copies to each of his students.[110] He also pushed to be named head of a new institute of German psychology in East Prussia, but his premature death in 1935 deprived him of further opportunity to serve the new political order.

In certain respects, Giese's allegiance to this new order had already been anticipated in his 1925 book *Girlkultur,* with its rather ambiguous vision of a regimented social type that Kracauer, for one, would later identify as "militarist." What Giese appreciated in the Tiller Girls was their embodiment of the highly drilled mechanical rhythms of "collective man," their "severe yet humane eugenic consciousness," which he identified as the key to national regeneration.[111] Between 1925 and 1933, however,

[105] Maier, "Between Taylorism and Technocracy" (cit. n. 12), 57; Nolan, *Visions of Modernity* (cit. n. 31), 234–5.
[106] Geuter, *Professionalization of German Psychology* (cit. n. 29), 181.
[107] Ibid., 161.
[108] Ibid., 90.
[109] Hinrichs and Peter, *Industrieller Friede?* (cit. n. 32), 75.
[110] Fritz Giese, *Nietzsche: Die Erfüllung* (Tübingen, Germany, 1934), 191. Giese died accidentally while undergoing an operation shortly before he was to be inducted into the SS in 1935. See Cornelius Borck, "Electricity as a Medium of Psychic Life," *Science in Context* 14(4) (2001): 12.
[111] Giese, *Girlkultur* (cit. n. 5), 141.

Giese's political views underwent a significant shift. In *Girlkultur* he had expressly favored American body-culture over its militarist German counterpart, which, with its early nineteenth-century roots in the nationalist gymnastics of Friedrich Ludwig Jahn, had been discredited by the general crisis into which Prussian militarism had delivered Germany in the Great War. In his writings of the early 1930s, however, with the American model tarnished, Giese embraced the new body-culture of the Nazis, for whom Jahn was a central figure. It was Nietzsche's *Übermensch,* according to Giese, that Hitler invoked when, at the Deutsche Turnfest held in 1933 in honor of Jahn, he conjured an image of the Germans as a people in whose "radiant" bodies and minds an image of a "higher unity" could be found.[112]

In this unity of body and mind, Giese saw the physiognomy of the National Socialist ideal of the self. It was now a question, for him, of preserving the "noncalculable character of all life." This shift had significant implications. In his early research, Giese had proposed, in effect, a kind of humanitarian intensification of work science, inscribing the principle of efficiency into the very core of the operator's being, albeit in a manner meant to win her consent. In his later work, amid increasing talk of the limits and human costs of rationalization, he began to reassess this project. Science, he opined, had to have "cultural" goals, if it was not to turn the worker into a purely instrumentalized being.[113] His writings of this period embraced metaphysical conceptions of work, stressing that the biological and technological components of labor had to be directed by a teleological impulse. This was evidently supplied by the leadership principle of the Nazis.[114]

Yet this shift did not lead to an outright repudiation of his earlier views nor to any minimizing of the role he envisioned for psychotechnics in the Third Reich. Giese resolved his reservations concerning the consequences of rationalization for the self by drawing a distinction between those attributes susceptible to psychotechnical calculation and those that lay beyond scientific analysis. Thus he exempted attributes such as courage, will, and nobility—the masculine virtues—from experimentation.[115] The implications of this gendering of the psychotechnical subject became clear with Giese's embrace of the National Socialist ideal. The promise of this ideal was its reconciliation of two seemingly opposed goals: on the one hand, to accomplish a (forced) "synthesis" of body and mind, labor and capital; on the other, to reinscribe the division between "body" and "mind" along gendered lines, by dividing the social body into the "masses" (coded female) and the masculine leadership principle. If Giese had earlier sought a new principle of social organization in the image of a humane, rational science whose aim was to win consent by replacing external with internal discipline, he now embraced a regime and a politics of the self that was based on a forced resolution of contradictions, recuperation of an authoritarian warrior ideal, and a gendered social order.[116]

[112] Giese, *Nietzsche* (cit. n. 110), 56.

[113] Fritz Giese, *Philosophie der Arbeit* (Halle, 1932), 26.

[114] On the metaphysical and National Socialist turn in German psychotechnics, see Rabinbach, *Human Motor* (cit. n. 1), 282–8.

[115] Fritz Giese, *Handbuch der Psychotechnische Eignungsprüfungen* (Halle, 1925), 209–10.

[116] On the gender politics of National Socialism, see Nienhaus, *Vater Staat und seine Gehilfinnen* (cit. n. 15); Claudia Koonz, *Mothers in the Fatherland* (New York, 1988); Jill Stephenson, *Women in Nazi Germany* (Harlow, England, 2001).

CONCLUSION

Does the story recounted here represent a case of the birth of fascism out of the spirit of social reform?[117] Was the project of social and personal transformation to which the human sciences made themselves handmaiden in the Weimar state one that, according to a fatal logic, led to the radicalism of the Nazi regime? The social critic Siegfried Kracauer, Giese's contemporary, certainly felt this to be the case. When in 1928 he bestowed the name "mass ornament" on the assemblage of bodies displayed in the precision dance team, he strove to capture in this term the wish image of a scientifically restructured, perfectly rationalized society that had banished conflict and transcended its own contradictions. It was a wish image that, by virtue of the powerful aura surrounding American-style social models, came to be internalized by a significant sector of Weimar society, including those modern types, the operators, whom Kracauer identified as the "natural objects" of psychotechnics.[118]

Kracauer construed this wish image in largely negative terms. "In the German view," wrote an industrial expert he cited in his study of Germany's salaried masses, "work must lead to an unfolding and realization of one's person. It must be viewed as service to the great tasks of the national community to which we belong."[119] In such statements, Kracauer discerned a form of false consciousness that failed to recognize the latent authoritarianism within the rationalization movement. The story I have told here, however, suggests a somewhat more complicated picture. Nikolas Rose, in his history of the sciences of "psy," suggests we need to understand how the language of scientific experts comes to be introjected by individuals as norms and standards both rewarding and persecuting.[120] The case of the operators offers an instructive illustration of this problem. It suggests that the scientifically mediated relation between state, industry, and labor in Weimar Germany rested on an intricate process of negotiation, one that only began to break down in the mid-1920s with the erosion of the postwar social consensus. Up to that point, however, the self-image of this new type of mental laborer was closely bound up with the scientific and organizational ethos embraced by the German state and industry after the war, under the spell of powerful models imported from America and then reworked to meet the requirements of the German workplace. Efficiency, rationalization, personal curve: like many others in Germany, the operators embraced these watchwords of the industrial experts out of a belief in both their aura of scientific objectivity and their promise of a new, distinctively modern form of selfhood.

For many Germans in the 1920s, rationalization meant not simply a streamlined work process but the production of a new kind of worker. At the very center of the rationalization project, indeed, stood the possibility of self-rationalization, a powerful, yet ambiguous, notion according to which workers would remake themselves in line with a more general ideal of social transformation. Postwar social restructuring entailed a simultaneous project of personal restructuring that coupled demands for dis-

---

[117] Detlev Peukert, "The Genesis of the 'Final Solution' from the Spirit of Science," in *Nazism and German Society,* ed. David Crew (New York, 1994), 274–99.

[118] Kracauer, *The Salaried Masses* (cit. n. 4); Kracauer, "The Mass Ornament" (cit. n. 6). For a rereading of Kracauer's analysis of the mass ornament that stresses the pleasure taken by the subject in the "drill" that accompanies the rationalization process, see Theweleit, *Male Fantasies* (cit. n. 39), 1:431–4.

[119] Kracauer, *The Salaried Masses* (cit. n. 4), 44.

[120] Rose, *Governing the Soul* (cit. n. 24), 152.

cipline and sacrifice with promises of emancipation. Psychotechnics was granted a key role in this dual undertaking, a role granted at least initially on the strength of its claim to move beyond the authoritarian social and political arrangements of prewar Germany and to embody a new social paradigm. As both a science of the workplace and a theory of the self, it was embraced by scientists, politicians, and workers alike as one of the keys to fashioning a democratic new order. Yet it never fully resolved the ambiguities inherent in this project, not least those concerning its claim to mediate between workers and employers and its ultimate consequences for the self. The new social order rationalization promised to erect in place of the old one remained elusive within the existing framework of Weimar society. Meanwhile, its subjects, initially receptive to the demand to remake themselves in line with the challenges of the postwar experiment in democracy, found themselves on the front lines of a wider debate about the limits and the human costs of rationalization, a debate in which productivist goals and political reliability were increasingly privileged over questions of job satisfaction and occupational identity.

Interpellated as a special workforce and as exemplars of a new kind of personhood, whose status was protected by certain rights and privileges guaranteed by the new republic, switchboard operators were eventually forced to confront how deeply their position had been compromised by the Weimar rationalization project. By the end of the 1920s, rationalization—once hailed as a magic formula of industrial relations—had itself become entangled in the deepest crisis of industrial civilization. Switchboard operators, who had tantalized experts with visions of a perfectly rationalized workforce, eventually withdrew their consent from this vision. The persisting conflicts of interest in German society could not be wished away; under National Socialism they would be resolved by force.[121]

---

[121] Peukert, *The Weimar Republic* (cit. n. 30), 115. For Peukert's own reflections on the mass ornament, see ibid., 161–3.

War Neurosis, Adjustment Problems in Veterans, and an Ill Nation:

The Disciplinary Project of American Psychiatry during and after World War II

By Hans Pols[*]

ABSTRACT

After World War II, the confidence of American psychiatrists was at an all-time high as a result of their successful participation in the war. When the incidence of mental breakdown in the American armed forces rose to unprecedented heights, new and effective psychotherapeutic methods were developed to treat the traumatic effects of the extraordinary stresses of warfare. At the same time, social scientists concluded that breakdown incidence was inversely related to morale, which led to the development of preventive measures aimed at specific groups. Both initiatives stimulated a number of psychiatrists to plan projects of social engineering after the war. They first focused on aiding the reintegration of returning veterans. Later, they addressed the poor mental health of the American population as a whole, which they considered to be the consequence of faulty child-rearing methods.

INTRODUCTION

In the air, passing time during the long journey on a soon to be decommissioned B-17 bomber between an unidentified major airport and the little airport of Boone City, Ohio, the three main characters of the 1946 film *The Best Years of Our Lives* exchanged thoughts about their years in the army and what life would be like now that the war was over. Air Force bombardier Fred Derry admitted that he was "just nervous out of the service, I guess." Army sergeant Al Stephenson responded: "The thing that scares me most is that everybody is gonna try to rehabilitate me."[1] Most contemporary viewers of this movie would have recognized the significance of these state-

* Unit for History and Philosophy of Science, Carslaw F07, University of Sydney, NSW 2006, Australia; hpols@science.usyd.edu.au.

The research for this article has been supported by a Discovery Project Grant of the Australian Research Council, "War, Trauma, and Rehabilitation: The Army, Psychiatry, and World War II" (DP0450751). The author is grateful to Greg Eghigian, Andreas Killen, and Christine Leuenberger for organizing the conference from which this volume resulted, the participants at the conference for their feedback, and Stephanie Oak for her detailed and insightful comments on earlier drafts.

[1] *The Best Years of Our Lives* (videorecording), screenplay by Robert E. Sherwood, produced by Samuel Goldwyn, directed by William Wyler, presented by Samuel Goldwyn Pictures Corporation, Santa Monica, Calif., MGM Home Entertainment, 2000 [1946]. In the original script for this Oscar-winning movie about the predicament of returning ex-servicemen, the character of Homer Parrish, the third man on the plane, was that of a veteran suffering from shell shock. In the final version, Parrish

ments immediately. In 1944, there began appearing an unusual number of magazine articles, newspaper features, lectures, radio broadcasts, and pamphlets, written by psychiatrists, psychologists, sociologists, priests, ministers, politicians, and social commentators, devoted to the veteran problem and detailing the psychological assistance necessary to transform former soldiers familiar with war, violence, and death into peaceful and productive postwar citizens. According to American psychiatrists, psychologists, educators, and opinion leaders, the main challenges in the transition from war to peace were psychological in nature. This made the psychology of the ex-serviceman one of the most discussed topics after the armistice—much to the chagrin of most returning veterans, who felt that the attention devoted to their mental lives was mistaken and misplaced.

Defining the principles for aiding the readjustment and the rehabilitation of returning veterans after World War II constituted the first major postwar political project of American psychiatry. These principles were based on the utopian beliefs of a number of reformist psychiatrists, inspired by prewar public health and mental hygiene ideas, that their discipline not only had the tools to treat mental illness and mental disorder in individuals but also could contribute to the social and cultural transformation of society as a whole. These beliefs were strengthened by the enormous success psychiatrists had had in treating mentally unwell soldiers during the war. Short-term methods of psychotherapy, administered near the battlefield, as soon as possible after a soldier had broken down, had proven to be very effective. In addition to developing new psychotherapeutic approaches for the treatment of individuals, American psychiatrists had also been involved in preventive activities directed at units, platoons, battalions, and the armed forces as a whole. By boosting morale and disseminating information among soldiers about the nature of the human mind and how to best manage emotions, psychiatrists hoped to reduce the incidence of mental breakdown. The perceived results of both initiatives—developing effective treatment methods for individuals and preventive programs for the army as a whole—increased the confidence of American psychiatrists in developing postwar disciplinary projects, which involved making psychiatric care available within the community on an out-patient basis as well as preventing a future world war and the maintenance of peace and democracy. As G. Brock Chisholm, former director-general of medical services for the Canadian Army and future founding director-general of the World Health Organization, stated: "We cannot afford to squander our best brains on psychotherapy . . . [but] should do [our] utmost to examine, compare, study, understand and treat the ills of our society."[2] According to Chisholm, not only did the transformation of society have a much higher priority than the treatment of individuals, but it was also imperative to prevent the annihilation of humankind in the near future.

The psychiatric discourse on the veteran problem held immediate appeal for Americans who wanted to organize the best possible welcome for returning soldiers, who

---

was a double amputee, a role played by real veteran (and amputee) Harold Russell. For an analysis of this film and other "reintegration dramas," see David A. Gerber, "Heroes and Misfits: The Troubled Social Reintegration of Disabled Veterans of World War II in *The Best Years of Our Lives*," in *Disabled Veterans in History,* ed. David A. Gerber (Ann Arbor, Mich., 2000); and Sonya Michel, "Danger on the Home Front: Motherhood, Sexuality, and Disabled Veterans in American Postwar Films," in *Gendering War Talk,* ed. Miriam Cooke and Angela Woollacott (Princeton, N.J., 1994).

[2] G. Brock Chisholm, "The Reestablishment of Peacetime Society" (William Alanson White Memorial Lectures), *Psychiatry* 9 (1946): 3–20, on 19. For Chisholm, see Allan Irving, *Brock Chisholm: Doctor to the World* (Markham, Ontario, 1998).

had prevailed, against all odds, over well-equipped and persistent enemies. The discourse also appealed to those individuals who were apprehensive about the return of more than 10 million veterans to a society that no longer benefited from the war economy. Many groups attempted to contribute to the reintegration of all returning veterans. However, many veterans hated all the fuss and wanted to have nothing to do with it. Some of them provided criticism of and ironical commentaries on the efforts of psychiatrists to turn every veteran into a neuropsychiatric problem in dire need of treatment and rehabilitation. The grand, and at times even utopian, ambitions of postwar American psychiatrists to contribute to the general political project of creating a peaceful, prosperous, and well-integrated society clashed with the needs, desires, and intentions of the individuals whose selves they aimed to shape and transform. The involvement of American psychiatrists in the rehabilitation of veterans would prove short lived. As psychiatry moved away from the clinical treatment of individuals with clearly identifiable diagnoses, veterans were less inclined to adopt its guidelines and perspectives. This did not deter psychiatrists from formulating even bolder goals for their postwar political project. They made the management of a peaceful, democratic, and open society in which aggression and hostility were handled with therapeutic prowess their aim for the cold war era.

In this article, I investigate the development of the involvement of American psychiatry with the armed forces during World War II and the broader social and political concerns of the discipline in the postwar world.[3] During the first stages of the war, psychiatrists developed extensive screening programs to weed out unfit individuals from the armed services. Despite the high rejection rates of inductees, the incidence of mental breakdown continued to climb during 1942, when military operations in the Mediterranean and Pacific Theaters of Operations faltered. As a response, new forms of psychotherapeutic intervention were developed, which were widely applied in the armed forces. Sociologists and psychologists investigating soldiers suggested preventive measures to improve morale. Both activities inspired the utopian ideas of American psychiatrists after World War II. I highlight the psychiatric discourse around the rehabilitation of returning veterans as the first attempt of psychiatrists to formulate a broader political project in the immediate aftermath of the war.

PSYCHIATRY AND SELECTION

There were three approaches in American psychiatry before 1940 that are relevant for understanding the development of American military psychiatry during World War II.[4] First, academic psychiatrists and superintendents of mental hospitals had

[3] For overviews of the involvement of American psychiatry with the armed forces during World War II, see Ellen Herman, *The Romance of American Psychology: Political Culture in the Age of Experts* (Berkeley, Calif., 1995), chap. 4, "Nervous in the Service"; Ben Shephard, *A War of Nerves: Soldiers and Psychiatrists, 1914–1994* (London, 2000); Hans Binnenveld, *From Shellshock to Combat Stress: A Comparative History of Military Psychiatry* (Amsterdam, 1997); and Edgar Jones and Simon Wesseley, *Shell Shock to PTSD: Military Psychiatry from 1900 to the Gulf War,* Maudsley Monographs 47 (New York, 2005); Paul Wanke, "American Military Psychiatry and Its Role among Ground Forces in World War II," *Journal of Military History* 63 (1999): 127–46.

[4] For an overview of psychiatric ideas in the 1930s, with an emphasis on mental hygiene, see Hans Pols, "Divergences in American Psychiatry during the Depression: Somatic Psychiatry, Community Mental Hygiene, and Social Reconstruction," *Journal of the History of the Behavioral Sciences* 37 (2001): 369–88.

experimented with somatic treatment methods (Metrazol shock, electroconvulsive therapy, and lobotomy) that had been introduced in the 1930s and hoped to identify the biological causes of mental illness. This approach had been dominant within American psychiatry during the first part of the twentieth century and had been reinvigorated with the introduction of new treatment methods.[5] Second, through a variety of out-patient agencies, psychiatrists associated with the National Committee for Mental Hygiene had advocated increasing the number of individuals who could benefit from psychotherapy by arguing that those mental disorders that did not constitute mental illness, among them the neuroses, were worthy of their attention.[6] These psychiatrists were inspired by American interpretations of psychoanalysis and were eager to broaden the array of conditions they considered suitable for professional intervention. A third strand in psychiatric thinking was represented by mental hygiene psychiatrists, who sought to realize the ideal of prevention, the guiding philosophy of the public health movement, within psychiatry. They hoped to develop preventive methods that could reduce the incidence of mental illness and mental disorder in the population as a whole. Their methods included educating the public on health issues, reducing economic instability, reforming education, and making mental health care available to those who needed it through general practitioners, teachers, and others.[7] Mental hygiene psychiatrists articulated projects of social and cultural transformation aimed at preventing mental disorders and fostering mental health. In their perspective, healthy, happy, and well-adjusted individuals constituted a peaceful and prosperous society. As Ellen Herman has commented, psychiatrists and psychologists blurred the lines between the individual and the collective, which enabled them to use their expertise in treating the mental disorders in the former to develop politically motivated intervention strategies aimed at the latter.[8]

Even before the official declaration of war, American psychiatrists had been contemplating how they could aid the war effort.[9] Initially, they focused their attention on processes of selection and screening by emphasizing the need to bar individuals with a predisposition to mental illness from the armed forces. As a consequence, they expected that the number of psychiatric casualties during war would be reduced significantly.[10] In December 1940, the psychoanalytic psychiatrist Harry Stack Sullivan

[5] For the history of somatic psychiatry before World War II, see Joel T. Braslow, *Mental Ills and Bodily Cures: Psychiatric Treatment in the First Half of the Twentieth Century* (Berkeley, Calif., 1997); Edward Shorter, *A History of Psychiatry: From the Era of the Asylum to the Age of Prozac* (New York, 1997); and Jack D. Pressman, *Last Resort: Psychosurgery and the Limits of Medicine* (New York, 1998).

[6] For an overview of these initiatives, see Kathleen Jones, *Taming the Troublesome Child: American Families, Child Guidance, and the Limits of Psychiatric Authority* (Cambridge, Mass., 1999); Theresa Richardson, *The Century of the Child: The Mental Hygiene Movement and Social Policy in the United States and Canada* (Albany, N.Y., 1989); and Elizabeth Lunbeck, *The Psychiatric Persuasion: Knowledge, Gender, and Power in Modern America* (Princeton, N.J., 1994).

[7] See, e.g., Pols, "Divergences in American Psychiatry" (cit. n. 4).

[8] Herman, *Romance of American Psychology* (cit. n. 3), 22.

[9] For an overview, see Albert Deutsch, "Military Psychiatry: World War II," in *One Hundred Years of American Psychiatry,* ed. J. K. Hall (New York, 1944); Harry Stack Sullivan, "Psychiatry and the National Defense," *Psychiatry* 4 (1941): 201–17; and [Sullivan], "Selective Service Psychiatry," *Psychiatry* 4 (1941): 440–66.

[10] See, e.g., William C. Porter, "Military Psychiatry and the Selective Service," *War Medicine* 1 (1941): 364–71; and other articles in the special issue on psychiatric aspects of military medicine published in the second issue of *War Medicine* in 1941.

was appointed by the Selective Service System to organize psychiatric selection for inductees. He developed a screening program that was much more extensive than previous ones had been.[11] In his opinion, psychiatrists needed to detect not only mental illness but also maladjustment in general. Individuals who had demonstrated that they lacked the ability to adjust to the changing challenges of everyday life could not be expected to adjust to the much more demanding and arduous conditions of army life. With these broadened criteria, it is not surprising that a high percentage, at times up to 25 percent, of inductees were considered unfit to serve.

The American armed forces eagerly accepted the advice of psychiatrists on organizing the screening as part of the Selective Service System. Partly because of the promise that thorough selection would obviate the need for psychiatric treatment on the battlefield, there were very few psychiatrists in the field. Virtually all psychiatrists working for the selective service and in the overseas theaters of operation were recruited from mental hospitals; and most held somatic theories on the nature of mental illness. Army policy dictated that soldiers diagnosed with mental illness had to be discharged and repatriated as they were suffering from preexisting conditions for which there were no effective treatments. Psychiatrists argued that the stresses of warfare might enable these conditions to come to the surface, but those stresses could not be considered to be the causes: in this way, "the Army can well be called the proving ground of man."[12] During the first few months of battle, the incidence of mental illness in the army was roughly equal to that in the population at large. Thus Roy D. Halloran, the first head of the neuropsychiatric branch of the Office of the Surgeon General of the U.S. Army and former medical superintendent of the Metropolitan State Hospital near Boston, Massachusetts, concluded that the war did not pose any unexpected challenges for psychiatry. Yet despite this assessment, he did report on a number of puzzling schizophrenic-like disturbances with sudden onset that disappeared rather quickly when soldiers were hospitalized.[13]

Despite the fact that Sullivan's psychoanalytic ideas were very different from Halloran's somatic approach, their ideas and approaches toward screening fitted together very well. The two psychiatrists agreed that stringent selection would reduce the incidence of mental illness in the army. Halloran viewed mental illness as the result of a probably inherited predisposition, which made it pointless to treat the condition near the battlefield. According to Sullivan, a predisposition to mental breakdown was the outcome of early childhood experiences and therefore essentially stable and predictable. Both men trusted that, on the basis of a short interview, psychiatrists could predict the extent to which individuals could adjust to new, unexpected, and demanding circumstances. Halloran advocated the removal of those individuals whose mental illness could have been detected if selection had been more thorough. Both psychiatrists intended to develop techniques to detect instability and predisposition in

[11] Harry Stack Sullivan, "Mental Hygiene and National Defense: A Year of Selective-Service Psychiatry," *Mental Hygiene* 26 (1942): 7–14; Sullivan, "Psychiatry and the National Defense" (cit. n. 9).
[12] Roy D. Halloran and Malcolm J. Farrell, "The Function of Neuropsychiatry in the Army," *American Journal of Psychiatry* 100 (1943): 14–20, on 17.
[13] Roy D. Halloran and Paul I. Yakovlev, *Military Neuropsychiatry,* vol. 7 of *Postgraduate Seminar in Neurology and Psychiatry* (Waltham, Mass., 1942); Halloran and Farrell, "Function of Neuropsychiatry in the Army" (cit. n. 12); Roy D. Halloran (chairman), "Proceedings of the Military Session of the American Society for Research in Psychosomatic Medicine on the Unfit: How to Exclude Them and How to Use Them," *Psychosomatic Medicine* 5 (1943): 323–63.

human beings to remove the problem of mental illness from the American army and thus increase its efficiency and fighting power.

WAR NEUROSIS AND BATTLE FATIGUE

Late in 1942, the number of soldiers who could not continue fighting because they were suffering from neuropsychiatric syndromes increased dramatically. The battles in the Pacific Theater of Operations (specifically, those around Guadalcanal) and in the African Theater of Operations (in particular, those in Tunisia) were unusually fierce, leading to great numbers of casualties on both sides. Many soldiers broke down and suffered from debilitating anxiety attacks, repetitive nightmares, tremors, stuttering, mutism, and amnesia. During the 1943 Tunisian campaign, when several catastrophic engagements with the German field marshal Erwin Rommel led to a number of strategic setbacks, 20 to 34 percent of all casualties were neuropsychiatric in nature. To U.S. Army physicians, these patients were as baffling as they were irritating because their symptoms changed rapidly, no physiological cause could be found, and none of the treatments offered proved effective. War neuroses, like shell shock during the Great War, successfully mimicked the symptoms of organic disease, confusing and frustrating physicians.

Military officials realized that the policy of evacuating soldiers with a neuropsychiatric diagnosis could not be maintained as more men were leaving the African Theater of Operations than were entering it.[14] As a consequence, they became receptive to the arguments of a small but outspoken group of psychoanalytically oriented psychiatrists who proposed providing psychotherapeutic treatment to soldiers suffering from war neuroses close to the front lines (a practice known as forward psychiatry during World War I). The armed forces started to enlist psychiatrists and commenced training army physicians in the principles of military psychiatry. In January 1943, Colonel William C. Porter, a leader in military psychiatry, started teaching a four-week course in military psychiatry for general physicians at the Lawson General Hospital in Atlanta, Georgia.[15] The U.S. Air Force engaged Roy G. Grinker and John P. Spiegel, two psychiatrists interested in psychosomatic medicine before the war, who developed techniques for short-term psychotherapy that could be administered near the front lines.[16] The authors of a widely used manual for medical officers on the treatment of war neuroses, Grinker and Spiegel relied on basic psychoanalytic notions by arguing that, under severe battle conditions, even the strongest and most mature soldiers regressed to the condition of a helpless child. "The ego reacts with the anxiety and helplessness of a child and abandons the scene altogether (stupor), or refuses to listen to it (deafness), or to talk about it (mutism), or to know anything about it

---

[14] In this respect, an analysis of the situation by military officials and the aspirations of a number of psychoanalytically oriented psychiatrists converged. For the military perspective on the incidence of war neuroses, see Elliot Duncan Cooke, *All But Me and Thee: Psychiatry at the Foxhole Level* (Washington, D.C., 1946). The chapters of this book had been previously published in the *Infantry Journal*.

[15] William C. Porter, "The School of Military Neuropsychiatry," *Amer. J. Psychiatry* 100 (1943): 25–7.

[16] Roy Grinker had been in analysis with Sigmund Freud under a stipend of the Rockefeller Foundation. For his career, see Daniel Offer and Daniel X. Freedman, *Modern Psychiatry and Clinical Research: Essays in Honor of Roy R. Grinker, Sr.* (New York, 1972). For an excellent overview of the role of psychoanalysts during and after World War II, see Nathan G. Hale, *Freud and the Americans, 1917–1985: The Rise and Crisis of Psychoanalysis in the United States* (New York, 1995), chap. 11, "World War II: Psychoanalytic Warriors and Theories."

(amnesia)."[17] In this perspective, war neuroses represented a flight into the comforts and dependence of childhood and constituted a retreat from the desired masculine behavior of soldiers, which could be defined as the ability to stand up under stress and not be overwhelmed by one's emotions.

Grinker and Spiegel argued that individuals suffering from war neuroses were neither cowards nor weaklings. Quite the contrary: the reactions of traumatized soldiers were entirely normal given the horrific conditions at the front line. According to them, "[I]t would seem to be a more rational question to ask why the soldier does *not* succumb to anxiety, rather than why he does."[18] Army psychiatrists were treating not pathological conditions in abnormal individuals but normal reactions of perfectly healthy and previously well-adjusted individuals who had been exposed to extraordinarily stressful situations. Thus treating soldiers suffering from neuropsychiatric conditions near the front lines required a reorientation in the perspective of military psychiatry. As two leading psychiatrists commented: "it was necessary to shift attention from problems of the abnormal mind in normal times to problems of the normal mind in abnormal times."[19] Rather than explaining mental breakdown as the expression of preexisting conditions, military psychiatrists came to acknowledge the importance of environmental stresses and that every man had his breaking point; estimates of when this point was reached varied between 100 days and a full year of battle exposure. The endurance of even the best-trained and hardiest of men was limited.

The treatment proposed by Grinker and Spiegel involved offering soldiers an opportunity to work through their trauma by reexperiencing it in psychotherapy.[20] With a sodium pentothal injection, the psychiatrist would induce a state of twilight sleep that aided in bringing repressed traumatic memories to the surface. The effects of this treatment, Grinker and Spiegel reported, were dramatic: "The stuporous become alert, the mute can talk, the deaf can hear, the paralyzed can move, and the terror-stricken psychotics become well-organized individuals."[21] Military psychiatrists now emphasized the importance of treating soldiers near the front line because treatment was often more successful when provided as soon as possible after the onset of symptoms. In addition, the bonds between soldiers, which were of essential importance for morale, were thus maintained. In these respects, the abilities of psychoanalytically inspired psychotherapy met the demands of military discipline: psychiatrists were able to return men to the front lines and salvage them for further duty. The astute management of soldiers' emotions made them more efficient fighting men. Several programs emphasizing the provision of rest, good food, and treatment near the front lines were developed, including that of neurologist Frederick R. Hanson, who was working in

---

[17] Roy R. Grinker and John P. Spiegel, *War Neuroses* (Philadelphia, 1945), 93. This book had been published as a confidential document in 1943 as Roy Grinker and John P. Spiegel, *War Neuroses in North Africa: The Tunisian Campaign* (New York, 1943).

[18] Ibid., 115.

[19] Malcolm J. Farrell and John W. Appel, "Current Trends in Military Neuropsychiatry," *Amer. J. Psychiatry* 101 (1944): 12–19, on 19. The stress concept became more important in Grinker and Spiegel's reinterpretation of their war experiences in Roy R. Grinker and John P. Spiegel, *Men under Stress* (Philadelphia, 1945). According to Russell Viner, this work was instrumental in making the stress concept generally known in the medical profession. See Russell Viner, "Putting Stress in Life: Hans Selye and the Making of Stress Theory," *Social Studies of Science* 29 (1999): 391–410.

[20] See the chapter on treatment in Grinker and Spiegel, *War Neuroses* (cit. n. 17), 75–114.

[21] Ibid., 82.

Algeria. Hanson introduced simple and straightforward treatments (rest, good food, hot showers, and sedation), which he claimed were successful in returning men to the fighting line in just a few days.[22]

In December 1943, William C. Menninger, a tireless advocate of psychoanalysis and the modernization of psychiatry, was appointed chief of the Division of Neuropsychiatry in the Surgeon General's Office of the U.S. Army. He informed all military medical officers of the principles of forward psychiatry.[23] Psychiatrists claimed that they were able to return up to 80 percent of neuropsychiatric cases to duty within a week.[24] After the war, these figures had to be adjusted downward, when it was acknowledged that the percentage of personnel able to return to the front lines had been disappointingly low (generally, they were only able to function in supporting roles).[25] Yet although the success of psychotherapeutic treatment near the front lines had been greatly overstated, its perceived success was tremendous, boosting the confidence of psychiatrists in their ability to treat mental disorder. In his presidential address to the American Psychiatric Association in 1944, Edward A. Strecker noted that "[p]ractically every member not barred by age, disability or ear-marked as essential for civilian psychiatry is on active duty."[26] About one-third of all American psychiatrists were serving in the armed forces. Not surprisingly, their experiences greatly influenced postwar psychiatry.

Managing the emotions was a task not left exclusively to the experts; the participation of soldiers themselves was essential. A great number of pamphlets and booklets informing soldiers about the latest findings of psychology and psychiatry were printed and distributed among the armed forces. One of the better-known examples was *Psychology for the Fighting Man,* compiled by a team of fifty-nine experts under the chairmanship of esteemed Harvard psychologist Edwin G. Boring and written by the science journalist Marjorie van de Water (see illustration 1).[27] More than 400,000 copies of the booklet, which was printed in a small format that could easily be taken anywhere, were sold during the war. *Psychology for the Fighting Man* discussed topics ranging from interacting with foreigners, seeing in the dark, and color and camouflage to food and sex as military problems, the soldier's personal adjustment, and the management of fear. With respect to the last topic, soldiers were informed that fear and panic were entirely normal and that, with some effort, these emotions could be brought under rational control. The authors advised their readers to develop a cool and detached attitude toward their inner selves and to discuss their feelings with their fellow soldiers or officers.

---

[22] Frederick Hanson, ed., *Combat Psychiatry: Experiences in the North African and Mediterranean Theaters of Operation, American Ground Forces, World War II* (Honolulu, 2005), originally published in *Bulletin of the U.S. Army Medical Department* 9, suppl. (Nov. 1949).

[23] Deutsch, "Military Psychiatry" (cit. n. 9).

[24] Leo H. Bartemeier et al., "Combat Exhaustion," *Journal of Nervous and Mental Disease* 104 (1946): 358–89, 489–525.

[25] Normal Q. Brill, Mildred C. Tate, and William C. Menninger, "Enlisted Men Discharged from the Army Because of Psychoneurosis: A Follow-up Study," *Journal of the American Medical Association* 128, 30 June 1945, 633–7; Normal Q. Brill and Gilbert W. Beebe, *A Follow-up Study of War Neuroses* (Washington, D.C., 1955).

[26] Edward A. Strecker, "Presidential Address," *Amer. J. Psychiatry* 101(1) (1944): 1–8, on 1.

[27] Edwin G. Boring and Marjorie Van de Water, *Psychology for the Fighting Man: Prepared for the Fighting Man Himself* (Washington, D.C., 1943).

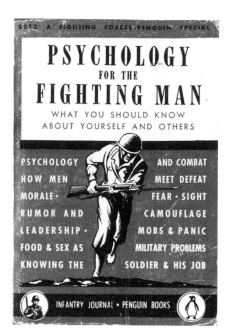

The cover of Psychology for the Fighting Man. *More than 400,000 copies were sold to soldiers during World War II.*

THE AMERICAN SOLDIER: A SOCIOLOGICAL PERSPECTIVE

Herbert Spiegel was one of the first psychiatrists to observe the development of war neurosis in Tunisia.[28] He was interested in the factors that contributed to the incidence of mental breakdown and made a number of interesting observations on the topic. He concluded that most soldiers were not motivated by a hatred of the enemy or an admiration for American values such as peace and democracy. Instead, soldiers kept on fighting out of loyalty to their buddies, not wanting to let them down. They were motivated not only by regard for their comrades but also by respect for those in positions of leadership, concern for the reputation of their units, and an urge to contribute to the success of their units.[29] Men often broke down after they had lost one or more of their comrades, when they received unwelcome news from the home front, or when they perceived conditions in the army to be bad. The incidence of war neurosis was low when morale was high. Spiegel argued that insight into the factors that contributed to a high morale could greatly aid the military in maintaining its strength.

American soldiers during World War II were probably the most thoroughly investigated soldiers in history. In October 1941, the Research Branch of the U.S. Army's Morale Division was established to conduct research into the attitudes and opinions of American soldiers using the most modern research techniques developed by the social sciences. The research, later published in four impressive volumes titled *The American Soldier,* was directed by Samuel Stouffer, a sociologist from the University

---

[28] Herbert X. Spiegel, "Silver Linings in the Clouds of War: A Five-Decade Retrospective," in *American Psychiatry after World War II,* ed. Roy W. Menninger and John C. Nemiah (Washington, D.C., 2000); Spiegel, interview with author, New York City, 19 March 1997.

[29] Herbert X. Spiegel, "Preventive Psychiatry with Combat Troops," *Amer. J. Psychiatry* 101 (1944): 310–5.

of Chicago.[30] Stouffer confirmed Spiegel's findings and stated that morale was inversely correlated to breakdown incidence. According to him, morale was intimately bound up with the trust soldiers had in their officers, their training, their outfits, their weapons, and their fellow soldiers. It was also related to the level of perceived support from the home front and a sense of fairness in the army. Lastly, it was related to the nature of the emotional bonds among soldiers and between soldiers and their commanders. It was somewhat disturbing to investigators to discover that most soldiers had no real idea, and did not care to find out, what the war was about or why they were fighting in the first place. For many soldiers, the war was a dirty job that needed to be finished as quickly as possible so they could return home.[31] Several morale-boosting initiatives were undertaken, such as a series of instructional movies titled *Why We Fight.* Nevertheless, it became increasingly difficult to maintain morale, in particular when the end of the war remained elusive and dissatisfaction with army conditions rose.

Grinker and Spiegel advocated a view inspired by psychoanalysis that characterized mental breakdown as a normal reaction of normal individuals under extraordinarily stressful conditions. The treatment methods they developed were perceived as very successful, which increased the confidence of psychiatrists in their therapeutic abilities. Because mental breakdown was seen as a normal response, *all* enlisted men and the ways in which they dealt with stress and adversity became of interest to psychiatrists. In the sociological perspective developed by social scientists, mental breakdown was considered to be related to the quality of intragroup relationships and morale rather than to individual characteristics. Constructive and preventive intervention, therefore, had to be directed at the level of groups, which inspired psychiatrists who, before the war, had been involved with the mental hygiene movement. In the American armed forces, the minds of soldiers were investigated by social scientists, received ministrations from morale boosters, were informed by psychological writings, and if they ceased working properly, would be treated by psychiatrists. Mobilization and engaging in a war against several countries on a global scale truly required the coordinated efforts of politicians, military officers, and scientists.

Soldiers adopted the language of psychiatry very quickly and commonly spoke about eight-balls, psychos, NPs, and psychoneurotics.[32] At first glance, this illustrates Ian Hacking's thesis on how the language of psychiatry provides the vocabulary to articulate complaints, illnesses, and identities and thereby constructs them. In other

---

[30] Samuel A. Stouffer et al., *The American Soldier: Adjustment during Army Life,* vol. 1 of *Studies in Social Psychology in World War II* (Princeton, N.J., 1949); Stouffer et al., *The American Soldier: Combat and Its Aftermath,* vol. 2 of *Studies in Social Psychology in World War II;* Carl I. Hovland, Arthur A. Lumsdaine, and Fred D. Sheffield. *The American Soldier: Experiments on Mass Communication,* vol. 3 of *Studies in Social Psychology in World War II;* Samuel A. Stouffer et al., *The American Soldier: Measurement and Prediction,* vol. 4 of *Studies in Social Psychology in World War II.*

[31] These findings indicate a general disenchantment with the war among the majority of soldiers, which was mostly forgotten after the war, when American society attempted to construct memories of valor and sacrifice about the just and glorious war. Recently, historians have embarked on a reappraisal of World War II, in which the attitudes expressed by American soldiers appear less surprising. See Paul Fussell, *Wartime: Understanding and Behavior in the Second World War* (New York, 1989); Fussel, *The Boys' Crusade: American GIs in Europe—Chaos and Fear in World War Two* (London, 2003); and Richard Polenberg, "The Good War? A Reappraisal of How World War II Affected American Society," *Virginia Magazine of History and Biography* 100 (1992): 295–322.

[32] Eight-balls (later, "odd-balls") refers to a Section 8 discharge ("unsuitable for Army service"); NP to "discharged because of neuropsychiatric reasons."

words, the language of psychiatry provides the means for "making up people."[33] Because of soldiers' writings, their newspapers, and their access to the media, their voices can be heard very clearly—and these voices provide a different perspective on the supposed success of psychiatry. Although soldiers appropriated the language of psychiatry, they subverted its meaning and turned the intentions of psychiatrists upside down. Psychiatry became the language in which soldiers expressed their own concerns. In this way, soldiers offered a very interesting counter-narrative to the narrative of the psychiatrists, not augmenting but delineating and limiting the powers of psychiatry. Soldiers related how the army made them into psychoneurotics by its hopeless inefficiency, pointless and tedious assignments, work assignments that did not fit an individual's ability, administrative errors that led to having to do the same basic training course repeatedly, and failure to listen to suggestions for improvements (improvements that could not be accommodated in any case).[34]

In the perspective of a number of vocal and outspoken soldiers, the armed forces, rather than battle conditions or the separation from home, were the cause of their trauma. The high, lofty, and patriotic ideals celebrated at home had no meaning for the average soldier. As one soldier stated: "The plain, unpublicized fact of the matter was that nine out of ten servicemen wanted nothing more to do with wars after their first week of training. . . . [The soldier's] special gripes, however, were reserved for the undemocratic, stupefying, favor-ridden totalitarian nature of military life itself." Many soldiers viewed the army as inefficient and often corrupt, while its policies were seen as arbitrary at best and plain stupid and unjust at worst. The initial American policy of not providing alcohol to soldiers, for instance, was thoroughly hated (although the tobacco companies provided cigarettes free of charge). The strict separation between officers, who received very good treatment, and the great majority of enlisted men was particularly galling to most soldiers: "[The soldier] had no use for a system in which one class got the best of everything, and the other class got less than what was left."[35] On several occasions, soldiers returned a day after they had conquered a town at the cost of the lives of several of their fellow soldiers only to find notices everywhere in the town stating, "For Officers Only" and "No Enlisted Men." African American soldiers, who were placed in separate units, suffered discrimination, and were generally assigned to insignificant and tedious tasks, had high rates of nervous breakdown.[36]

THE VETERAN PROBLEM

When the war came to an end, most Americans wanted to provide their soldiers the best possible welcome. After all, those soldiers had triumphed over two well-equipped and highly motivated enemies who had also excelled in cruelty and destruction. The return of veterans was widely celebrated. The GI Bill of Rights (officially, The Servicemen's Readjustment Act of 1944) aimed to provide access to home and business loans

---

[33] Ian Hacking, "Making Up People," in *Reconstructing Individualism: Autonomy, Individuality, and the Self in Western Thought,* ed. Thomas C. Heller (Stanford, Calif., 1986), 222–36.

[34] See, e.g., "The Neutral Soldiers," *Common Sense* 14(10) (1945): 25–6, providing the account of a soldier who received a neuropsychiatric discharge because he could not stand army culture.

[35] Edgar L. Jones, "One War Is Enough," *Atlantic Monthly,* Feb. 1946, 48–53.

[36] Ellen Dwyer, "Psychiatry and Race during World War II," *Journal of the History of Medicine and Allied Sciences* 61 (2006): 117–43.

and expand educational opportunities. Apart from solutions such as the GI Bill of Rights, many Americans were eager to make the readjustment of veterans as easy as possible. Yet there was also uneasiness and apprehension about the process of demobilization. During the war, a number of wounded and disabled veterans had returned home, among them soldiers discharged for neuropsychiatric reasons.[37] American business had thrived because of the war economy; many were concerned about what would happen when the war came to an end. When more than 12 million veterans came home, where would they find work? Moreover, many soldiers had enlisted as young men; the army had changed them into rough-talking, demanding, and potentially violent individuals. Army training had taught them how to kill efficiently; although these skills would not be helpful in civilian life, they might not disappear.

Starting in 1944, an impressive number of advice books, articles, and pamphlets on the readjustment problems of veterans was published. It was believed that with tact and understanding by the home front, in addition to some counseling and advice from experts, the intemperate emotions of returning ex-servicemen could be readjusted to postwar peace. One of the more ominous warnings about the coming veteran problem was presented by the educational sociologist and World War I veteran Willard Waller. Using several examples, he stated that "veterans have written many a bloody page of history, and those pages have stood forever as a record of their days of anger."[38] Calling veterans America's gravest social problem and citing the activities of southern veterans after the Civil War, the roots of the German Nazi party in the activities of German ex-soldiers, and the 1932 Bonus March in Washington, D.C., by 20,000 disgruntled Great War veterans, Waller concluded: "The veteran is, and always has been, a problematic element in society, an unfortunate, misused, and pitiable man, and, like others whom society has mistreated, a threat to existing institutions." His conclusion was clear: "Unless and until he can be denaturalized into his native land, the veteran is a threat to society."[39] The pleasant, obedient, and equable boys who were enlisted at the beginning of the war would return as hardened, bitter, and angry men. To counter this problem that threatened to overwhelm society, a new art of rehabilitation had to be invented and diligently applied.

Willard explored the reasons why so many veterans might face problems adjusting to peace. He related how the army had annihilated the individual soldier's personality and will to make him function like a cog in a machine. After the inductee had been handed his outfit, he was forced to discard all individuality. The lives of soldiers were completely regimented and regulated by superiors; obedience rather than initiative was valued. There was no privacy in the army: soldiers spent their time in cramped living quarters, continuously exposed to one another. Many soldiers had a hard time getting used to the fact that they were, in principle, expendable. Finally, in the army, prohibitions of peacetime society were set aside. Army training was a training in violence and killing. It was expected that a great number of soldiers would have been hardened from their experiences on the front lines. Many others could have acquired behavior patterns that would make them lastingly unsuitable for living under conditions of

---

[37] Dorothy Paull, "Psychiatric Rehabilitation of Rejectees and Men Discharged from the Armed Forces: The Wisconsin Service for Rejectees," *Mental Hygiene* 29 (1945): 248–54.

[38] Willard Waller, *The Veteran Comes Back* (New York, 1944), 5. This was not the only book bringing the veteran problem to the attention of the American public. See also Dixon Wecter, *When Johnny Comes Marching Home* (Cambridge, Mass., 1944).

[39] Waller, *Veteran Comes Back* (cit. n. 38), 13.

peace. Of particular worry were those soldiers who had received a neuropsychiatric discharge. Not only were they "probably poor marriage risks," but in general, the "odds were against them." For jobs in which interpersonal skills were important or that might be stressful, "psychoneurotics are suspect."[40] Yet it could be expected that almost *all* soldiers would suffer from psychological problems after their return. "The problem of helping these psychologically battered veterans to regain their balance is very great and most urgent, but not insoluble. If we are to solve it, we must begin now to train thousands of technicians to deal with such cases."[41]

There were many positive elements of army life that soldiers could be expected to miss after they had returned home. First, many soldiers had made deep and lasting friendships in the army; they would find civilian life cold and harsh without those. Second, many soldiers had grown up in small towns; after joining the army, they had traveled for the first time, seen different countries and large cities, and become friends with individuals they would never have met otherwise. They were unlikely to be content to return to their small and mundane hometowns. Third, many soldiers who had been out of work or in dead-end jobs during the Depression years had been given positions of power and responsibility. Very few soldiers would be happy to merely take up the jobs they had left behind. They expected more out of the postwar world and made their desires known. Because most of them had spent their formative years in the army, they had been completely alienated from civilian life; many of them might prefer army over civilian life.

Waller's opinions were repeated in the popular press many times.[42] As a result, numerous articles, pamphlets, lectures, speeches, radio talks, and sermons were devoted to the veteran problem, giving advice to friends and family of veterans on how they could assist the returnees' readjustment. Although the committee that had published *Psychology for the Fighting Man* also produced *Psychology for the Returning Serviceman,*[43] most books and articles were directed not at the soldiers, but at women back home—the mothers, wives, and sweethearts of returning soldiers.[44] After all, as Waller wrote in the *Ladies' Home Journal,* "[t]he personal side of reconstruction is woman's work." Waller advised women to help returning veterans readjust with the most modern advice: "[A wife] should begin by making a thorough study of the psychology of the veteran, his habits, attitudes, beliefs, desires and capacities, the ideals and values by which he lives. . . . The wife must study her own veteran in order to learn his needs and problems, his strengths and weaknesses."[45] Several psychiatrists

[40] Ibid., 169.

[41] Ibid., 55. See also Willard Waller, "Why Veterans Are Bitter," *American Mercury* 61 (Aug. 1945): 147–54; Waller, "The Veteran's Attitudes," *Annals of the American Academy of Political and Social Science* 238 (March 1945): 174–9.

[42] See, e.g., Agnes E. Meyer, "The Veterans Say—or Else!" *Collier's* 118, 12 Oct. 1946, 16, 17, 115–9; Cristopher La Farge, "Soldier into Civilian," *Harper's Magazine* 190 (March 1945): 339–46.

[43] Irvin L. Child and Marjorie Van de Water, *Psychology for the Returning Serviceman* (Washington, D.C., 1945). Directed also at returning soldiers were Dorothy W. Baruch and Lee Edward Travis, *You're out of the Service Now: The Veteran's Guide to Civilian Life* (New York, 1946); Maxwell Droke, *Good-by to GI: How to Be a Successful Civilian* (New York, 1945); Benjamin C. Bowker, *Out of Uniform* (New York, 1946); Leo M. Cherne, *The Rest of Your Life* (Garden City, N.Y., 1944).

[44] There is some analysis of this genre. See, e.g., Sonya Michel, "American Women and the Discourse of the Democratic Family in World War II," in *Behind the Lines: Gender and the Two World Wars,* ed. Margaret Randolph Higonnet et al. (New Haven, Conn., 1987), 154–67.

[45] Willard Waller, "What You Can Do to Help the Returning Veteran," *Ladies' Home Journal,* Feb. 1945, 26, 27, 92–8, on 26 and 95.

also wrote advice books and articles for the popular press on the veteran problem.[46] One of the most popular, written by George K. Pratt, a psychiatrist connected to the National Committee for Mental Hygiene, interpreted army life as a state of regression from which veterans had to be weaned through the efforts of the home front.[47] Several examples were given of readjustment gone wrong—not because of long-lasting war traumas or unbearable army conditions, but because of the lack of warmth, tact, and understanding of mothers, wives, and partners in encouraging returning veterans to take up their responsibilities.[48] Journalists, social commentators, critics, clerics, social workers, teachers, and others produced an even greater number of publications, thus establishing a new genre of rehabilitation literature, focused on the psychological aspects of readjustment, which would aid ex-servicemen to fit into a normal, peaceful society.

Spokespersons for veterans and outspoken veterans themselves generally reacted negatively to the concerns about their rehabilitation and loathed the endless stream of literature that portrayed them as psychoneurotics, disgruntled killers, potential communists, alcoholics, and louts bound to disturb the social order. They wanted to be treated like normal human beings who had had a number of unusual experiences and who were a few years older. They distrusted the motives of psychiatrists, journalists, and opinion makers who portrayed returning soldiers as threats to the social order to be managed with emotional finesse and psychological tactics. They knew that copious amounts of advice had been eagerly read by the home front in anticipation of the veterans' return. Most veterans were insulted by the advice and refused to view the difficulties they might encounter on their return as psychological in nature. They demanded that soldiers who were suffering from psychiatric symptoms receive the very best of care but rejected any psychological approach to the problems and issues facing normal veterans on their return home. In this respect, the attempt of psychiatrists and psychologists to implement their social project for the realization of a peaceful and prosperous postwar world appeared to be a failure.

Many soldiers did admit that veterans would be changed men, first, because they were a few years older and, second, because they had spent a significant amount of time away from home. There were other reasons as well. Many of them did not want to return to life as they had known it from before the war. They did not want to take up their old, meaningless jobs or return to their own small, provincial towns. And many of them had strong convictions that there should never be a war again—and supported political initiatives such as the establishment of the United Nations. Many were restless after their return. They could not settle down, were not happy with their jobs, and were dissatisfied with the uneventful life of civilians. As a group of sociologists

---

[46] See, e.g., Thomas A. C. Rennie and Luther E. Woodward, *When He Comes Back; If He Comes Back Nervous: Two Talks to Families of Returning Servicemen* (New York, 1944); Alexander G. Dumas and Grace Keen, *A Psychiatric Primer for the Veteran's Family and Friends* (Minneapolis, Minn., 1945); Herbert I. Kupper, *Back to Life: The Emotional Adjustment of Our Veterans* (New York, 1945). For articles in the popular press, see Carl R. Rogers, "Wartime Issues in Family Counseling," *Journal of Home Economics* 36 (Sept. 1944): 390–3; and Norman Q. Brill, "Veterans with Problems," *Journal of Home Economics* 38 (June 1946): 325–8.

[47] George K. Pratt, *Soldier to Civilian: Problems of Readjustment* (New York, 1944).

[48] For a case history from Pratt's *Soldier to Civilian,* in which a veteran survived a mortar attack well, only to succumb to the pressures of his parents, see Hans Pols, "The Repression of War Trauma in American Psychiatry after World War II," in *Medicine and Modern Warfare,* ed. Roger Cooter, Mark Harrison, and Steve Sturdy (Amsterdam, 1999), 251–76, at 264–5.

"How old is your problem child, Madam?"

observed about veterans in the months after their discharge: "This quality of restless-ness, of impatience, of irritability, was evident in many of the average veteran's early post-service behavior."[49] They liked to meet with other veterans, drink, and go over old and cherished memories. In the memories of war of veterans, two diametrically opposed images emerged. On the one hand, they saw the years in the army as the best and the most exciting years in their lives. On the other, they saw the war as an endless series of frustrations: the army must have been the most inefficient and undemocratic institution on earth. The two repertoires are presented as mutually exclusive, but dur-ing their drinking bouts, soldiers easily switched between them.

Many people commented on the sheer amount of writing on the veteran problem. As a teacher observed: "Pick up almost any current periodical and the chances are you will find an article on how to deal with the returning veteran."[50] One veteran turned journalist complained: "Apart from politicians and movie personalities, few individ-uals in America were so publicized as veterans." He felt that the public only became interested in American soldiers after their return home because of the perceived threat they posed to society: "It was only *after* victory that the invasion of America became

[49] Robert J. Havighurst et al., *The American Veteran Back Home: A Study of Veteran Readjustment* (New York, 1951), 74. Similar observations were made by journalists and casual observers.
[50] Walter R. Goetsch, "The G.I. in Civvies," *School and Society,* 21 July 1945, 45–6, on 45.

a reality."[51] Often, veterans, or at least their spokesmen, showed resentment about the unwanted and misdirected attention given to them. As Charles G. Bolte, organizer of the American Veterans Committee, stated:

> GI Joe, mythical darling of the advertising copy writers, has had more nonsense written about him than has ever been set down concerning any other American folk hero . . . We are daily flooded with speeches, articles, editorials, books, pamphlets and confidential reports telling us what this legendary character thinks, feels, wants, needs and plans.[52]

Many veterans were suspicious of these efforts, which they viewed as attempts to sell particular viewpoints and products to them. To continue Bolte's statements:

> Even more oddly, the courses of action they recommend to deal with what they call "the veteran problem" always parallel their own professional bents: the educators urge more schooling for Joe, the psychiatrists want mental clinics, the businessmen say he needs free enterprise in big doses, the labor leaders demand strong unions to receive him, and so on. Everybody is out to help Joe; one half the country is telling the other half how to treat him; thousands of "experts" are interpreting him.[53]

After 1944, politicians, businessmen, and the clergy all attempted to frame the veteran problem in a way that furthered their own interests. As one critic implored: "No one speaks for the soldiers but the soldiers. . . . It may be a good idea to try to get accustomed to their voices as early as possible."[54]

American psychiatrists were unusually successful in convincing the public that their expertise was needed in managing the readjustment of returning veterans. One veteran decried how psychiatrists "had found their opportunity to catapult into public consciousness" as members of the Selective Service System.[55] Now that the war was over, "the more vocal and zealous psychiatrists made it sound as though most veterans would return as mental problems. . . . Like the other special-pleaders, the psychiatrists used the veteran as a package for displaying the same wares to which the public had previously been apathetic."[56] Psychiatrists had been able to convince the public that not only was the wounded or neurotic veteran in need of their special attention but also that every veteran was a potential psychological problem. Soldiers writing in *Stars and Stripes* indicated that many ex-servicemen felt a strong disgust for the "'psychoneurotic fad' which is sweeping the country, and which is based on the belief that every homecoming veteran is 'maladjusted',", or "half hero and half problem."[57] They begrudged these ideas, which they thought were disseminated "by amateur psychiatrists writing in the Sunday supplements and the ladies magazines who saw a psychopath or a neurotic beneath every uniform."[58] Bolte described Pratt's manual as

[51] Bowker, *Out of Uniform* (cit. n. 43), 25.
[52] Charles G. Bolte, *The New Veteran* (New York, 1945), 2. The American Veterans Committee aimed to be an alternative veterans organization apart from the American Legion and was motivated by left-wing political ideals.
[53] Ibid.
[54] "No One Speaks for the GI," *Common Sense* 14 (1945): 27–8, on 28.
[55] Bowker, *Out of Uniform* (cit. n. 43), 32.
[56] Ibid., 34.
[57] Kupper, *Back to Life* (cit. n. 46), 19; Bolte, *New Veteran* (cit. n. 52), 6.
[58] Bolte, *New Veteran* (cit. n. 52), 108.

one that "would make the average civilian a little jumpy about going down a dark alley with a veteran," while a literary critic claimed to be unable to escape the disappointing conclusion that the manual was "essentially an elementary text in psychiatry—with a contemporary cover."[59]

Publications on the veteran problem also appeared overseas and were read by soldiers who wanted nothing more than to return home. Soldiers had mixed reactions. One psychologist reported how a "young officer suffering from fatigue and battle strain [had] come upon one of these direly distressing pieces in a popular magazine." This poor patient was upset for a week and had to be assured time and again "that he was 'all right' and that the war hadn't transformed him into a beast or a bogeyman."[60] A veteran claimed that if soldiers had paid attention to these articles, they had probably developed "a terrific inferiority complex" because they were portrayed as "tamed dogs gone wild who must pause on the road back to normalcy in order to be rehabilitated." He concluded that it would be "a shock to some of our sons and brothers to find themselves the subjects of highly academic psychological study."[61] Other veterans were merely bemused or irritated. One of them reported that he felt compelled to "act nervous so as not to disappoint the barrage of relatives who hovered about me like homing pigeons!"[62] Another even reported how his mother, on the basis on an advice article, pretended that he had never been away. Fortunately, he guessed what she had been trying to do; otherwise he "might have thought his mother a good case for a psychiatrist."[63]

Some veterans expressed their annoyance with the widespread concern about their supposed problems by writing their own manuals about a strange and outlandish group of people with rather peculiar habits: the Civilians. A former journalist wrote one of the better-known advice books, which contained extensive guidelines on, among other matters, restrooms (latrines for Civilian personnel); eating habits ("Civilians have an odd custom of serving food in individual dishes or in neat and separate piles on the same plate"); manners ("Don't be misled by the Civilian tendency to be soft-spoken and polite"); or even Civilian speech ("Don't make fun of Civilian speech or accents").[64] Some booklets even contain a dictionary translating GI to Civilian. The same symbolic reversal is played out in explaining the overly concerned reaction to the returning veteran:

> [Civilians] are apt to treat as a mental case any returned soldier who has just been paroled to the community at large. It's not entirely their fault. They've been . . . bombarded with learned articles on how to treat a returned soldier, how to handle him and humor him and what to say and what not to say. . . . You'll find, especially on your first day, that Civilians will wear strained expressions when they talk to you, and start agreeing with you before you open your mouth. Some of them will try to get you to talk about your experiences.[65]

---

[59] Ibid., 140; Ralph Peterson, "Problems in Reconversion [Review of *Soldier to Civilian*]," *Saturday Review of Literature* 28, 13 Jan. 1945, 12.

[60] Droke, *Good-by to G.I.* (cit. n. 43), 44.

[61] William Best, "They Won't All Be Psychoneurotics," *Saturday Evening Post* 216, 14 April 1945, 112.

[62] Kupper, *Back to Life* (cit. n. 46), 20.

[63] Best, "They Won't All Be Psychoneurotics" (cit. n. 61), 112.

[64] Morton Thompson, *How to Be a Civilian* (Garden City, N.Y., 1946), 37, 40, 41, 75.

[65] Ibid., 31–2.

As for dealing with those overly concerned Civilians, the following advice sufficed: "You have fought a more demanding war than the Civilian. Don't rub it in. Play fair with him. He can be a pal in need."[66]

Most veterans were afraid that all the advice literature could turn mothers and wives "into kitchen psychologists determined to 'cure' the veteran—even at the cost of his sanity."[67] One journalist reported how veterans had implored him to convey this advice: "stop trying to practice amateur psychiatry on them" so that new veterans did not have to "undergo the patronizing, over sympathetic, kid-glove treatment they had encountered on their return." Instead, they wanted "only to be treated like normal human beings without any of the pampering advocated in most 'When-He-Comes-Home' articles."[68] The great majority of veterans had rather conventional ideas of what would ease their readjustment to peace. They would "be happy to sit in front of our fireplaces and let our wives and sweethearts fetch us Old Fashioneds."[69] A natural approach by their wives, mothers, and sweethearts would be the best: "A wife can make her husband's adjustment easy, not by 'psyching' him, but with a little pre-Freudian love and understanding."[70]

Veterans generally resisted the attempts of psychiatrists to transform them into individuals who needed special attention and care. They conceded that veterans with psychiatric problems were entitled to the best professional psychiatric care available but vigorously resisted the expansion of the psychiatric domain to include normal soldiers. The first postwar project of psychiatry, to extend its domain beyond the mentally ill and soldiers who suffered from war-related psychiatric syndromes to include guiding the readjustment of average veterans to postwar society, had failed. The attempt to provide the language and tools for the shaping of the postwar veteran self had found an unappreciative audience.

MOTHERS

Despite the veterans' reaction, the popular esteem of American psychiatrists rose to unprecedented heights immediately after World War II. Their war efforts had been widely publicized and praised by the general public. William C. Menninger appeared on the cover of *Time* magazine (October 1948), and psychiatric themes appeared in several movies (one only needs to think of *Spellbound* [1945], directed by Alfred Hitchcock, and *Home of the Brave* [1949]). The trust in their abilities had been greatly increased by their wartime successes. Immediately after the war, psychiatrists sought to translate their wartime successes into peacetime gains. A number of leading psychiatrists were eager to participate in the building of America's postwar society, characterized by the restoration of a traditional domestic order. The management of environmental stresses had proven to be the key in the management of war neurosis and in boosting morale in the army; psychiatrists were optimistic that these lessons could be applied in the postwar world as well. Many former military psychiatrists banded together and formed the Group for the Advancement of Psychiatry in an attempt to

[66] Ibid., 76.
[67] David Dempsey, "Veterans Are Not Problem Children," *American Mercury,* Sept. 1945, 326–31, on 326.
[68] Don Wharton, "The Veteran Is No Problem Child," *Common Sense* 14 (1945): 17–18, on 17.
[69] Kupper, *Back to Life* (cit. n. 46), 19.
[70] Dempsey, "Veterans Are Not Problem Children" (cit. n. 67), 331.

reform their discipline by adopting psychoanalytic ideas and prewar mental hygiene ideals.[71] Consequently, after the war, the interests of psychiatrists shifted away from treating soldiers scarred by battle and aiding veterans to readjust to issues of social engineering, cultural change, fortifying peacetime society, and preventing a third world war. Their interests in veterans had only been a station stop on the road to expanding their views to projects for postwar society that would encompass all citizens.

By the end of the war, nearly 1,000 psychiatrists, almost one-third of all American psychiatrists, had been enlisted in the armed forces.[72] After the war, many of them expressed concern about the mental health crisis that was facing postwar America and cited the high neuropsychiatric rejection and discharge rates as an indication of the seriousness of the problem. They shared these concerns with those psychiatrists who had stayed at home during the war. Psychiatric attention then shifted from those soldiers who had broken down after heroic service to the high numbers of inductees who had been rejected because of mental imbalance and the high number of soldiers who had broken down without having been exposed to battle. In 1946, the conservative Philadelphia psychiatrist Edward A. Strecker published *Their Mothers' Sons,* which explicitly dealt with these soldiers.[73] Strecker excluded those soldiers—true war heroes—who had broken down after prolonged exposure to battle, claiming they were entitled to the best possible psychiatric care available. His treatise was about the soldiers who had broken down without having faced any unusual adversity or simply in reaction to uncomfortable barracks, tasteless army food, and separation from their families. Strecker contrasted the behavior of the heroes, who, upon their first phone call to their parents after having been wounded in action, made "light of the loss of an arm or a leg" with patients in a ward of soldiers discharged from army service whose complaints had no physiological basis whatsoever: "It was like trying to push back a wall of water. There was nothing solid to grasp."[74]

Strecker indicted the American "Mom" for raising sons of weak and questionable character who broke down for the slightest reasons—the undeserving psychiatric casualties of war. Invariably, he claimed, behind each broken-down soldier stood an overprotective Mom who had not untied the emotional apron strings with which she bound her children to her. Strecker decried "the emotional satisfaction, almost repletion, she derives from keeping her children paddling about in a kind of psychological amniotic fluid rather than letting them swim away with the bold and decisive strokes of maturity from the emotional maternal womb."[75] By her overprotective attitude and

[71] Gerald N. Grob, "The Forging of Mental Health Policy in America: World War II to New Frontier," *J. Hist. Med. Allied Sci.* 42 (1987): 410–46; Rebecca Jo Plant, "William Menninger and American Psychoanalysis, 1946–1948," *History of Psychiatry* 16 (2005): 181–202.

[72] William C. Menninger, "Psychiatric Experience in the War, 1941–1946," *Amer. J. Psychiatry* 103 (1947): 577–86, 578.

[73] Edward A. Strecker, *Their Mothers' Sons: The Psychiatrist Examines an American Problem* (Philadelphia, 1946). The most important chapters of this book also appeared in the conservative *Saturday Evening Post* as Strecker, "What's Wrong with American Mothers?" *Saturday Evening Post* 219, 26 Oct. 1946, 14–5, 85–104. These thoughts were initially expressed in Strecker, "Psychiatry Speaks Democracy," *Mental Hygiene* 29 (1945): 591–605, which was based on a lecture given April 27, 1945. Strecker's use of the phrase "Mom" was, as he acknowledged, derived from the strongly misogynistic Philip Wylie, *A Generation of Vipers* (New York, 1942).

[74] Strecker, *Their Mothers' Sons* (cit. n. 73), 15, 14.

[75] Ibid., 31. The gendered metaphors are almost too obvious to point out. For the equation of femininity with fluidity, fluids, floods, and water, see Klaus Theweleit, *Male Fantasies,* vol. 1, *Women, Floods, Bodies, History* (Minneapolis, Minn., 1987).

because of her own emotional needs, the American Mom had produced a generation of psychiatric casualties, rejectees from army service, and draft dodgers "who resorted to any device, however shameful, even to the wearing of female clothing."[76] Everything a Mom did was unconsciously designed "exclusively to absorb her children emotionally and to bind them to her securely."[77] The good mother, by contrast, produced a proper balance of give-and-take in her children, which enabled them to attain social maturity and independence later in life.

According to a number of American psychiatrists speaking after World War II, the war had done the nation a favor by exposing a mental health problem that had existed for a long time. In Strecker's words: "War is a great leveler. It strips the soft, protective swathing of civilian life from millions of young men and exposes them to a threat to survival. It brings out their strengths and reveals their weaknesses both physical and mental."[78] War conditions only triggered traumas that had been established in peacetime and would, naturally, continue to find expression in grown men now that the war had been won. Life at the front lines and life under peacetime conditions were essentially the same: both could be stressful at times, but nothing a mentally robust and well-adjusted man would not be able to handle.

Strecker's account is striking for many reasons. In the first place, the importance of a stress-inducing environment had virtually disappeared, to be replaced by a renewed emphasis on the importance of predisposition—albeit a radically different conception as compared with the one that had been generally accepted among prewar psychiatrists. Second, Strecker, who had been an army psychiatrist during World War I and an adviser to various government bodies during the previous four years, hardly mentioned the horrific nature of battle, which was repeatedly highlighted by Grinker and Spiegel and other psychiatrists in the armed forces. The latter had justified their intervention by emphasizing the extraordinarily traumatizing nature of warfare, while Strecker considered the stresses of peace and war essentially the same. In Strecker's perspective, the war had been useful in revealing America's tremendous mental health problem. In his view, the main contribution his discipline could make to maintaining peace and prosperity was by reforming child-rearing methods or roping in American Moms. The 1950s would become an unusually domestic decade in which prosperous families in wealthy suburbs focused on raising families.[79] It was also the era of Dr. Spock, during which mother blaming became one of the main characteristics of psychiatry.[80] These tendencies were inspired and reinforced by the cold war. Psychiatrists and others became concerned about child-rearing methods contributing to fostering authoritarian personalities and how families could best manage the expression

[76] Strecker, *Their Mothers' Sons* (cit. n. 73), 18.

[77] Ibid., 36.

[78] Ibid., 17.

[79] Elaine Tyler May, *Homeward Bound: American Families in the Cold War Era* (New York, 1988); Fred Matthews, "The Utopia of Human Relations: The Conflict-Free Family in American Social Thought, 1930–1960," *Journal for the History of the Social and Behavioral Sciences* 24 (1983): 343–62.

[80] The first edition of Benjamin Spock's famous child-rearing book was published in 1946: Spock, *The Common Sense Book of Baby and Child Care* (New York, 1946). For an analysis of the presence of Dr. Spock's manual in American families, see Nancy Pottishman Weiss, "Mother, the Invention of Necessity: Dr. Benjamin Spock's *Care for Infant and Child*," in *Growing up in America: Children in Historical Perspective,* ed. N. Ray Hiner and Joseph M. Hawes (Urbana, Ill., 1985). See also Molly Ladd-Taylor and Lauri Umansky, eds., *"Bad" Mothers: The Politics of Blame in Twentieth-Century America* (New York, 1998), 519–46.

of frustration to prevent aggression from building up.[81] A consumerist society might aid the formation of mass man, which had lost its critical capacities and could potentially follow a strong leader when one was able to appeal to the masses.[82]

CONCLUSION

During World War II, American psychiatrists had been eager to expand their domain and implement utopian ideas of social engineering that they had developed before World War II. During the first years of the war, they were actively involved in the Selective Service System and were successful in implementing a broad definition of conditions, based on the concept of adjustment, which would justify rejection from army service. These efforts were based on the idea that the lack of a predisposition to mental illness was the most reliable indicator of the likelihood that a person would be able to stand the challenges of army life. In 1943, this conception was replaced by one emphasizing the effects of environmental stresses on essentially normal individuals. Successful methods of short-term psychotherapy were introduced near the front lines. At the same time, social scientists investigated breakdown incidence and related it to the level of morale. Both approaches provided new stimulants for psychiatrists to develop a postwar political project: their methods of psychotherapy had been unusually effective and could be applied on a mass scale.

The first postwar project of American psychiatrists was to guide the reintegration and rehabilitation of returning ex-servicemen. In this attempt, they were not successful. As psychiatry moved away from the immediate clinical care of individuals with clearly identifiable diagnoses who appeared to benefit from well-circumscribed and defined treatment interventions, individuals were less inclined to follow its mandate or to adopt its vocabulary and techniques. This, nevertheless, did not discourage psychiatrists, who considered the reform of child-rearing methods as an essential condition of changing the national character of Americans. As far as the veterans were concerned, they had rather conventional ideas of what would ease their return home. According to them, jobs, job security, financial support for further education and housing, and family life were essential ingredients for a happy postwar life.

On the flight home in *The Best Years of Our Lives,* after stating that he is scared that everybody is going to rehabilitate him, Al Stephenson expresses his own opinion about what it will take to get used to postwar life: "All I want's a good job, a mild future, a little house big enough for me and my wife. Give me that much and I'm rehabilitated [snaps his fingers] like that."[83]

---

[81] Theodore W. Adorno, *The Authoritarian Personality* (New York, 1950); John Dollard et al., *Frustration and Aggression* (New Haven, Conn., 1939).

[82] David Riesman. *The Lonely Crowd: A Study of the Changing American Character* (New Haven, Conn., 1950); Erich Fromm, *The Sane Society* (New York, 1955). See also Hans Pols, "Anomie in the Metropolis: The City in American Sociology and Psychiatry," *Science and the City, Osiris* 18 (2003): 194–211.

[83] *The Best Years of Our Lives* (cit. n. 1).

Sick Heil:
Self and Illness in Nazi Germany

by Geoffrey Cocks[*]

ABSTRACT

Illness in Nazi Germany was a site of contestation around the existing modern self. The Nazis mobilized the professions of medicine and psychology, two disciplines built around self, to exploit physical and mental capacity. Nazi projects thus instrumentalized the individual and essentialized a self of race and will. A cruel and anxious obsession with health as a means of racial exclusion was a monstrous form of the modern turn inward to agency of body and mind. The Nazis regulated the individual through family and factory (social control), areas of ordinary life in which modernity located human activity and meaning, and propagandized traditional values the populace internalized (social discipline). A Nazi premodern warrior ethos was served by a liberal ethic of productivity and an absolutist tradition of state control. Medicalization and commodification of health was continuous with modern trends and became a wartime site of attempted well-being of the self at the expense of the Nazi ethnic community.

INTRODUCTION

One of the hallmarks of Nazism was its ruthless emphasis on the subordination of the individual to the ethnic community, or *Volksgemeinschaft*. The common Nazi slogan for this was *Gemeinnutz geht vor Eigennutz,* or "collective need goes before individual greed."[1] This ideal and its enforcement shaped life and policy in Nazi Germany between 1933 and 1945. For some time, of course, historians have known that the very vagueness of Nazi ideology and the competing and conflicting interests within Hitler's state and society made the Nazi order anything but seamless. Still, given the Nazi emphasis on collective duty to an ethnic community, it would appear that the individual self played little or no role in Nazi policy or in daily life during the Third Reich. However, a study of the central place assumed by health and illness in Nazi Germany reveals that dynamics of the self developed over the course of the modern era in Western history were a vital constituent of both policy and popular experience during the twelve years of Hitler's rule.

The Nazis, because of their racism, literally saw the nation as a "body politic" (*Volkskörper*) in the organic sense of the term. It was the aim of the regime to ensure

[*] Department of History, Albion College, Albion, MI 49224; gcocks@albion.edu.
I am grateful to Greg Eghigian, Andreas Killen, and Christine Leuenberger for their scholarly and editorial assistance; all translations, unless otherwise noted, are mine.
[1] Quoted in Claudia Koonz, *The Nazi Conscience* (Cambridge, Mass., 2003), 145.

the survival and advancement of the race and the duty of every individual member of the ethnic community to demonstrate and develop the physical and mental characteristics that were both the proof of racial superiority and the means of racial struggle. The Nazis instrumentalized the individual in service to the state while essentializing an "Aryan" self of race and will, emphasizing internal qualities of alleged racial character and culture—"healthy ethnic sensibility" (*gesundes Volksempfinden*).[2] The Nazis were thus constrained by their own policies, as well as by modern realities, to channel their commands and exhortations through millions of individuals who had been acculturated and socialized into the modern conception and practice of the largely autonomous self. This furthered a Western tradition of thinking in terms of individual abilities and capacities that was present in Germany because of the industrialization and the commercialization that had been spreading through central Europe since the nineteenth century. Although the Nazis inveighed against degenerate liberalism and individualism to the bitter end, the regime increasingly opted to propagandize "great personalities," to portray in word, picture, and film model individuals of Nazi character and performance, and even to indulge—and indulge in—advertising that catered to individual needs, desires, and demands.[3] Some Nazi agencies, from the same racist view of nature that exalted the family for purposes of procreation, even promoted a liberalized ethic of sexual pleasure that reflected not only the modern utilitarian ethic of the individual pursuit of happiness but also a culture of consumption in Germany dating to the late nineteenth century.[4]

The Nazi slogan concerning need and greed itself displays tension between the individual and the collective. Unlike in the propaganda slogan "You are nothing. Your *Volk* is everything!" in the more common iteration of the same idea "collective need" only "goes before" (*geht vor*) "individual need." This is not a textual or rhetorical quibble, particularly in the modern light of Hegel's explication of the dialectical nature of human thought, Freud's analysis of the contradictions of human motivation, and postmodern inquiry into the ambiguities of language. This rhetorical tension reminds us that National Socialism was not *sui generis* but a monstrous muddle of extant traditions, trends, and conflicts coursing through the modern history of the West. The tension in the Third Reich between self and society betrays, in particular, the strong cultural and social presence of the modern self in Germany during the first half of the twentieth century. This presence had been marked by widespread political discourse in Germany on the need for national solidarity in the wake of the disastrous effects and outcome of the First World War as well as the severe economic, political, and

---

[2] Ibid., 140, 197–8.

[3] Moritz Föllmer, "Nazism and the Politics of the Self in Berlin" (paper presented at the conference "The Self as Scientific and Political Project in the Twentieth Century: The Human Sciences between Utopia and Reform," Pennsylvania State Univ., 10–11 Oct. 2003); Ray Cowdery and Josephine Cowdery, *German Print Advertising, 1933–1945* (Rapid City, S. Dak., 2004), 23, 32, 73; Hartmut Berghoff, "Enticement and Deprivation: The Regulation of Consumption in Pre-War Nazi Germany," in *The Politics of Consumption: Material Culture and Citizenship in Europe and America*, ed. Martin Daunton and Matthew Hilton (Oxford, 2001), 169–71.

[4] Dagmar Herzog, *Sex after Fascism: Memory and Morality in Twentieth-Century Germany* (Princeton, N.J., 2005), 16; Charles Taylor, *Sources of the Self: The Making of the Modern Identity* (Cambridge, Mass., 1989), 169; Koonz, *Nazi Conscience* (cit. n. 1), 103; Matthew Jeffries, "For a Genuine and Noble Nakedness? German Naturism in the Third Reich," *German History* 24 (2006): 62–84; David Hamlin, "Romanticism, Spectacle, and a Critique of Wilhelmine Consumer Capitalism," *Central European History* 38 (2005): 250–62, at 250–55; and Wolfgang König, *Volkswagen, Volksempfänger, Volksgemeinschaft* (Paderborn, Germany, 2004).

social problems of the postwar Weimar Republic. As Moritz Föllmer has argued, the surging "rhetoric of *Volksgemeinschaft*" after 1918 only underscored the pervasiveness of self-interest even—or especially—among those protesting it.[5] Even after 1933, Victor Klemperer, a German-Jewish professor of Romance languages and literature in Dresden, could note in the Nazi regime's "ever greater tyranny a sign of ever greater uncertainty," thereby generalizing just this sort of tension.[6]

The Nazis also mobilized the relatively new disciplinary field of medicine and the still newer field of psychology to help shape individual Germans to the mold of National Socialist duty and sacrifice. This recruitment was due, in part, to the incomplete Nazi reformation of German institutions and ideas and a "polycratic" governing system of rivalry, duplication, patronage, and self-promotion.[7] Nazi policy was thus continuous with the modern medicalization and psychologization of the experience of health and illness as well as with the construction of a modern "therapeutic" state of official control and concern. As Greg Eghigian has shown, such "psychologization of human subjectivity," in particular, was continuous under authoritarianism in Germany after 1945 in the German Democratic Republic.[8] In the Third Reich, this marshalling of expertise served the Nazi political and scientific project to define, exploit, and enhance the imputed superior human qualities of members of the "master race" (*Herrenvolk*). However, these now newly ambitious professions under Nazism had been created in the modern era for the purpose of the study, protection, cultivation, and disciplining of the individual. The growing emphasis in institutionalized psychotherapy, for one, on integration of the individual into society eased collaboration with both progressive and fascist regimes. This came at the expense of an orientation within Freud's psychoanalysis, in particular of some advocacy for individual desire in conflict with society. In all these iterations, however, the individual self remained the object of scientific and political attention.[9]

As this new racialized medical project collided with established social and cultural realities as well as with events occasioned by Nazi actions, the result was not the elimination of the space in which the self could operate but rather its restriction and redefinition. Industrial mobilization around 1936 and the outbreak of war in 1939 worked not only to advantage professionally the fields of medicine and psychology but also to perpetuate and sharpen habits having to do with the self at the expense of the Nazi *Volksgemeinschaft,* forging continuities with powerful social and cultural dynamics in Germany and the West before 1933 and after 1945. The practice and discourse of

---

[5] Moritz Föllmer, "The Problem of National Solidarity in Interwar Germany," *Germ. Hist.* 23 (2005): 202–31, on 215; cf. Peter Fritzsche, "Cities Forget, Nations Remember: Berlin and Germany and the Shock of Modernity," in *Pain and Prosperity: Reconsidering Twentieth-Century German History,* ed. Paul Betts and Greg Eghigian (Palo Alto, Calif., 2003), 35–60; Fritzsche, *Germans into Nazis* (Cambridge, Mass., 1998), 163, 178, 184, 192, 198, 201–2, 204, 208–9, 211, 228–31, 233.

[6] Victor Klemperer, *I Will Bear Witness: A Diary of the Nazi Years, 1933–1941,* trans. Martin Chalmers (New York, 1998), 70.

[7] Peter Hüttenberger, "Nationalsozialistische Polykratie," *Geschichte und Gesellschaft* 2 (1976): 419–42.

[8] Greg Eghigian, "The Psychologization of the Socialist Self: East German Forensic Psychology and Its Deviants," *Germ. Hist.* 22 (2004): 181–205, on 181.

[9] Philip Rieff, *Freud: The Mind of the Moralist* (Chicago, 1959), 161, 247; Russell Jacoby, *The Repression of Psychoanalysis: Otto Fenichel and the Political Freudians* (New York, 1983), 77; Paul Roazen, *Erik H. Erikson: The Power and Limits of a Vision* (New York, 1976), 172; Geoffrey Cocks, *Psychotherapy in the Third Reich: The Göring Institute,* 2nd ed. (New Brunswick, N.J., 1997); Ulfried Geuter, *The Professionalization of Psychology in Nazi Germany,* trans. Richard J. Holmes (Cambridge, 1992).

politicized human sciences under National Socialism were therefore not only compromised morally through their collaboration with the Nazi regime but also promoted and hobbled by the especially chaotic conditions set by theory, policy, and event for most of the brief existence of the Third Reich.

Both practice and limitations on practice magnified the importance of the various responses of Germans to the conditions surrounding issues of individual well-being under Hitler, that is, responses to *problems* as well as to projects. A focus on subjects is germane because social and cultural response to scientific expertise is particularly important in medicine, which not only was a field of special importance to the Nazis but also is the most powerful modern discipline intimately involved with the ways in which individuals live their lives and construct their identities. In the end, the disastrous events that constituted the brief history of the Third Reich perpetuated habits of the self characteristic of the modern era. The self in Nazi Germany would become both servant of the regime and sanctuary for the individual, experiences that exemplify the central place that the objects—or, better, subjects—of projects must take in any study of policy and science in the history of the modern era.

ME AND WE

The Nazi obsession with health was itself one outcome of the long modern Western turn from the ancient preoccupation with the soul as intimation of cosmic order to the body as the location not only of pleasure, pain, demand, and desire but also of personal identity. In this, Nazi ideology was reflecting an even more general feature of Western history in the modern era—the conception and construction of the self. Charles Taylor has argued that the modern self constitutes a sense of "inwardness" related to the ineluctable human search for meaningfulness and morality. Augustine, the Reformation, and the Enlightenment had turned Plato's contemplation of external rational order into individual faith and reason as internal sources of knowledge, obligation, and identity. Coupled with this was a modern notion of natural and subjective rights, inherent in every person regardless of status, that are emblematic of individual autonomy. This idea was complemented by the utilitarian principle of the avoidance of suffering for oneself and others as well as by the "affirmation of [the] ordinary . . . life of production and the family."[10] This affirmation of ordinary life stood in marked contrast to the archaic ethos of honor held by warriors and public citizens of rank in Greece and Rome. Such an ethos was opposed by philosophers from Plato to Descartes who celebrated reason as internalized "domination of passion by thought."[11]

The Nazis, of course, essentially reverted to a warrior ethic. But they did so by means of an imposed order that not only dealt with the extant praxis of the self in modern society but also embodied features of the modern society that was both cause and effect of the development of modern selfhood. From the eighteenth-century Enlightenment, the Nazis inherited the instrumentalizing reason that underlay the science, technology, and industry the Nazis themselves used to build everything from Messerschmitts to gas chambers. From nineteenth-century Romanticism they crudely adapted a collectivist, racist "appeal to expressive integrity" that was a crabbed and

---

[10] Taylor, *Sources of the Self* (cit. n. 4), 13, 285; Fritz Breithaupt, "Goethe and the Ego," *Goethe Jahrbuch* 11 (2002): 77–109.
[11] Taylor, *Sources of the Self* (cit. n. 4), 153.

cancerous mutation of the Romantic "power of expressive self-articulation."[12] These two sets of ideals had originally been nurtured and provoked, respectively, by the rise of a bourgeois commercialism that valued and rewarded the individual purposeful-ness and discipline that comes with self-mastery. Such ideals had also been embod-ied, advanced, and compromised by the discipline and order imposed by the modern state and its bureaucracy. This system of centralized institutional rewards and pun-ishments was based upon "the growing ideal of a human agent who is able to remake himself by methodical and disciplined action" through mastery of powers within the self.[13] Such agency in a different modern context took the form of unbridled will in place of reason or faith and thereby served as another source for the brutal Nazi im-poverishment of modern ideas.

More troublingly even than this, the Nazi state also embodied what Taylor calls "hypergoods," an inherent inclination of human beings toward "spiritual aspirations." This is not to adopt the ultimate moral relativism that assigns inherent ethical value even to Nazi ideology. It is to say that Hitler and the Nazis had aspirations to what they conceived of as the "good," an ideal of active earthly benevolence Taylor identifies as an essential element of the modern human identity. Hitler and the Nazis, of course, did not share in the ideal of *universal* benevolence that underlies modern liberalism and humanism. They restricted consideration to a specific group of humans regarded as "racially" more valuable than all others. However irrational any of their underlying motives were, Nazi true believers were convinced that the ultimate end of their march into the future was the scientific and rational creation of one perfect race inhabiting the entire world. This was a more brutal form of a tradition of earlier eugenic think-ing among many Germans from right to left on the political spectrum. However, even original liberalism and humanism had imposed their own hierarchal "meta-narrative" schemes of "improvement" upon humanity, schemes that represented blindness, prej-udice, and cruelty toward any number of "lesser" beings.[14]

In pursuit of their aims, however, the Nazis had to rely not only on social control effected by modern state propaganda and terror but also on the complementary dy-namic of social discipline, the internalization of mores and dictates by the population. This in the case of Hitler's Germany is what Claudia Koonz has recently defined as the "Nazi conscience." Koonz begins by noting that, in opposing what they saw as the degenerate and materialistic self-interest of modern society, "Nazi ideals embedded the individual within the collective well-being of the nation."[15] Hitler promised heav-ily masculinized honor and dignity, a reflection, as we have noted, of an ancient war-rior and early modern aristocratic ethos. However, the Nazis were not able to extirpate an idea and a practice of self that had firmly embedded themselves in modern society, and the effects of which, as we have seen, became a considerable site of contestation in Germany after the First World War. Hitler was in power for only twelve years, and his regime was shot through by inefficiency, corruption, and self-interest. The national

---

[12] Ibid., 415, 390; see also 372, 384, 388.

[13] Ibid., 159, 276–77; David Sabean, *Power in the Blood: Popular Culture and Village Discourse in Early Modern Germany* (Cambridge, 1984), 211.

[14] Young-sun Hong, "Neither Singular nor Alternative: Narratives of Modernity and Welfare in Ger-many, 1870–1945," *Social History* 30 (2005): 133–53, on 138; Taylor, *Sources of the Self* (cit. n. 4), 85, 88, 97, 100, 101, 345; Johannes Vossen, *Gesundheitsämter im Nationalsozialismus: Rassen-hygiene und offene Gesundheitsfürsorge in Westfalen, 1900–1950* (Essen, Germany, 2001).

[15] Koonz, *Nazi Conscience* (cit. n. 1), 2; Hong, "Neither Singular nor Alternative" (cit. n. 14), 137, 145, 148–49.

and international events that followed upon the Nazi assumption and use of state and military power created conditions in which the dissolution of the self into the mass was complicated and compromised before collapsing almost completely by the end of the Second World War.

The Nazi *Volksgemeinschaft* also undermined itself in a variety of ways, for instance through fear and frequent denunciations that set "Aryans" against each other. The ideal of ethnic community itself was also inherently divisive precisely because it was based "on self-love and other hate."[16] Being propagandized to as a member of a superior race almost always enhanced an individual German's own sense of self-worth. Such propaganda helped produce the Nazi conscience, the internalization of standards by which one's own self is judged and measured against others. Beginning early in their regime, the Nazis sought to cement popular support through a positive message of ethnic solidarity and progress, having found that the strident and crude anti-Semitism of such vicious Jew-baiters as Julius Streicher tended to alienate the solid and respectable Germans the Nazis wished to co-opt. The promotion of virtue had a more positive effect on the public mood in Germany than the brutal rant and exercise of anti-Semitism. The one-day boycott of Jewish businesses in April 1933, for example, was a failure in terms of public opinion. This did not constitute popular admiration for Jews, however; rather, it was an indication of the population's preference to have the dirty work of the regime done in secret and as far away from daily life as possible. Klemperer's diaries for the entire period of the Third Reich describe the wide range of "Aryan" attitudes toward individual Jews, from sympathy and loyalty to indifference and hatred, all of which provided a *soupçon* of comfort and assistance for Jews and a ladleful of quiescence and thereby support for their persecution.

In their propaganda, the Nazis stressed virtues congruent with most Germans' sense of themselves as living up to high moral standards, particularly in forging an ethnic identity that could be measured more implicitly and subtly against others deemed inferior and threatening. Such "positive messaging" also diverted attention from the small and powerful groups of fanatics organized to persecute "racial enemies." Nazi terror was directed primarily against specific groups, most of them socially marginal, so the majority of Germans were free to exercise what Koonz calls "selective compliance" with the regime's policies and practices. The Nazis thus relied not simply on external pressure or on persuasion but on cultivating the internal conviction of many, if not most, Germans, an approach that exploited the internalization of motive and identity integral to the modern self. This meant that certain degrees of choice were available to the majority of Germans in Nazi Germany on the basis of the fact that they were, by and large, convinced of the virtue of their actions and attitudes and, designated as Aryans, spared by and large the threat of imprisonment, torture, or death at the hands of their regime. The latitude allowed individuals not identified as members of persecuted groups such as communists and Jews can be seen in the widespread phenomena in Nazi Germany—with moderating or radicalizing effects—of loyal or opportunistic "self-coordination" and the functionary initiative called "working towards the Führer."[17]

---

[16] Koonz, *Nazi Conscience* (cit. n. 1), 10; Robert Gellately, *The Gestapo and German Society: Enforcing Racial Policy, 1933–1945* (New York, 1990); Marion A. Kaplan, *Between Dignity and Despair: Jewish Life in Nazi Germany* (New York, 1998), 37.

[17] Koonz, *Nazi Conscience* (cit. n. 1), 5, 12–15, 75, 108, 141, 177; Ian Kershaw, "'Working Towards the Führer': Reflections on the Nature of the Hitler Dictatorship," *Contemporary European History* 2 (1992): 103–18.

All such actions and attitudes fit easily into an established German civic identity flatteringly racialized by the new government. Because one was now a member of a superior race, one's own identity was not extinguished but rather enhanced and reconfirmed by others' recognition and affirmation in the course of everyday existence, an existence that for most people continued (until the war) in many ways as it had before 1933—only better. In the early years of the Nazi regime, the common allusion among "self-coordinating" intellectuals, scholars, and experts to Friedrich Nietzsche's concept of the "transvaluation of all values" is typical of related convictions, confusions, and conceits of the era.[18] The creation of a Nazi conscience followed some of what Taylor describes as "the conflicts of modernity," by which what he defines as the essential human disposition to be or to do good serves evil. In this process, the conscience serves as a powerful source of identity for the self by way of guilt, self-satisfaction, hypocrisy, and—horribly notable in the Nazi record—"projection of evil outward" onto others.[19]

THE BODY POLITIC

The Nazis habitually deployed metaphors of health and illness to describe their political aims. That which was "healthy" in the German nation was to be preserved and enhanced, while that which was "diseased" was to be cut out of the German body politic. Hitler, more than anyone else, projected this concern in terms of aesthetics, making a direct correlation between what was beautiful and healthy and what was ugly and unhealthy. Health and illness were not just metaphors for Hitler and the Nazis. They were obsessive constituents of a political and scientific project designed to address what the Nazis believed were central crises of the modern world and which represented vital persistent problems and discourses in the history of modern Germany and the West on the health of individual body and mind.

In Germany as elsewhere in the West, concern with individual physical and mental well-being had been considerably heightened around the turn of the twentieth century by the growing medicalization of health and illness. While mortality rates from acute illness and especially from epidemics steadily declined due to better hygiene and medical care, morbidity rates for various chronic illnesses connected with poor living and working conditions increased. Furthermore, although health care was improving for the population in general, there was still significant social inequality when it came to health and thus a distressing degree of physical and mental suffering and dysfunction among the urban and rural lower classes.[20] Of course, doctors and governments could no more banish illness for anyone than they could death, and the increased official and popular attention paid to the differences between health and illness only served to raise disappointments as well as expectations. Everyday episodic illness, "being sick" (*Kranksein*), was the most common, if usually manageable, reminder of the disorders and dysfunctions of the morbid and mortal body. All three categories of illness—acute, chronic, and episodic—therefore, underscored the fragility of the

[18] Koonz, *Nazi Conscience* (cit. n. 1), 107, 110; Taylor, *Sources of the Self* (cit. n. 4), 65; for a case study of "self-coordination" among psychotherapists, see Kl. Erw. 762.2, Bundesarchiv Koblenz, Koblenz, Germany (hereafter cited as BAK).

[19] Taylor, *Sources of the Self* (cit. n. 4), 217, 503, 516.

[20] Reinhard Spree, *Health and Social Class in Imperial Germany: A Social History of Mortality, Morbidity, and Inequality,* trans. Stuart McKinnon and John Halliday (Oxford, 1988).

individual body at a time when "the sick person became a social entity with an actual and perceived identity."[21] Thus it is no surprise that in the early twentieth century a good deal of popular, professional, and official discourse and activity was devoted to issues of health and illness in Germany as it was elsewhere. The economic troubles and political uncertainties of the postwar Weimar Republic increased social discourse about the body, a discourse sharpened by the legion of physical and mental casualties of the First World War. Greg Eghigian argues that after 1918, in particular, there was widespread demand by organized interests for state intervention in the form of victim compensation, rendering one means of modern state control a means of individual assertion and protection as well.[22]

Alternative therapies, which became widespread in Germany after 1918, also represented a reaction against the impositions and limitations of scientific medical control over health and illness. According to Michael Hau, alternative therapies and practices, in particular, offered individual Germans before and after the First World War a way toward health and beauty of the body as a means of reasserting autonomy in the face of external forces over which, in relative contrast to their bodies, they had little or no understanding or control. Even before the war, and alongside the operation of professional medicine and government health programs, there was bustling commerce in and consumption of natural regimens and mechanical devices produced to relieve pain and restore physical and mental vigor.[23] Just as the Nazi regime coordinated and exploited the medical profession for its own purposes, many alternative therapies found a place, if a subordinate one, in the Third Reich. As we have already noted, neither the regime's pursuit of nazification nor blustery prattle about *Volksgemeinschaft* meant an end to individual struggles over issues of self and identity, especially with hysterical Nazi promises and threats concerning the health and illness of body and mind.

The flattering Nazi promotion of the "master race" was reinforced by gushes of "psychological capital" from apparent and loudly trumpeted Nazi economic, diplomatic, and military successes. Even in peacetime health, however, was the source of continuing, indeed heightened, individual anxiety. At the extreme, people worried about whether they or their loved ones might be declared unfit to procreate or even to live as a result of some genetic taint. Rumors gusted and swirled about the forced sterilization and, during the war, gassing of mental patients. One official in the Reich Interior Ministry wrote of the "almost psychotic fear" among the populace about a Nazi decree of 1933 calling for the prevention of hereditary illness that led to mass sterili-

[21] Geoffrey Cocks, "The Old as New: The Nuremberg Doctors' Trial and Medicine in Modern Germany," in *Medicine and Modernity: Public Health and Medical Care in Nineteenth- and Twentieth-Century Germany,* ed. Manfred Berg and Geoffrey Cocks (Cambridge, 1996), 180; and in Geoffrey Cocks, *Treating Mind and Body: Essays in the History of Science, Professions, and Society under Extreme Conditions* (New Brunswick, N.J., 1998), 199; Claudine Herzlich and Janine Pierret, *Illness and Self in Society,* trans. Elborg Forster (Baltimore, 1987); Edward Shorter, *Bedside Manners: The Troubled History of Doctors and Patients* (New York, 1985).

[22] Greg Eghigian, *Making Security Social: Disability, Insurance, and the Birth of the Social Entitlement State in Germany* (Ann Arbor, Mich., 2000); Hong, "Neither Singular nor Alternative" (cit. n. 14), 143; Geoffrey Cocks, "Modern Pain and Nazi Panic," in Betts and Eghigian, *Pain and Prosperity* (cit. n. 5), 96.

[23] Michael Hau, *The Cult of Health and Beauty in Germany: A Social History, 1890–1930* (Chicago, 2003), 3, 5, 15, 17–31, 208 n. 8; Pierre Bourdieu, *Distinction: A Social Critique of the Judgment of Taste* (Cambridge, Mass., 1984), 179, 190–93; Joachim Radkau, *Das Zeitalter der Nervosität: Deutschland zwischen Bismarck und Hitler* (Munich, 1998).

zation of mental patients.[24] Writer Christa Wolf strikingly recalls how an aunt struggled with knowledge of the murder of a mentally disabled sister and in the end "had a violent attack of migraine . . . and . . . when she recovered, her face had taken on a greater likeness to her dead twin sister."[25]

In the public realm of Nazi Germany, therefore, "images of disability and . . . illness constantly circulated."[26] Because the body was closely joined with modern conceptions of individual identity, this was a massive aggravation for individuals of the even more general modern fact of life that, as inevitably defined in terms of social context, "the stability of the self is always in question."[27] Even before the outbreak of war in 1939, the cuts in medical services because of rearmament and mobilization and the regime's insistence on individual fitness (*Leistungsfähigkeit*) for purposes of labor reinforced a growing widespread feeling of anxiety about the body (and the mind) that had long been a feature of social constructions around health and illness in Germany.[28] These concerns could also mobilize guilt among most Germans as they had long since internalized social and ethical standards having to do with the responsibilities of individuals to themselves, their families, their workplaces, and society. As we have seen, the Hitler regime exploited this in the construction of a "Nazi conscience." All such feelings would be further magnified—and problematized—by patriotism mobilized by Nazi rhetoric and the thrills and agonies of a war that for most Germans began very well, got very bad very fast, and ended very badly indeed.

THE BODY IN QUESTION

Nazi mobilization for war and the war itself had a massive impact on health and illness in Germany, not just in terms of the millions of people who suffered and died but also in terms of the effect of Nazi projects involving the regimentation of the energies of individual Germans for purposes of national and racial warfare. Most Germans were victimized in one way or another by Hitler's war, and narratives of the suffering of German civilians, particularly under the Allied bombing of German cities, have been a staple of discourse in Germany since 1945. Germans' health experiences during the war were an extension of those under Nazism at peace, which were in turn an elaboration on the modern phenomenon of the body and mind as location for assertions of and struggle over self and identity, which was most recently further aggravated by the Nazi regimen.

The suffering inflicted on others, of course, especially on the Jews of Germany and occupied Europe, far outweighed in gruesome quantity and quality that imposed by

---

[24] Reichsministerium des Innern, n.d., R 18, 5585, BAK.

[25] Christa Wolf, *Patterns of Childhood,* trans. Ursule Molinaro and Hedwig Rappolt (New York, 1980), 198; Koonz, *Nazi Conscience* (cit. n. 1), 119, 127, 262; Hau, *Cult of Health and Beauty* (cit. n. 23), 8, 200–203, 209 n. 15.

[26] Carol Poore, "Who Belongs? Disability and the German Nation in Postwar Literature and Film," *German Studies Review* 26 (2003): 21–42, on 22.

[27] Jerrold Siegel, "Problematizing the Self," in *Beyond the Cultural Turn: New Directions in the Study of Society and Culture,* ed. Victoria E. Bonnell and Lynn Hunt (Berkeley, Calif., 1999), 281–314, on 299; Ronald Hoffman, Mechal Sobel, and Frederika J. Teute, eds., *Through a Glass Darkly: Reflections on Personal Identity in Early America* (Chapel Hill, N.C., 1997), 9; Hong, "Neither Singular nor Alternative" (cit. n. 14), 138.

[28] *Deutschland-Berichte der Sozialdemokratischen Partei Deutschlands* (hereafter cited as *Sopade*) 4 (1937): 1320–1; Hau, *Cult of Health and Beauty* (cit. n. 23), 15.

the regime on the Aryan population of Germany.[29] Klemperer, who survived because of his Aryan wife, filled his diaries with accounts of such matters as heart and stomach problems from stress and inadequate food, colds, influenza, physical and psychological abuse by officials, and the politically aggravated complaints of aging. In the summer of 1942, he wrote: "Fear and hunger fill the day. Hunger alternates with nausea. When they are cold, I can only eat the old, bad potatoes in spite of myself, and they agree neither with my stomach nor with my bowels. But more than that, I am tormented by fear."[30] Medical care for Jews in Nazi Germany was almost nonexistent. As the medical assessor for the Jewish community in Dresden told Klemperer in 1943, "'Today a Jew must not fall ill.'"[31]

The mobilization of industry for war, beginning around 1936, marked a decisive turn in Nazi health policy. Medical practice now—and as accelerated by the outbreak of war in 1939—was not one devoted so much to racial improvement as to securing and maintaining healthy bodies for deployment as soldiers and workers. Doctors were massively conscripted into the armed forces and assigned by the Nazi Party's German Labor Front (Deutsche Arbeitsfront, DAF) to enforce the regime's demands for worker productivity. These moves further compromised a state health care system that had been purged of the many socialists working in its bureaucracy and raided for funds to assign to industrial and military projects. Such reallocation not only served Nazi war aims but also aggravated an ideological stress on healthy human productivity rather than on indulgent care of the weak and the sick. It also removed (although incompletely as we shall see) the modern tradition of health care as being principally the preserve of the family and the individual physician. While various party and state agencies strove to cultivate racially healthy families and individuals, Nazi alarums and excursions placed the body in ever greater question for the vast majority of Germans living under Hitler. More than ever, the regime imposed a hierarchy of care with the productive at the top and the less or nonproductive at the bottom while also using the war as cover and occasion for murderous intensification of the campaign against racial enemies.

The fields of medicine and psychology played key roles in Nazi preparations for war by increasing and defining official and popular attention to matters of the individual body and mind as practical resource rather than as idealized collective racial body. Psychology and psychotherapy, though newly and incompletely professionalizing in Germany during the 1930s and 1940s, took advantage of the general shaking up of German life by Nazism to establish themselves as competitors to traditional medicine and psychiatry. Part of this was mutually ideological and political. Psychotherapists in the Third Reich, for example, undertook the rhetorically and professionally advantageous promotion of a Romantic disciplinary heritage of concern for the collective expression of unique collective characteristics of peoples and nations.[32]

---

[29] Mary Nolan, "Germans as Victims during the Second World War," *Cent. Eur. Hist.* 38 (2005): 7–40; Richard J. Evans, " '*Facilis descensus averni est*': The Allied Bombing of Germany and the Issue of German Suffering," *Cent. Eur. Hist.* 38 (2005): 75–105.

[30] Victor Klemperer, *I Will Bear Witness: A Diary of the Nazi Years, 1942–1945,* trans. Martin Chalmers (New York, 1999), 84.

[31] Ibid., 208.

[32] Fritz Künkel, *Charakter, Einzelmensch und Gruppe* (Leipzig, 1933); and Künkel, *Das Wir: Die Grundbegriffe der Wir-Psychologie,* 3rd ed. (Schwerin, Germany, 1939); Ernst Kretschmer, "Konstitution und Leistung," *Westfälische Landeszeitung,* 20 Aug. 1944, T-78, reel 190, frames 1866–67, National Archives, College Park, Maryland (hereafter cited as NA).

Such an intellectual tradition had been strong in Germany since the nineteenth century and, as we have seen, influenced Nazi conceits and confusions. The girding for war would, however, increase Nazi demands for expertise and competence at the expense of earlier radical demands for professional projects based on "new" Germanic racial concepts. Psychotherapeutic "we-psychology," for instance, would serve pragmatically productive aims rather than exclusively racial ones. Academic psychologists, for their part, found employment in industry and the military (until 1942, when psychological selection became a luxury that could no longer be afforded by the armed forces) as part of an ongoing effort intensified to maximize human resources. As early as 1936, the DAF was supervising vocational training and developing occupational preference guides to direct young workers into areas in which there were shortages of skilled labor. The DAF's approach combined American technocratic thought with a disciplined and dutiful Nazi "sense of community" and "joy in one's creative and productive labor."[33] As one element in this project, at the end of September 1939 the DAF assumed formal supervision of the German Institute for Psychological Research and Psychotherapy, an institution founded in 1936 that, among other activities, promoted and practiced short-term psychotherapeutic methods designed to foster the healthy and happy integration of individuals into family and work.

As the deployment of psychologists and physicians within the German civilian population in the 1930s was limited and as their function was designed to serve the aims of the state more than the aims of their patients, it is not surprising to learn that many Germans, even before the war, found relatively little help in managing the difficult and stressful demands of life and work. Patients were increasingly left to their own devices in an environment of considerable self-reliance. Much like the wartime researchers in Great Britain whose work led to the Beveridge Report in 1942 and the construction of the National Health Service after the war, regional medical officials in new Nazi Germany found that health conditions, living circumstances, hygienic habits, and the resultant psychological state of much of the general population often left a great deal to be desired. This was especially apparent in the sections of Annual Health Reports (*Jahresgesundheitsberichte*) concerning "General Health Conditions" (*Allgemeine gesundheitliche Verhältnisse*) in the first years of the regime, as the Great Depression had for many people worsened an already marginal degree of security against health problems.[34] There were complaints in these reports even about physical toughening in Nazi auxiliary formations. Excessive exercise, such as barefoot marches by Hitler Youth brigades, was, according to officials of military recruiting districts, aggravating a host of existing physical conditions to the detriment of the armed services.[35]

There were other sources of concern and complaint as well. The German Social Democratic Party had fled into exile but had agents in Germany to report on conditions

[33] Cocks, *Psychotherapy in the Third Reich,* 256; Geuter, *Professionalization of Psychology,* 83–162. (Both cit. n. 9.)

[34] Jahresgesundheitsbericht, Teil A, Abschnitt II, Allgemeine gesundheitliche Verhältnisse (hereafter cited as Jgb) Gesundheitsamt Düsseldorf, 1932, Nordrheinwestfälisches Haupstaatsarchiv, Düsseldorf (hereafter cited as NWH).

[35] Wehrwirtschafts-Inspektion VII, Munich, 10 Sept. 1937, T-77, reel 248, frames 824–28, NA; Wolf, *Patterns of Childhood* (cit. n. 25), 190–1; Jgb, Sachsen/Glauchau, 1935, Sächsisches Hauptstaatsarchiv, Dresden, Germany (hereafter cited as SHD); Winfried Süss, *Der "Volkskörper" im Krieg: Gesundheitsverhältnisse und Krankenmord im nationalsozialistischen Deutschland 1939–1945* (Munich, 2003), 96–110; Hong, "Neither Singular Nor Alternative" (cit. n. 14), 151.

for the working class. There was much to report about the impositions placed upon workers in particular by the Nazi regime. Although unemployment declined markedly, especially with the onset of rearmament, the health of workers was compromised in a variety of ways, including overwork, often among women in particular.[36] There were no longer any unions to protect the interests of the working classes as the DAF served the interests of employers and the state. The Nazis substituted supervision for service in the reduction of resources for sick funds, welfare, compensation, and disability pensions. The new emphasis was on productivity, so doctors were directed to cut the number of sick days and sick leaves. Increasingly, company medical officers (*Betriebsärzte*) and independent medical examiners (*Vertrauensärzte*) displaced both the family and the family doctor as the locus of decisions concerning health, sickness, and work.[37] The job of the company medical officers was to limit sick days and expenses as much as possible, while the medical examiners checked private doctors' diagnoses for indulging the alleged "flight into sickness" of their patients. This represented a merciless sharpening, in the absence of organized working class opposition, of traditional business management mistrust of patient freedom to choose a doctor. This mistrust was based on the conviction that inexperienced doctors could be deceived by their patients and would indulge them for fear of losing them to competition.[38]

Although the overall health record of the Nazi regime during the early years of the Third Reich was mixed, the increase in the attention paid to matters of health and illness created significant new degrees of uncertainty for most Germans about personal expectations and prospects. Mobilization for war increased the pressure by the regime on its human resources in the tradition of the modern state's practice of supervision, discipline, and punishment. This pressure was met by various means of coping, strategies of self-protection, and even some (almost always nonpolitical) resistance on the part of the populace. What sociologist Michel de Certeau calls "antidiscipline" and "the polytheism of scattered practices" were especially acute and widespread because official concern with matters of health and illness intrudes more deeply into people's lives than in other realms. Although Germans have long been stereotyped as responding with particular alacrity to demands for discipline and obedience, there is also a long history in Germany of defiance of authority.[39] In October 1936, a Social Democratic agent recounted a meeting in Saxony at which an official spoke glowingly about disability pensions, then unwisely opened the floor for discussion. One woman who had been receiving disability payments for twenty years for a finger lost to blood poisoning had just had her payments cut off. She was so profanely vociferous—and

[36] Jgb, Mecklenburg/Wismar, 1935, Landeshauptarchiv Schwerin, Schwerin, Germany (hereafter cited as LHS); Jgb, Bayern/Regensburg, 1937, 1939, 1940, Staatsarchiv Amberg, Amberg, Germany (hereafter cited as SAm); Jgb, Württemberg/Ravensburg, 1941, Staatsarchiv Ludwigsburg, Ludwigsburg, Germany (hereafter cited as SL); Jgb, Bayern/Memmingen, 1941, Staatsarchiv Augsburg, Augsburg, Germany (hereafter cited as SAu).

[37] *Sopade* 1 (1934): 42; 2 (1935): 788, 839; 3 (1936): 632, 637, 1287–88; 4 (1937): 1314–15; 6 (1939): 348; and Süss, *"Volkskörper" im Krieg* (cit. n. 35), 242–68.

[38] Martin Geyer, *Die Reichsknappschaft: Versicherungsreformen und Sozialpolitik im Bergbau 1900–1945* (Munich, 1987), 237; Süss, *"Volkskörper" im Krieg* (cit. n. 35), 151–60.

[39] Michel de Certeau, *The Practice of Everyday Life,* trans. Steven Rendall (Berkeley, Calif., 1984), 47; Alfons Labisch, *Homo Hygienicus: Gesundheit und Medizin in der Neuzeit* (Frankfurt am Main, 1992), 133; Arthur Kleinman, *The Illness Narratives: Suffering, Healing, and the Human Condition* (New York, 1988); Richard J. Evans, "In Pursuit of the *Untertanengeist:* Crime, Law, and Social Order in German History," in *Rethinking German History,* ed. Richard J. Evans (London, 1987), 156–87.

received such support from the audience—that the chair of the meeting threatened to call the police.[40]

Such protests, of course, did not arrest Nazi martial momentum. From northwestern Germany came a report that even badly injured First World War veterans were being declared fit for work.[41] The ironic—and tragic—result of these demands was an increase in accidents, injuries, and illnesses that negatively affected production despite draconian measures to keep people at work and for more hours. Even before the war, there had been the growing prospect of a severe shortage of experienced doctors primarily because of military conscription but also because of the full exclusion from practice in 1938 of the large number of Jewish physicians. As a result, doctors who were available to civilians tended to be younger and less experienced than the doctors they replaced.[42] In the end, according to a 1937 *Sopade* report, the Nazis had done nothing less (or more) than reintroduce fundamental insecurity about health and disability into the lives of millions of German workers.[43]

BODY BLOWS

The war radicalized and rationalized Nazi health policy, rendering the war both occasion and cover for unprecedented exterminatory racism, on the one hand, and implementation of heartlessly militarized social-utilitarian triage, on the other.[44] This utilitarian turn placed more emphasis on productivity by individuals of "racial quality" and less on collective ethnic identity. This was a decisively modern standard of the inherent—that is, internal—productive characteristics of the self applied to the brutal and aggressive program of Nazi imperial and racial warfare. In the civilian realm, however, this utilitarian standard, even with its own rigors and outrages, allowed some space for individuals to operate on the basis of habits of the self. More than ever, the self became sanctuary for the individual as well as servant of the regime. Recourse to this space more often constituted a cultivated habit of self-protection than articulated dissatisfaction with the regime or its ideology, although such discontent (in the form of disappointment or disillusion) was increasingly common with the deterioration of life during wartime. Such recourse was especially the case in the newly problematized realm of personal health and illness, which—particularly with all of the disruptions of Nazi rule that came with escalating military and social disaster—took the place of the Nazi fantasy of a happy, healthy ethnic community. It was also the case, as we shall see, that medical care at this time was, even under the body blows of war, a sphere of some patient agency and choice in spite of all the coercion exercised by agencies of the state. During the war, the *Volksgemeinschaft* suffered from official neglect at the very top, too, as Hitler and the Nazi elite were now focused on conquest, extermination, and survival rather than on half-baked Nazi prewar plans for the creation of a new German racial community.[45]

Pragmatic standards, though ruthless in their own way, were applied even to the

---

[40] *Sopade* 3 (1936): 1288–9.

[41] Ibid., 1292.

[42] *Sopade* 3 (1936): 1320; 4 (1937): 1333–4; 5 (1938): 546, 1082; and Süss, *"Volkskörper" im Krieg* (cit. n. 35), 181–92.

[43] *Sopade* 4 (1937): 1320.

[44] Süss, *"Volkskörper" im Krieg* (cit. n. 35), 32–40, 69, 213, 405–16.

[45] Hong, "Neither Singular nor Alternative," (cit. n. 14), 150–51; Michael Thad Allen, *The Business of Genocide: The SS, Slave Labor, and the Concentration Camps* (Chapel Hill, N.C., 2002).

huge numbers of wounded soldiers. Wartime Nazi propaganda consistently ethereal-
ized death and propagandized pain and suffering as a sign of noble heroic sacrifice
and proof of racial superiority.[46] The war wounded were to be integrated productively
into society as "war sufferers" (*Kriegsversehrten*) rather than uselessly pensioned as
the heroic "disabled" (*Kriegsbeschädtigten*).[47] The Nazis once again had to acknowledge the limits of social control and worry as well about the limits of social discipline
brought about by popular internalization of standards and values. One result of this
increasing national sensitivity to suffering and death was debate within the armed
forces about public relations with regard to wounded soldiers. Although some officials, especially earlier in the war, argued that the wounded served as heroic inspiration for the populace, others increasingly pointed out that the sight of sick and
wounded soldiers reassigned to work in Germany had a depressing effect on civilian
morale.[48]

Although there was generalized suffering from illness and disease in Nazi Germany
at war, not everyone, of course, was equally vulnerable. Apart from those millions targeted by the regime for extermination and enslavement, there emerged among Aryans
a hierarchy of medical care in wartime Nazi Germany, with soldiers and armaments
workers at the top along with still racially protected mothers, children, and infants.
Eva Klemperer could observe as late as 1943 that "the very little ones are blooming"
as "children's food and . . . full-cream milk are provided . . . up to the age of six."[49]
At the bottom were Jews, mental patients, forced laborers from eastern Europe, and
those, usually older, patients with severe chronic illnesses. Neither the Nazi regime
nor the bulk of the German population was concerned with the suffering of those regarded as outside the ethnic community. Victor Klemperer wrote of this dichotomy in
spare and desperate eloquence from the perspective of one denied any consideration
or choice when, in July 1942, he noted the large number of proudly pregnant Aryan
women bearing "their stomachs like a Party badge" and concluded that all of Germany was "a meat factory and butcher shop."[50]

Within the wartime meat factory of Aryan life there circulated a mix of reactions to
the demand for physical and mental sacrifice in service to the state, the nation, and the
race. Unlike those automatically consigned to the Nazi butcher shop, most Germans
experienced a clash of loyalty and concern for the self. Patriotism, even or especially
in the face of a long and destructive war, could largely be relied upon by the regime.
Christa Wolf recalls, early in the war, an "autosuggestion" of health through collective commitment.[51] Suffering in their own butcher shop, whether in terms of the death

---

[46] Cocks, "Modern Pain and Nazi Panic" (cit. n. 22), 97, 103, 106.

[47] "Das Ziel: Vollwertige Arbeitskraft," *Dresdner Zeitung,* 16 Aug. 1944, T-78, reel 190, frame 1853;
reel 191, frames 4571–659; Arbeitstelle der Reichsgruppe Industrie für Wiedereinschulung von
Kriegsversehrten, Richtlinien für den Einsatz von Versehrten in Industriebetrieben, T-73, reel 109,
frames 7553–56; Heeres-Sanitätsinspekteur, Die Wiedereinsatzfähigkeit nach Verwundungen, Er-
frierungen, Erkrankungen, 30 Jan. 1944, T-78, reel 1865, frames 6301–25, NA; Cocks, "Modern Pain
and Nazi Panic" (cit n. 22), 105–7.

[48] Dr. Wolff, "Die Betreuung unserer Schwerverletzten," *Deutsches Ärzteblatt* 47/48 (1941), T-78,
reel 191, frame 680; "Sanitätsdienst und Propaganda," T-78, reel 189, frames 1133–49; Kommandeur,
Sanitäts-Abteilung Chemnitz, "Arbeitsbehandlung in den Res. Lazaretten," 6 March 1943, T-78, reel
189, frames 1401–30, NA; Süss, *"Volkskörper" im Krieg* (cit. n. 35), 286–7.

[49] Klemperer, *Bear Witness, 1942–1945* (cit. n. 30), 190.

[50] Ibid., 111.

[51] Wolf, *Patterns of Childhood* (cit. n. 25), 200; Evans, "'*Facilis descensus averni est*'" (cit. n. 29),
99–101; Nolan, "Germans as Victims" (cit. n. 29), 33, 38–39.

and injury to friends and loved ones fighting at the front or their own or others' suffering from enemy action in the skies over Germany, was often experienced as being shared and was thus another source of comforting and encouraging solidarity. Such feelings were bolstered by the long-standing internalization of social and cultural norms regarding discipline, obligation, and the expectations of one's peers. These were intensified by propaganda and its internalization as Nazi conscience, along with growing anger and fear directed at Germany's approaching enemies. Although absence from work—or at least the attempt at it—was widespread, levels of German morale, production, and fighting ability remained high until the very end of the war.

There is considerable evidence, however, that on the German home front atomization as well as solidarity existed within a society—hardly a racial community anymore, if ever—under great stress. One can speak of a "community of fate" under such conditions, but it is clear that individuals and groups were increasingly thrown upon their own resources—as multiple and competing communities of fate—amid the escalating chaos in Germany. That, under these conditions, one's needs were ever more related to basic survival and any possibilities for relief from physical and mental suffering augmented the tendency to think in terms of threats and opportunities having to do with the embodied self. Commitment to an idealized, and still heavily propagandized, Nazi ethnic community proved hollow and brittle under the body blows of a war spawned by a Nazi society that had always favored the "strong" over the "weak," even within the ethnic community.[52]

Officially at least, illness and disability were no longer private matters. The great mass of Germans were expected by the regime to work in spite of illness, what in the military had long been called managed illness (*schonungskrank*), a policy enforced by various medical controls over patients and physicians.[53] By 1942, the right to choose a doctor was done away with by law; in the new city built around the Volkswagen factory in Lower Saxony, for one, doctors from the plant assumed medical supervision over the entire population and the city's hospital.[54] However, just as the "people's car" served only the military, these actions had nothing to do with an idealized "people's community." The discourse and practice of medicine and psychology reflected a turn toward the immediate, the practical, and the multiplying individual cases of difficulty and dysfunction within the context of militant order and disorder. References to racial biology, which had been rare in the Annual Health Reports even in the early years of the regime, now all but disappeared from them.[55] In the armed services, physicians and psychotherapists propounded a holistic interaction of body and mind in treating casualties that conflated Nazi enthusiasm for the will with modern medical theory. Yet this, too, was less in service to Nazi ideological projects than it was continuous with professional thinking and pursuant to the immediate demands for treatment of catastrophic numbers of physical and mental casualties. The jettisoning of psychological selection in the military was likewise prelude and accompaniment to numerous

[52] Tilla Siegel, "Wage Policy in Nazi Germany," *Politics and Society* 14 (1985): 37; Ulrich Herbert, "Arbeiterschaft im 'Dritten Reich': Zwischenbilanz und offene Fragen," *Geschichte und Gesellschaft* 15 (1989): 320–60, on 356; Detlev J. K. Peukert, *Inside Nazi Germany: Conformity, Opposition, and Racism in Everyday Life*, trans. Richard Deveson (New Haven, Conn., 1987), 63, 77, 113, 117; Fritzsche, *Germans into Nazis* (cit. n. 5), 228.

[53] *Sanitätsbericht über das Reichsheer (1933–1935)*, ed. Heeres-Sanitätsinspekteur im O.K.H. (Berlin, 1940), reel 187, frame 189, NA.

[54] Süss, *"Volkskörper" im Krieg* (cit. n. 35), 181–241, 254–68.

[55] Jgb, Sachsen/Glauchau, 1935, SHD; Jgb, Mecklenburg/Ludwigslust, 1937, LHS.

Wehrmacht and SS executions of deserters and defeatists carried out not on the basis of racial selection but in an attempt to stem and preempt the loss of manpower.[56]

The official and professional focus, too, was now on problems rather than on projects. There was a growing, if regionally varied, shortage of doctors (particularly specialists, especially surgeons), drugs, and hospital beds in the civilian sector, leading to the greater employment of female and foreign physicians on the German home front and even prompting the brief consideration by some medical officials of employing Jewish doctors to treat Aryans.[57] The prevention of epidemics of the greatly feared typhus was largely successful, particularly in comparison with efforts during the First World War, due to the strict isolation of those, mostly slave laborers, afflicted with it in camps.[58] The incidence of common illnesses, however, grew exponentially amid wartime dislocations and difficulties, especially those caused by the British and American bombing of German cities. Although Nazi Germany did not suffer the widespread malnutrition and starvation that occurred during the First World War, food shortages (along with shortages of vitamins, clothing, and housing) led in many areas to loss of weight and physical strength, underdevelopment in children and youth, and a decline in psychological well-being (*subjektives Wohlbefinden*) among all ages. Reich Health Leader Leonardo Conti himself observed the loss of weight among boys attending the funeral in Berlin of Security Service Chief Reinhard Heydrich in 1942. A year later, in Dresden, Eva Klemperer noted "how unhealthy the schoolchildren look."[59] In 1944, infant mortality in urban areas suddenly increased, while the incidence of tuberculosis, influenza, and colds rose steadily due to the amount of time people were spending in damp air raid shelters and cellars. Together with colds, the most common diseases were gastrointestinal ailments and bacterial skin inflammations. Widespread stomach problems were the result of psychosomatic factors as well as the declining quality of meal for bread. Skin problems, particularly among children, came primarily from poor living conditions and the shortage of soap. Outbreaks of diphtheria and scarlet fever were not uncommon. There was a general decline in physical vigor and the strength of the immune system and thus an increase in recovery time from illness. There was an overall increase in nervousness, particularly in reaction to air raids and to losses at the front. One public health report from the first year of the war noted that, simply when spoken to, women would "often break into tears."[60]

---

[56] Geuter, *Professionalization of Psychology* (cit. n. 9), 233–58.

[57] Geoffrey Cocks, "Partners and Pariahs: Jews and Medicine in Modern German Society," *Leo Baeck Institute Yearbook* 38 (1991): 191–205, on 191 and 197–200; Cocks, *Treating Mind and Body* (cit. n. 21), 173, 179–81; *Sopade* 6 (1939): 910; Süss, *"Volkskörper" im Krieg* (cit. n. 35), 201–4.

[58] Jgb, Württemberg/Neuenstadt, 1943, SL.

[59] Klemperer, *Bear Witness, 1942–1945* (cit. n. 30), 190; Leonardo Conti, *Stand der Volksgesundheit im 5. Kriegsjahr* (Berlin, 1944); Gesundheitsamt Düsseldorf, Reg. Aachen 16486, NWH; Jgb, Preussen/Solingen, 1943, NWH; Jgb, Mecklenburg/Parchim, 1942, LHS; Jgb, Württemberg/Backnang, 1942, SL; Jgb, Preussen/Celle, 1943, Niedersächsisches Hauptstaatsarchiv Hannover, Hannover, Germany (hereafter cited as NHH); Leonardo Conti to Martin Bormann, 3 July 1942, T-175, reel 68, frames 206–7; SD-Berichte zu Inlandsfragen, 13 July 1942, 9, T-175, reel 263, frame 7104, NA; Süss, *"Volkskörper" im Krieg* (cit. n. 35), 390–3.

[60] Jgb, Preussen/Jerichow II, 1939, 1940, Landesarchiv Magdeburg, Magdeburg, Germany (hereafter cited as LM); Jgb, Preussen/Lüneburg, 1941, NHH; Jgb, Bayern/Kirchheimbolanden, 1942, Landesarchiv Speyer, Speyer, Germany (hereafter cited as LS); Jgb, Preussen/Celle, 1943, NHH; Jgb, Württemberg/Neuenstadt, 1941, LS; Jgb, Mecklenburg/Schwerin, 1943, LHS; Jgb, Bayern/Nördlingen, 1942, SAu.

These personal burdens regarding health, illness, and injury increased popular concern about the possible consequences stemming from official expectations of ability or willingness to work. The regime worried about this, as evidenced by the conditions surrounding Hitler's order in September 1941 to end the systematic gassing of mental patients that had begun just after the outbreak of the war, in 1939. It seems that this decision was based largely on widespread popular anxiety about possible expansion of extermination to other categories of the "disabled." By summer 1941, the population had become worried about a long war and about the tremendous numbers of German dead and wounded in the invasion of Russia. It was also in the spring and summer of 1941 that the Royal Air Force began the large-scale bombing of cities in western Germany, raids that destroyed many large urban hospitals. When Catholic bishops in the spring and the summer publicly protested the "euthanasia" of the mentally ill, a secret program that nevertheless was widely suspected and feared, they made a point of raising the more general issues of who would be next and how any doctor could be trusted in an environment in which the "unproductive" were eliminated. What then as well, Bishop Clemens August Graf von Galen asked, about severely wounded and thus "unproductive" soldiers returning home from the front? Outrage was effectively and universally adulterated by fear. Nazi racist terror could indeed radicalize the hard social-utilitarian ethic that dominated wartime policy toward the working and fighting populace, although the desperate war situation prevented action along such lines. In 1942, a company medical officer in Bremen, with the appropriately chilling name of Warning, recommended to the Reich Justice Ministry the "physical elimination" of all Germans, regardless of ethnic pedigree, who did not measure up to high standards of discipline and productivity.[61] Resultant public fear of such an ethos was the reason that Hitler ordered a halt to the so-called T4 Program (medical killing on a wide scale) in August 1941. Mass support of body and mind for the war and for the regime was at the time more important than pursuing betterment of the race. Mental patients would be allowed to die by starvation and injection on regional initiative starting in 1942, when the institutions throughout Germany that housed them were increasingly needed for wounded soldiers and for civilians bombed out of their houses and patients out of their hospitals in cities crumbling under the Allied air assault.[62]

Wartime pressures also prompted individuals to exploit gaps in the systems of control over health and illness. There was sanctuary in the self as well as in service. There continued to be institutions that could offer ways and means of coping—loyally and productively, to be sure—with the problems of individual bodies and minds, but many of these institutions, particularly as the British and American bombing of Germany intensified, were lamed or destroyed. For example, in 1940 a psychotherapist in Düsseldorf by his own account successfully counseled a young mother who was having difficulty breastfeeding her infant after witnessing the deaths of several children in an air raid. The devastation caused by bombing in the succeeding years, however, made a mockery of the work that could be done by the small number of psychotherapists. In a situation indicative of the widespread disruption—though not total destruction—of the urban medical infrastructure in Germany by bombing, as early as January 1941

[61] H. Warning, Das Asozialproblem im Betriebe, 23 Feb. 1942, R 22, 1932, BAK.
[62] Süss, *"Volkskörper" im Krieg* (cit. n. 35), 127–51, 301–2, 311–50.

the psychotherapeutic society in Düsseldorf had to cancel its evening meetings due to air raids.[63]

At the extreme, illness could even serve as a way for those opposed to the regime to avoid service, as when a young Heinrich Böll evaded a Hitler Youth camp "by more or less extorting an impressive medical certificate from our family doctor."[64] Even racial enemies could exploit the regime's general (and sexist) fear of disease and contagion, such as female forced laborers from eastern Europe who, to escape work, painfully faked or aggravated symptoms of serious illness by introducing irritating substances onto or into their bodies.[65]

Furthermore, although many doctors had responded enthusiastically to recruitment by the Nazis and medicine remained a critical element in Nazi racial policy, most physicians in private practice serving Aryans remained more or less committed to the traditional model of the individual doctor-patient relationship. Doctors on the home front, even though increasingly scarce, too young, too old, or too authoritarian and thus often unpopular, had become ever more important sources for the provision of various scarce goods. In an attempt at social control, the regime required medical certificates (*Atteste*) for extra food rations, vacations, transportation, shoes, and clothing.[66] Such control, however, was imperfect and even troublesome. For example, in 1942 the SS Security Service (*Sicherheitsdienst,* SD) complained that doctors' valuable time was being taken up by hordes of patients seeking such certificates. Many of these individuals, the report asserted, had been referred by government officials trying to get rid of petitioners or by shopkeepers who told customers they needed a doctor's note for such items as raincoats and hot-water bottles.[67] Such individual and organizational recourse did not often represent resistance to the regime on political or moral grounds. Rather it was a matter of serving individual or corporate interests in ways that, while causing friction that compromised the Nazi system, also maintained an ongoing functional role in service to a regime and a nation at war. Individuals could also justify or rationalize such actions as looking out for family or friends, a selfless instance of "community feeling" (*Gemeinschaftsgefühl*) that could serve to assuage an internalized Nazi conscience. This might have been the case, particularly as much of the Nazi propaganda battle having to do with individual obligations to the ethnic community was fought on the ground of the familiar and the familial.[68]

---

[63] Fritz Mohr, "Brief an eine durch Fliegerangriff stillunfähig gewordene Mutter," *Hippokrates* 11 (1940): 1016; Gustav Richard Heyer, *Menschen in Not: Ärztebriefe aus der psychotherapeutischen Praxis* (Stuttgart, 1943), 101–3; Cocks, *Psychotherapy in the Third Reich* (cit. n. 9), 228; Otto Curtius to Matthias Heinrich Göring, 1 Jan. 1941, Kl. Erw. 762.2, BAK; Süss, *"Volkskörper"im Krieg* (cit. n. 35), 269–91, 403.

[64] Heinrich Böll, *What's to Become of the Boy? Or, Something to Do with Books,* trans. Leila Vennewitz (New York, 1984), 68.

[65] Hermann Köhler to Kurt Schössler, 14 Oct. 1942, NSDAP Kreisleitung Eisenach, folder 000142a, Myers Collection, University of Michigan, Ann Arbor, Michigan; Süss, *"Volkskörper" im Krieg* (cit. n. 35), 376.

[66] Süss, *"Volkskörper" im Krieg* (cit. n. 35), 370–80; Marie Vassiltchikov, *Berlin Diaries, 1940–1945* (New York, 1987), 62; Jgb, Preussen/Kreuznach, 1939, 1940, Landeshauptarchiv Koblenz, Koblenz, Germany; Michael H. Kater, *Doctors under Hitler* (Chapel Hill, N.C., 1989); Robert N. Proctor, *Racial Hygiene: Medicine under the Nazis* (Cambridge, Mass., 1988).

[67] SD-Berichte zu Inlandsfragen, 29 Jan. 1942, 14–15, T-175, reel 262, frames 5728–29; Leonardo Conti to all Gauleiter, 30 April 1942, T-81, reel 663, frame 584, NA.

[68] Lisa Pine, *Nazi Family Policy, 1933–1945* (Oxford, 1997); Claudia Koonz, *Mothers in the Fatherland: Women, the Family, and Nazi Politics* (New York, 1987).

Individual Germans avoided doctors when it was in their interests to do so. Drugs, in particular, had been an important instance of both approach and avoidance with respect to physicians since the 1930s, when pharmaceuticals became a significant means of addressing and controlling physical and mental pain. During the war, drugs also served as a substitute for increasingly unavailable alcohol and watered-down beer.[69] The Nazis attempted to make physicians the gatekeepers for drugs by requiring that more drugs be available by prescription only (*rezeptpflichtig*). As a result, there were reports that patients ranked doctors in terms of desirability on the basis of how many prescriptions they wrote.[70] To save money, the regime also limited the number of prescriptions a doctor working under the state health insurance system could issue. The response to this by the public, according to SS Security Service reports on public attitude and behavior, was to avoid such doctors. Wealthier Germans visited private physicians, the SD averred in 1940, while poorer Germans went directly to a pharmacy.[71] A year later, the SD was grumbling that a flood of advertising for headache remedies had increased the number of chronic stomach problems and that many of the products contained caffeine. The fact that these products were available in grocery stores, hair salons, and DAF consumer clubs made the problem all the greater.[72] This last concern was compounded by the fact that drugstores (*Drogerien*) were emerging as competitors to pharmacies (*Apotheken*) through drugstores' capacity to sell packaged drugs at lower prices than those made up on order by pharmacists. As early as 1937, the German Pharmacists Association had objected to a court's decision that "natural health stores" could sell aspirin for "nonpathological headaches."[73] Strategies of popular negotiation dealing with drugs represented a decentralization of medical service that was also a strong indication of the fracturing of the idealized Nazi ethnic community through individual and corporate pursuit of relief from physical and mental pain and stress.

The Nazi regime itself had a history of drugs. It had inherited both a growing market for pharmaceuticals and the practice, since the 1920s, of state and private health insurance drug coverage. Although alcohol remained the most popular painkiller, by the 1930s a variety of analgesics and barbiturates had become widely available. Drug companies advertised aggressively while popular and medical demand for drugs— what one medical official in 1936 labeled among patients *Arzneihunger* (medicine hunger) and among avaricious doctors *Arzneischatz* (medicine treasure)—was strong.[74] The Nazis themselves added official encouragement of the use of amphetamines to

[69] Jgb, Mecklenburg/Stargard, 1940, LHS; Jgb, Mecklenburg/Güstrow, 1941, LHS; Jgb, Preussen/Magdeburg, 1940, LM.

[70] Landrat des Landkreises Aachen to Deutscher Gemeindetag, 30 March 1936, RW 53, 455, NWH; Jgb, Preussen/Düsseldorf-Mettmann, 1941, NWH.

[71] SD-Berichte zu Inlandsfragen, 3 April 1940, 10, T-175, reel 258, frame 1171, NA.

[72] SD-Berichte zu Inlandsfragen, 20 Feb. 1941, 14–15, T-175, reel 260, frames 3285–86, NA; Reichsministerium des Innern to Reichswirtschaftsministerium, 18 May 1943, R 86, 3900, BAK; cf. Jgb, Württemberg/Backnang, 1939, SL; and Jgb, Preussen/Fallingbostel, 1942, NHH.

[73] Deutsche Apothekerschaft to Reichsgesundheitsamt, 18 June 1937, R 86, 3900, BAK.

[74] Auszug aus der Niederschrift über die 7. Sitzung der Rheinischen Arbeitsgemeinschaft für Wohlfahrtspflege, 4 July 1936, RW 53, 455, NWH; Peter Reichel, *Der schöne Schein des 3. Reiches: Faszination und Gewalt des Faschismus* (Frankfurt am Main, 1993), 374–75; Carl Fervers, *Schmerzbetäubung und seelische Schonung* (Stuttgart, 1940), 152–53, 155; Paul Ridder, *Im Spiegel der Arznei: Sozialgeschichte der Medizin* (Stuttgart, 1990), 11–25; Ludwig Sievers, *Handbuch für Kassenärzte* (Hanover, Germany, 1925), 37–45.

raise industrial and military productivity, a policy called by one rather bold expert, "the chemical whip."[75] The frequent popular recourse to stimulants (and painkillers) had the significant negative effects of addiction and declining effectiveness. Drugs served as a means of individual escape and avoidance, usually temporary, nonpolitical, and thus in overall functional support of the system. Heinrich Böll (who was compelled to serve the regime by being drafted into the Wehrmacht) and members of his family became addicted to the most popular methamphetamine, Pervitin, which they got from a hospital where it was put "into the coffee of malingerers to encourage them to leave voluntarily."[76] Damage from air raids to the highly centralized drug industry and, subsequently, to the German transportation network that caused growing shortages of drugs and vitamins only increased demand. Individuals and groups engaged in a countrywide scramble to secure supplies of painkillers, cough medicines, sleep aids, and stimulants. Factories hoarded drugs for their workers while the army manufactured its own. Although doctors were still sources for pharmaceuticals, it is clear that even with official controls on prescriptions, the avoidance, as well as the shortage, of physicians contributed to what was almost an orgy of self-medication.[77]

The widespread recourse to pharmaceuticals in Nazi Germany was itself also indicative of the persistence, and not just wartime reemergence, of personal comfort and pleasure as individual social and cultural goals. Before the war, the Nazis had not only indulged but also actively promoted consumption of what material goods were available in the wake of the Depression and in the bow wave of mobilization. This was designed to manufacture support for the concomitant demands of rearmament and, subsequently, the war itself. Yet it was also an appeal to the hedonism central to the modern utilitarian ethic of the pursuit of pleasure and the avoidance of pain that had been recently reinforced by the first stirrings of a consumer society in Germany before and after the First World War.[78] So although, for example, the Nazis on racial-collective grounds discouraged smoking, especially among women, smoking generally increased during the 1930s and 1940s. More than one regional public health report in 1939 expressed concern about the link between smoking and high rates of heart and circulatory disease. (There was scant mention of cancer in these reports, because, as Robert Proctor has documented, the links between tobacco smoking and lung cancer, in particular, were only just being established by German research between 1939 and 1944.[79]) The Nazi antismoking campaign, although tied to a reactionary social program, was thus in line with modern progressive initiatives also consistent with

[75] F. Eichholtz, "Ermüdungsbekämpfung: Über Stimulanten," *Deutsche medizinische Wochenschrift* 67 (1941): 1356; Heeres-Sanitäts-Inspektion, Richtlinien zur Erkennung und Bekämpfung der Ermüdung (Anwendung und Wirkung des Pervitins), Nr. 120, 18 June 1942, T-78, reel 198, frames 2511–14, NA; Leonardo Conti, Rundschreiben Vg. 9/41, n.d., T81, reel 663, frames 711–14, NA; Cocks, *Psychotherapy in the Third Reich* (cit. n. 9), 242, 312–13; Vicki Baum, *Hotel Berlin '43* (Garden City, N.Y., 1944), 11, 24, 26, 82, 85; Jgb, Preussen/Wuppertal, 1940, NWH.

[76] Böll, *What's to Become of the Boy?* (cit. n. 64), 73.

[77] Reichsministerium des Innern to Reichswirtschaftsministerium, 22 July 1942, Reg. Düsseldorf, 54364 III, NWH; Manfred Fischer, Die Eigenfertigung und die Betriebsleistung—Vergleich der Wehrkreis-Sanitätsparke, 1 April 1944, T-77, reel 297, frames 5712–14, NA; SD-Berichte zu Inlandsfragen, 9 May 1940, T-175, reel 259, frame 1485, and 2 Oct. 1941, reel 261, frames 4928–31, NA; Süss, *"Volkskörper" im Krieg* (cit. n. 35), 204–8, 397–8; United States Strategic Bombing Survey, *The Effect of Bombing on Health and Medical Care in Germany* (Washington, D.C., 1945), 294–339.

[78] Berghoff, "Enticement and Deprivation" (cit. n. 3), 165–84.

[79] Jgb, Württemberg/Biberach, 1939, SL; Jgb, Bayern/Eichstätt, 1939, Staatsarchiv Nürnberg, Nuremberg, Germany; Robert N. Proctor, *The Nazi War on Cancer* (Princeton, N.J., 1999), 21, 180–98, 214–18, 284 n. 8, 320 n. 94, 330 n. 51.

modern concern for the physical and mental preservation and enhancement of the self.[80] The largely negative—or indifferent—public reaction to the campaign, too, was based on habits built around consumable personal pleasure and the resources—and threats to individual indulgence—that prompted such consumption. The Nazis further undermined their own antismoking campaign by worrying about the effects on morale of wartime shortages of tobacco. Merchandising played a large role as well in the increase in smoking in Nazi Germany. Print advertising was ubiquitous, and in 1938 a public health officer in northeastern Germany complained about the spread of cigarette vending machines, particularly as a means of attracting young smokers.[81]

There was a similar, if less insistent, Nazi offensive against drinking. Public health officers in the 1930s worried about excessive drinking in times of unemployment, when need for "spiritual" comfort was greater, but they also complained about the increased demand for alcohol that came with economic recovery.[82] The Third Reich did see a decline in official reports of alcoholism, coincident with a 1934 law mandating sterilization for alcoholism and the practice of putting alcoholics in concentration camps, but general consumption of alcohol was not significantly affected.[83] During the war, of course, almost everyone had more reason and, sometimes in spite of shortages, more opportunity to indulge in drinking as well as smoking and in legal and—for some—illegal drugs. Before alcohol became especially scarce in urban Germany during the war, it often replaced the coffee that was almost impossible to get after 1939.[84]

Sex, too, attracted official and popular attention. The Nazis condemned modern sexuality and promiscuity as one more sign of a degenerate and mongrelized materialistic culture. However, drawing selectively from the Romantic heritage that was strong in Germany, some propaganda contrasted the shameless eroticism and perversions of modern society with the "natural" beauty of open and healthy Aryan heterosexuality. Like official encouragement and subsidy of marriage and family, the promotion of extramarital sex was designed to increase the German birthrate and thus secure a future supply of workers and soldiers for the Reich. It was also designed to bind the populace closer to the regime. But here, too, the Third Reich was constrained to tolerate personal satisfaction "in the context of the broader modernization of consumer culture under Nazism."[85] During the war, the vastly increased mobility of large groups of people, such as soldiers, forced laborers, and evacuees, radically increased the incidence of sexual activity, activity that literally reached depths of the grotesque with the frantic and open sex deep inside Hitler's bunker during the last days in besieged Berlin.[86] Of course, the war was anything but one long party for the Germans.

[80] Robert N. Proctor, "The Nazi War on Tobacco: Ideology, Evidence, and Possible Cancer Consequences," *Bulletin of the History of Medicine* 71 (1997): 435–88.

[81] Jgb, Mecklenburg/Wismar, 1938, NWH; Jgb, Preussen/Moers, 1938, NWH; Jgb, Mecklenburg/Parchim, 1940, LHS; Jgb, Preussen/Erfurt, 1934, LM; Jgb, Bayern/Kirchheimbolanden, 1939, LS; Proctor, *Nazi War on Cancer* (cit. n. 79), 228–42.

[82] Jgb, Bayern/Weilheim, 1933, 1934, Staatsarchiv München, Munich, Germany; Jgb, Mecklenburg/Ludwigslust, 1935, LHS; Jgb, Mecklenburg/Waren-Müritz, 1935, LHS; Jgb, Thüringen, 1935, Hauptstaatsarchiv Weimar, Weimar, Germany; Jgb, Bayern/Regensburg, 1939, SAm; Jgb, Bayern/Amberg, 1940, SAm; Jgb, Württemberg/Ravensburg, 1938, 1939, 1941, SL.

[83] Jgb, Württemberg/Tübingen, 1935, SL; Jgb, Württemberg/Biberach, 1936, SL; Jgb, Bayern/Kempten, 1938, SAu; Jgb, Bayern/Regensburg, 1939, SAm.

[84] Jgb, Mecklenburg/Waren-Müritz, 1939, LHS; Jgb, Württemberg/Biberach, 1942, SL.

[85] Herzog, *Sex after Fascism* (cit. n. 4), 16.

[86] Anthony Beevor, *The Fall of Berlin 1945* (New York, 2002), 344; Süss, *"Volkskörper" im Krieg* (cit. n. 35), 396–97.

In the spring of 1942, for example, a report from Mecklenburg noted a steep decline in reports of venereal disease due to "the catastrophe," that is, the three devastating nights of air raids on the city of Rostock near the end of April that killed and injured thousands and led to the evacuation of thousands more into surrounding rural areas. The decline in reports of infection could also indicate official failure to report or individual inability to seek treatment for sexually transmitted diseases. It is clear, however, that such events could only in the short term diminish the opportunity and the inclination for sexual self-indulgence and diversion; the very next year a rise in the incidence of venereal disease in the region was ascribed to further evacuations and the arrival of more eastern workers (*Ostarbeiter*).[87]

As with smoking and drinking, medical officials adopted an increasingly pragmatic problem-oriented discourse and practice concerning the increased sexual activity as a result of war and the mobilization for war. As early as 1935, Annual Health Reports from around Germany were reporting rises in the rates of gonorrhea, most often in connection with the establishment or expansion of military bases. Such behavioral and attitudinal changes were also the result of recent loosening of traditional social inhibitions that both complemented and contradicted Nazi standards and practices of ethnic and national order and discipline. Widespread sexual activity was endemic even—perhaps especially—within the Nazi youth organizations, in which instances of homosexuality prompted efforts at psychotherapeutic treatment in place of punishment.[88] There was equally widespread medical condemnation of apparent sexual promiscuity among young girls. One report from Regensburg in 1940 attributed the rise in the rate of gonorrhea infection not so much, if at all, to activity by prostitutes as to "the wild sexual intercourse of young girls, many of whom are adolescents."[89] Sexual pleasure in wartime, too, was a source for the exercise of the modern self in Nazi Germany as well as a source of popular and official anxiety over the already compromised and endangered body.[90]

SELF-SERVICE

Hitler's war elevated individual productivity over racial community and further problematized individual well-being in ways that reinforced the modern ethic of serving the self. The modern Western self that had developed in Germany had already been manifest as challenge and tension within and around Nazi peacetime attempts at the construction of a *Volksgemeinschaft*. By the end of the Second World War, any actual Nazi *Volksgemeinschaft* was in tatters largely as a result of social dynamics privileging the individual body, mind, and self. The modern tradition of identifying self with nation, too, had been severely compromised. Now, in contradiction to still strident Nazi appeals for racial solidarity, the *Volk* had become nothing—or at least hardly "everything"—in the face of individual self-preservation. This ongoing trend was par-

---

[87] Jgb, Mecklenburg/Rostock-Land, 1942, 1943, LHS.

[88] Jgb, Sachsen/Glauchau, 1935, SHD; Jgb, Bayern/Regensburg, 1937, SAm; Adelheid von Saldern, "Victims or Perpetrators? Controversies about the Role of Women in the Nazi State," in *Nazism and German Society, 1933–1945,* ed. David F. Crew (London, 1994), 141–65, 151, 156–7; Michael H. Kater, *Hitler Youth* (Cambridge, Mass., 2004), 58, 70, 77, 105–12, 135–8, 143, 149–50, 159–60, 164, 186–7, 234–5; Cocks, *Psychotherapy in the Third Reich* (cit. n. 9), 286–8, 294.

[89] Jgb, Bayern/Regensburg, 1940, SAm; Jgb, Preussen/Lüneburg, 1939, 1940, NHH; Jgb, Mecklenburg/Stargard, 1938, LHS; Jgb, Württemberg/Ravensburg, 1941, 1942, 1943, SL.

[90] Conti, *Stand der Volksgesundheit* (cit. n. 59); Süss, *"Volkskörper" im Krieg* (cit. n. 35), 394–5.

ticularly the case in the realm of health and illness, wherein the "Nazi version of the modern Western material culture of purchasable individual well-being—at least via doctors and the brave new world of drugs—perpetuated certain continuities of structure, appetite, and expectation in the development of a consumer culture."[91] More and more, life in collapsing Nazi Germany had become an individual, as well as a traditionally German corporate, war of all against all in a frenzied landscape of battle and work, discipline and diversion, suffering and survival, fear and flight, injury and death. Similar difficulties persisted and even multiplied during the immediate postwar period with the almost complete breakdown of civilian authority,

The social and cultural habits surrounding the self in the Third Reich in six years of peace and six years of war constituted distinct continuities in the relationship between self and society not only with the history of Germany and the West before 1933 but also with the German successor states, Europe, the West, and eventually much of the world after 1945. The modern self, along with its social, cultural, and scientific support system, survived—and had itself supported—the Third Reich. In this way as in others, the Nazi years in Germany were firmly, if ambiguously and obscenely, lodged into the history of Europe and the West in the modern era.

[91] Cocks, "Modern Pain and Nazi Panic" (cit. n. 22), 109; Fritzsche, *Germans into Nazis* (cit. n. 5), 121–2; Cocks, *Psychotherapy in the Third Reich* (cit. n. 9), 407–12; Christian Goeschel, "Suicide at the End of the Third Reich," *Journal of Contemporary History* 41 (Jan. 2006): 153–73; Richard Bessel, "'Leben nach dem Tod'—Vom Zweiten Weltkrieg zur zweiten Nachkriegszeit," in *Wie Kriege enden: Wege zum Frieden von der Antike bis in die Gegenwart,* ed. Bernd Wegner (Paderborn, Germany, 2002), 239–58.

From the Inside Out:

Therapeutic Penology and Political Liberalism in Postwar California

By Volker Janssen*

ABSTRACT

In the wake of World War II, California emerged as leader in modern therapeutic corrections thanks to the structure of its military welfare state. By the late 1950s, however, prisoners, staff, the public, and the political leadership questioned the effectiveness of group counseling as the path back to citizenship, prompting the creation of a Correctional Research Division. Replacing case file evaluations with rigorous statistical prediction tools, the division built the necessary political legitimacy for a series of bold therapeutic community programs. Promoting personality change through democratic community life upset institutional discipline, industrial work order, and professional standards, but it also legitimized prisoners' political awareness and group identities incompatible with reintegration into white middle-class society. In the wake of the Watts riots, a conservative administration not only withdrew political support for rehabilitative research but also converted statistical prediction tools from an argument for bold liberal experimentation into an argument for incarceration.

INTRODUCTION

On an ordinary Monday morning in 1964, clinical psychologist Dennie Briggs arrived at his workplace in Southern California's Chino prison. He had just positioned himself at a large table when an inmate approached him. Addressing the psychologist with a short, "You first," he dumped a huge pile of white terry cloth towels in front of Briggs and directed him to start folding. Briggs worked in the institutional laundry that day, and inmate T. J. was his foreman. It was part of California's therapeutic community experiments that ran from 1960 to 1965, for which Dennie Briggs served as project coordinator.

These projects were the culmination of several years of research in rehabilitative group strategies. In their most radical stage, they turned life in prison inside out, subverting the industrial work order, institutional discipline, and professional authority. More important, correctional efforts at building new personalities as the foundation for good citizenship validated prisoners' personhood. Encouraging the prisoners'

---

* California State University Fullerton, PO Box 6846, Fullerton, CA 92834-6846; vjanssen@fullerton.edu.

I thank Michael A. Bernstein, Greg Eghigian, Michael Meranze, David Miller, and the editors of this journal for their help and suggestions.

agency in the rehabilitative process, these efforts opened the door for the prisoners' claims to rights as citizens. Moreover, by relying on the group as a site of social learning, they legitimized the search by prisoners for their own community and inadvertently fostered a political sense of self among prisoners that proved central to the crisis in American prisons in the late 1960s.

In the years following World War II, California's innovations in correctional rehabilitation represented the vanguard of what has often been called "therapeutic" or "clinical" penology in the United States. Most pronounced in the Golden State but recognized among sociologists as a nationwide phenomenon, the era of therapeutic corrections between World War II and the 1970s has not drawn the attention of historians.[1] Penal slavery in the postbellum South has seemed to hold more pertinent historical lessons for a nation with a rapidly expanding prison population, three-strike rules, and mandatory sentences.[2] The few students of the liberal prison regimes of the postwar years have typically identified their rehabilitative programs as ineffective, insignificant, or insincere, or as insidious forms of mind control. Instead of therapeutic and disciplinary technologies, their political origins, and their unexpected outcomes, the focus of historical research has been the prisoners' agency and racial divisions, often with fascinating results.[3] Nonetheless, as the political bankruptcy of mass incarceration becomes increasingly apparent, the history of California's efforts at a scientific reconstruction of democratic citizens within authoritarian institutions of confinement deserves our attention.

In many respects, the rehabilitative practices in postwar corrections fit squarely within the structure of the military welfare state and the ways in which World War II altered the relationships between state, science, and individuals.[4] The cooperation and public service of American prisoners in civil defense projects and prison industries

---

[1] Notable exceptions are Charles Bright, *The Powers That Punish: Prison and Politics in the Era of the "Big House," 1920–1955* (Ann Arbor, Mich., 1996); Jonathan Simon, *Poor Discipline: Parole and the Social Control of the Underclass, 1890–1990* (Chicago, 1993); Daniel Glaser, *Preparing Convicts for Law-Abiding Lives: The Pioneering Penology of Richard McGee* (New York, 1995).

[2] Emphasizing the continuity between slavery and imprisonment are Loïc J. D. Wacquant, *Les prisons de la misère* (Paris, 1999); Wacquant, "From Slavery to Mass Incarceration: Rethinking the 'Race Question' in the U.S.," *New Left Review* 13 (2002): 41–60; Christian Parenti, "The 'New' Criminal Justice System: State Repression from 1868 to 2001," *Monthly Review* 53(3) (2001): 19; Kim Gilmore, "Slavery and Prison—Understanding the Connections," in *Social Justice* 27(3) (2000): 195. This continuity is also apparent in the recent scholarship on prisons: Robert Perkinson, "'Between the Worst of the Past and the Worst of the Future': Reconsidering Convict Leasing in the South," *Radical History Review* 71 (1998): 207–16; Mary Ellen Curtin, *Black Prisoners and Their World: Alabama, 1865–1900* (Charlottesville, Va., 2000); Paul Michael Lucko, "Prison Farms, Walls, and Society: Punishment and Politics in Texas, 1848–1910" (PhD diss., Univ. of Texas, Austin, 1999); Norbert Finzsch, "'The Obsession With Work': Criminology, Labor, Convict Labor, and Social Control in Nineteenth- and Twentieth-Century America" (unpublished paper, Univ. of Hamburg, Germany, 1998); Martha A. Myers, "Race, Labor, and Punishment in the New South," *Social Problems* 38 (May 1991): 267–87; Matthew J. Mancini, *One Dies, Get Another: Convict Leasing in the American South, 1866–1928* (Columbia, S.C., 1996). Most remarkably, Alex Lichtenstein, *Twice the Work of Free Labor: The Political Economy of Convict Labor in the New South* (New York, 1996), has connected southern penal slavery at the turn of the century with progressive politics.

[3] See, e.g., Eric Cummins, *The Rise and Fall of California's Radical Prison Movement* (Palo Alto, Calif., 1994); Robert Perkinson, "'Scrap Iron, Broken Glass, and Barbed Wire': Convict Writing and the Rise and Fall of Convict Leasing in Texas," in *Prison Writing: Probing the Boundaries,* ed. Diana Mendlicott (Amsterdam, 2003).

[4] Consider, in particular, the work of Michael Bernstein, Ellen Herman, and Alice O'Connor on how the changing political circumstances of the war and postwar years shaped the relationship between experts, their science, and statecraft. Michael A. Bernstein and Allen Hunter, eds., *The Cold War and Expert Knowledge: New Essays on the History of the National Security State, Rad. Hist. Rev.* 63 (Fall

during the war had brought home the importance of civic loyalty over plain obedience. In fact, noted a member of the War Industries Board with some concern, "[w]hat made a good convict" was exactly "what ma[de] a poor citizen." Training prisoners "for the duties and responsibilities of citizenship" required not just industrial education, vocational training, and civics classes but also practice in "simple prison democracy" so as to foster a democratic personality.[5] California's celebrated introduction of group counseling seemed to meet this challenge.

Employing a psycho-scientific method with a political rationale, California's counseling program constituted both therapeutic practice and statecraft. The entire correctional program was, indeed, clinical in its design to the extent that it focused on the scientific classification of the individual, identified patterns and psychological dynamics, and prescribed a "treatment." The public discourse over delinquency and social order on which correctional practice depended for public support, however, reflected particular postwar concerns about youth, family order, the status of women, racial integration, full employment, and democracy. This discourse identified white young men as the chief victims of social circumstances and worthy of the state's very best efforts at restoring them to good citizenship. Here, California's new rehabilitative practices in vocational training, psychotherapy, and group counseling benefited from similar social projects in the armed forces and the Veterans Administration, which legitimized the public investment in professional expertise and correctional experimentation.[6]

By the late 1950s, however, correctional practice in California found the legitimacy of its counseling program challenged by prisoners, staff, and the public as well as by the state's political leadership. Pressured to increase its scientific rigor and innovate its treatment methods on better scientific grounds, the California Department of Corrections (CDC) created a Correctional Research Division. Drawing on research and experiments they had conducted for the armed forces, a young cadre of professionals developed a new community-based method for the transformation of personalities and replaced the old method of case file evaluation with a statistical prediction tool. Combined, these two innovations enabled a series of therapeutic community projects, arguably the most radical experiments in the history of American prisons. Reflecting a liberal imagination of the reconstructive powers of incarceration, these experiments

---

1995); Michael A. Bernstein, *A Perilous Progress: Economists and Public Purpose in Twentieth-Century America* (Princeton, N.J., 2001); Ellen Herman, *The Romance of American Psychology: Political Culture in the Age of Experts, 1940–1970* (Berkeley, Calif., 1995); Alice O'Connor, *Poverty Knowledge: Social Science, Social Policy, and the Poor in Twentieth-Century U.S. History* (Princeton, N.J., 2001).

[5] Harry Elmer Barnes, *Report on the Progress of the State Prison War Program under the Government Division of the War Production Board* (Washington, D.C., 1944), 86. Theodor W. Adorno had revealed the disturbingly undemocratic and fascist tendencies among prisoners in San Quentin in his seminal study, Adorno et al., *The Authoritarian Personality* (New York, 1950). However, the horrors of carceral exclusion and destruction of prisoners in German concentration camps and Soviet gulags had highlighted the totalitarian roots of imprisonment. James B. Jacobs, "Macrosociology and Imprisonment," in *Corrections and Punishment,* ed. David F. Greenberg (Beverly Hills, Calif., 1977), 89–107, on 92.

[6] The chief designer of California's therapeutic corrections initiatives, Norman Fenton, described his charges as fellow veterans with "mental wounds," and his narrative of family life, delinquency, institutionalization, and redemption closely followed the model provided by army psychologists. For more details, see Volker Janssen, "Convict Labor Civic Welfare: Rehabilitation in California's Prisons, 1941–1971" (PhD diss., Univ. of California, San Diego, 2005).

also bore the seeds of their own failure. The nurture of the prisoners' group identity translated into a political consciousness that prompted a conservative backlash when it emerged among prisoners of color. Ironically, when Ronald Reagan's conservative administration withdrew California's support for rehabilitative research, it converted the statistical prediction tools from an argument for bold liberal experimentation into an argument for incarceration.

BEYOND THE LIMITS OF PROFESSIONAL COUNSELING

When sociologist John Irwin served time as an inmate in Central California's Soledad prison in the early 1950s, he was impressed by the "optimistic, tolerant, and agreeable mood," even "enthusiasm for the new penal routine." He wrote: "We were bombarded with sophisticated tests administered by young, congenial, 'college types'. . . .We were examined thoroughly by dentists and physicians. For six weeks we attended daily three-hour sessions with one of the college types. . . . [W]e became convinced that the staff members were sincere and were trying to help us. We believed they were going to make new people out of us."[7]

Only a few years later, however, prisoners expressed disappointment with the state's group counseling programs. Few held out any hope for a personal adjustment that would benefit them upon their return to citizenship. Pressured to attend the weekly sessions, most of them participated only to get an early parole date. Having little trust in either their keepers or their fellow prisoners, few were willing to discuss sensitive, personal issues in prison, and they disparaged others for doing so. Prisoners quickly learned not to set their hopes in counseling programs. They participated routinely and cynically. A follow-up interview with seventy parolees in 1966 confirmed that more than half of the prisoners thought correctional treatment programs ineffective. Most agreed that correctional programming had become a self-sustaining bureaucracy that existed only to "get more money from the state for more prisons." A substantial share of the interviewees recognized counseling sessions as yet another tool of authority and found questions about their personal lives and thoughts as probing and unwelcome as a cell search. Many considered the suggestion that they were "emotionally disturbed" more offensive and humiliating than the allegation of "moral unworthiness."[8]

Prisoners quickly learned how to co-opt the system and take advantage of the opportunities the medical treatment model offered.[9] "If convicts were in prison to be

---

[7] John Irwin, *Prisons in Turmoil* (Boston, 1980), 56, 61.

[8] A prisoner in San Quentin during the heyday of therapeutic corrections, Malcolm Braly, recalled the deadening boredom and growing cynicism of counseling sessions in his novel *On the Yard* (Boston, 1967), 103–4; John Irwin, *The Felon* (Englewood Cliffs, N.J., 1970), 52–3.

[9] By placing conditions for a return to citizenship and freedom into the realm of the self, some have argued, the practice of therapeutic corrections disenfranchised prisoners for whom the specialists' psychological terrain of selfhood remained unfamiliar. Most accounts suggest that prisoners found the expectations of therapists and guards confusing and contradictory, and many resented being considered "mad" rather than "bad." Rebecca McLennan, "Citizens and Criminals: The Rise of the American Carceral State, 1890–1935" (PhD diss., Columbia University, New York, 1999); McLennan, "Punishment's Square Deal: Prisoners and Their Keepers in 1920s New York," in *Journal of Urban History* 29(5) (2003): 597–619, on 608; Cummins, *Rise and Fall of California's Radical Prison Movement* (cit. n. 3), 14; Bright, *The Powers That Punish* (cit. n. 1), 278. A 1966 survey among recent parolees suggested, however, that a large majority—72 percent—did not understand counseling as a form of social control but one of boredom and irritation. Irwin, *The Felon* (cit. n. 8), 52–3. Few understood

cured," sociologist Ronald Berkman observed, "it made no sense to continue to punish them with beatings, solitary confinement, and severe deprivation. Once their cure had been effected, it made no sense to keep them in prison one minute longer." Prisoners thus played the therapeutic game and quickly familiarized themselves with the language and terminologies of basic psychology. Having studied the new penological paradigm, prisoners conned their way through counseling and turned the experts' tour to self-discovery into hollow morality skits.[10]

As counseling failed to change prisoners' personalities and make them better citizens, so it failed to produce institutional peace. The introduction of college-educated social workers and mental health professionals in charge of new classification methods and therapeutic programming split corrections personnel into custodians and caregivers. On the one side stood the guards—overworked, underpaid, typically retired military personnel or poor white southern migrants, in charge of maintaining the prisoners' captivity. On the other side stood the treatment staff—enthusiastic college graduates with salaries and comfortable work schedules, eager to practice their craft and knowledge by turning prisoners into free citizens. A "cold war between individual treatment and collective security" emerged.[11]

It did not help that custodial staff had to provide many of the counseling services, because the state could not afford trained counselors for its entire prison population. As a result, "staff persons with no formal training in psychology led many, if not most, groups" in the state's correctional institutions.[12] There are sufficient indications that custodial personnel resented such assignments. Forced not just to guard but also to act as if they cared about "self-understanding" and "emotional release," rank-and-file officers were made uncomfortable by counseling, both as guards and as men. Working

---

counseling as a coordinated effort at mind control as some critics of the New Left school of criminology suggested in the 1970s. See, e.g., Richard Speiglman, "Prison Psychiatrists and Drugs: A Case Study," Stephen J. Pohl, "Deciding on Dangerousness: Predictions of Violence as Social Control," and Bob Martin, "The Massachusetts Correctional System: Treatment as an Ideology for Control," in *Punishment and Penal Discipline: Essays on the Prison and the Prisoners' Movement,* ed. Tony Platt and Paul Takagi (Berkeley, Calif., 1980).

[10] Ronald M. Berkman, *Opening the Gates: The Rise of the Prisoners Movement* (Princeton, N.J., 1977), 61; Cummins, *Rise and Fall of California's Radical Prison Movement* (cit. n. 3), 13; Bright, *The Powers That Punish* (cit. n. 1), 188. Robert M. Harrison and Richard A. McGee to Group Counseling Program Supervisors: Script of Role Playing Situation at San Quentin, memo, 19 Nov. 1959, F3717:581 Corrections, Correctional Program Services, Projects and Programs, Group Counseling 1955–1960, California State Archives Records (hereafter cited as CSA).

[11] Sociologists David Powelson and Reinhard Bendix noted the basic tension between custody and care in 1951, notes Irwin, *The Felon* (cit. n. 8), 52; James B. Jacobs, *New Perspectives on Prisons and Imprisonment* (Ithaca, N.Y., 1983), 140; R. Theodore Davidson, *Chicano Prisoners: The Key to San Quentin* (New York, 1974), 29–30; Bright, *The Powers That Punish* (cit. n. 1), 276, 279. For a gendered interpretation of the class divide among prison staff, see Dennie Briggs, Stuart Whiteley, and Merfyn Turner, *Dealing with Deviants: The Treatment of Antisocial Behavior* (New York, 1973), 99–100. A gendered division between welfare and discipline is the theme of Velia Garcia, "My Momma the State: A Socio-Cultural Study of the Criminalization of Chicanos" (PhD diss., Univ. of California, Berkeley, 1990).

[12] In 1953, only 100 of the 8,000 psychiatrists in the United States were employed by correctional institutions, with less than one-third of them employed full time. In 1957, that number had grown somewhat to 133 psychiatrists, 90 psychologists, and 162 social workers, more than half of them in California. Irwin, *Prisons in Turmoil* (cit. n. 7), 44; Gene Kassebaum and David Ward, *Prison Treatment and Parole Survival* (New York, 1971); Joseph C. Finney, Ph.D., M.D., "A Report on Mental Health in Corrections in California," With Recommendations for Hawaii, 31, F3717:931 Corrections—Medical Services Division, Central Files Psychiatric Services, 1961–62, CSA; F3717:1587 Corrections Administration Manuals, Group Counseling Manuals, 1962–1965, Chapter I GC-I-00, CSA.

in an archetypical "good ole boys" environment, correctional officers were as unlikely as prisoners to sit in a circle and talk.[13]

Since the prison riots of 1953, which had occurred in most modern state correctional systems across the nation with the exception of California, advocates of therapeutic corrections also faced challenges from critics on the outside, such as public intellectual John Bartlow Martin. "We have improved food and buildings and other appurtenances considerably," Martin acknowledged. He noted as well that the inmate was no longer considered a slave of the state afflicted with *civiliter mortuus*. Certainly, the prisoner was no longer "just an animal in the pen." Yet the reconstruction of citizenship, speculated Martin, still eluded Americans. "[B]edazzled by the myth of rehabilitation," he concluded, society had in fact reached its lowest point. "Prison is not just the enemy of the prisoner. It is the enemy of society."[14] The correctional "big house," it seemed, provided welfare and security but had lost the struggle for the individualism and creativity that made good citizenship.

Much more important than such public challenges, however, were challenges on the political level. The lack of any specific evidence for the effectiveness of group counseling as a rehabilitative technique raised concerns in the state legislature. When the CDC announced two costly intensive counseling projects in its 1955–1956 report to Governor Goodwin Knight, claiming that "counseling has already proved its value," it caused considerable concern. In an effort to rein in the department's apparent spending spree, the legislature and California legislative analyst Fred Lewie requested the creation of a research division in Corrections to determine whether counseling made a difference in the recidivism rates of parolees. Departing from a long history of research in penology and criminology, this agency focused on state correctional activities rather than on its subjects. Not the criminal, nor his neighborhood or family, became the subject of scientific scrutiny, but the practices of state officials in charge of his correction. A novelty in the field, California's Research Division became an international role model in corrections research. Within five years, the division was employing a staff of forty professionals and contracted with several specialists at the University of California, Berkeley, and other universities. By the mid-1960s, California had become the world's leading social scientific, psychological laboratory in corrections, spending as much on the scientific research of correctional practices as the rest of the country combined.[15]

[13] James Park, a corrections psychologist and later associate warden of San Quentin 1964–1972, thought that Fenton's trust in the guards "was due to some scholarly misreading of the prison employees' blue collar culture." Quoted in Joseph W. Eaton, *Stone Walls Not a Prison Make: The Anatomy of Planned Administrative Change* (Springfield, Ill., 1962), 172.

[14] Kenyon J. Scudder, *Prisoners Are People* (Garden City, N.Y., 1952); Glaser, *Preparing Convicts for Law-Abiding Lives* (cit. n. 1); Norman Fenton, *What Will Be Your Life?* (Sacramento, Calif., 1955); *Coffin v. Reichard,* 143 F.2d 443, 445 (6th Cir. 1944); John Bartlow Martin, "Prison: The Enemy of Society," *Harper's Magazine,* April 1954, 29–38, on 36.

[15] The two projects were the Pilot Intensive Counseling Organization (PICO) and the Intensive Treatment Project (ITP). Both efforts involved the intensified use of professional therapeutic staff. California Department of Corrections, *Biennial Report, 1955–1956* (Sacramento, Calif., 1957), 4; Dennie Briggs, interview with author, 1 Nov. 2003, Berkeley, Calif.; Intensive Treatment Program, First Annual Report, Oct. 1957, Institutional Treatment Programs 1958, F3717:350 Corrections Administration, Reports & Studies, CSA; Douglas J. Grant, Joan Grant, and Dennie Briggs, "Personality Integration and Delinquency Rehabilitation: An Occasional Paper" (unpublished paper, in author's possession), 8; Glaser, *Preparing Convicts for Law-Abiding Lives* (cit. n. 1), 128; Glaser, *The Effectiveness of a Prison and Parole System* (Indianapolis, 1964), 7; Briggs, Whiteley, and Turner, *Dealing with Deviants* (cit. n. 11), 102.

Under the leadership of its new director, Douglas J. Grant, the division, by 1961, had developed an evaluative tool that could measure the results of correctional treatment experiments, that is, changes in the rate of recidivism. Previously, social workers had used voluminous case files and personal experience to render "expert judgments" on a prisoner's propensity toward a return to delinquency. Grant's new base expectancy score (BES) employed multivariate regression analysis to derive a more reliable prediction formula from the aggregate data of paroled prisoners. Direct comparisons between caseworker judgments and formula predictions showed that, indeed, "experience was not enough" and that the BES predictions correlated 48 percent with parole outcome compared to the experts' meager 21 percent. Even a combination of professional expertise and a mathematical formula could not improve the prediction score of the BES alone. For modern correctional research, Grant announced, it was "time to start counting."[16]

Corrections professionals were aware that identifying different prisoner types with a statistical failure rate could support a conservative and restrictive application of parole.[17] However, it could also provide scientific legitimacy for the expansion of correctional experimentation. In other words, the BES not only gave a numerical value to the effect of experimental treatment. It also made the risk behind correctional research projects—which so often meant lower security levels—calculable, literally. Indeed, Grant used the BES to prove the accountability and scientific competence of the California Department of Corrections, stating that "[a]ny correctional agency not using a prediction procedure to study the effectiveness of its decisions and operations is perpetrating a crime against the taxpayer." The fact that the state's parole board was under no obligation to draw on the base expectancy score for its decisions underscores the central role of the BES as a source of scientific legitimacy for correctional experiments, which were subject to scrutiny in the state's political and public spheres.[18]

Just like the counseling approach introduced after World War II, California's BES drew heavily on research conducted on behalf of the armed forces.[19] Prior to his di-

[16] Don M. Gottfredson and Kelley B. Ballard, *Offender Classification and Parole Prediction* (Sacramento, Calif., 1966). In developing the base expectancy score, Grant drew on the work of Hermann Mannheim and Leslie Wilkins in England. Douglas J. Grant, "It's Time to Start Counting," *Crime and Delinquency* 8 (1962): 261; Hermann Mannheim and Leslie T. Wilkins, *Prediction Methods in Relations to Borstal Training* (London, 1955). The base expectancy states the expected parole violation rate for a given group made on the basis of past experience with such groups. Maxwell Jones, Dennie Briggs, and Joy Tuxford, "What Has Psychiatry to Learn from Penology?" *British Journal of Criminology* 4 (1966): 227–38, on 228; Michael Hakeem, "Prediction of Parole Outcome from Summaries of Case Histories," *Journal of Criminal Law, Criminology, and Political Science* (July-Aug. 1961): 145–55; E. Savides, "A Parole Success Prediction Study," and Don M. Gottfredson, "Comparing and Combining Subjective and Objective Parole Predictions," *Research Newsletter of the California Department of Corrections* 3–4 (Sept.-Dec. 1961).

[17] Simon, *Poor Discipline* (cit. n. 1), 173. For a debate among corrections professionals over the strengths and weaknesses of statistical prediction tools, see Victor H. Evjen, "Current Thinking on Parole Prediction Tables," *Crime and Delinquency* 8 (July 1962): 215–38.

[18] Douglas J. Grant, "It's Time to Start Counting" (cit. n. 16), 259–64; Simon, *Poor Discipline* (cit. n. 1), 173.

[19] Statistical prediction tools had their own history, as Jonathan Simon has pointed out, beginning with the work of criminologists such as Ernest Burgess at the University of Chicago in the 1920s. Sociologists Sheldon and Eleanor Glueck subsequently refined the Burgess method with the help of multivariate regression analysis. Social scientists further honed this technique for military applications during World War II and subsequently in the Department of Defense. Simon, *Poor Discipline* (cit. n. 1), 172. See also Jonathan Simon and Malcolm M. Feeley, "True Crime: The New Penology and

rectorship in the Research Division, Douglas Grant had spent his entire career in the Veterans Administration and the U.S. Naval Reserves and had served as the chief clinical psychologist at the U.S. Navy Retraining Command at Camp Elliott, near San Diego. Grant developed the basis expectancy score at this military prison. His "rehabilitation research" also included intensive interviews with and psychological testing of military prisoners to derive a series of sets that defined the interpersonal relationships of individuals.[20] The result of these efforts was a new taxonomy of interpersonal competency. Grant's interpersonal maturity levels—or I-levels—sorted prisoners on a scale of seven levels of sociopsychological development.[21] Drawing on the interpersonal relations approach that military psychologists developed to determine soldiers' ability to operate effectively in a military command, Grant's maturity levels described progressive stages of social capabilities, which roughly corresponded to successive stages in developmental psychology.

I-levels ranged from the interactions of a newborn infant to "an ideal of social maturity which is seldom or never reached in our present culture." Level 2 personalities only recognized others "as givers or withholders" and showed no interest in things outside themselves. Level 3 personalities tried to "manipulate [their] environment" but still categorized others solely by their usefulness. Level 4 personalities had adopted a set of standards by which to judge themselves and others and recognized others' influences and expectations. Their inability to conform to their own standards, however, often produced feelings of guilt that in turn produced antisocial behavior. Personalities on Level 5 and beyond were sufficiently aware of the complexity of human relations. Their delinquency, explained Grant, was merely situational. However, lower maturity levels, according to Grant, meant a propensity toward delinquency and antisocial behavior. In fact, Grant compared level 4 subjects to the authoritarian personality type in Theodor Adorno's celebrated study "The Authoritarian Personality," which, incidentally, had surveyed San Quentin prisoners in 1946.[22] Like his predecessors, then, Grant considered a prisoner's psychological make-up an important foundation for democratic citizenship.

Grant's I-level theory had important implications for correctional treatment strategies because it identified particular prisoner types as unsuitable for traditional psychotherapy. Designed to unlock conflicts deeply internalized and translated into neuroses, conventional counseling did little for delinquents with personalities on levels 3 and 4, Grant argued. These men periodically dissipated their anxieties "by running away, striking out at someone, or having an affair," thus releasing any pressure that might produce the need for personality change. The delinquent, Grant explained,

Public Discourse on Crime," in *Punishment and Social Control: Essays in Honor of Sheldon L. Messinger,* ed. Stanley Cohen, Sheldon L. Messinger, and Thomas G. Blomberg (New York 1995). Although the connection between research in the military-scientific complex and the social sciences has become a thriving field in the history of science and postwar politics, the entanglement of liberal therapeutic corrections in the structures of the military welfare state has not been recognized.

[20] Douglas Grant and Marguerite Q. Grant, "A Group Dynamics Approach to the Treatment of Nonconformists in the Navy," *Annals of the American Academy of Political and Social Science* 322 (March 1959): 126–35, on 126. Dennie Briggs, *In Prison: Transitional Therapeutic Communities. Research and Demonstration Project Conducted by the California Department of Corrections, 1958–1965* (London, 2000), 8, http://www.pettarchiv.org.uk.

[21] Briggs, Whiteley, and Turner, *Dealing with Deviants* (cit. n. 11), 103.

[22] Grant and Grant, "A Group Dynamics Approach" (cit. n. 20), 128–9.

"would run away from any therapy relationship which made him feel anxious." The delinquent's behavior needed to become the subject of constant therapeutic intervention, "leaving the man no space to escape from facing his issues."[23]

Even before correctional research confirmed the limited impact of counseling on recidivism rates, a new form of group treatment started to garner the enthusiasm of young mental health professionals—milieu therapy. As had been the case with group counseling, milieu therapy originated as an experimental response to the exigencies of war and industrial production before becoming a psychotechnology in the correctional repertoire. Called "therapeutic community" as well, this practice first emerged in wartime England under the leadership of British psychiatrist Maxwell Jones. Conceived as a new rehabilitative technique for veterans, Jones's therapeutic community focused on men dislocated by war from family, home, community, and labor. Instead of costly one-on-one counseling, Jones simply had his patients work together in jobs resembling "as closely as possible the environment of unskilled factory work." In daily discussion groups, patients were to build a culture for the community that would provide them with the techniques necessary to enrich and control their social relationships. Indeed, after only four months, Jones noted that "isolated, depressed, bitter, resentful, suspicious, uncooperative, unproductive personalities respond[ed] to a fairly healthy social situation by accepting responsibility and learning active cooperative participation."[24]

American psychologists first experimented with therapeutic communities in the Veterans Administration and the Office for Naval Research. In 1952, psychologist Harry Wilmer brought the lessons learned at Belmont Hospital to the Veterans Hospital in Oakland, California, to treat psychological casualties from the Korean War. For this first American therapeutic community project, he recruited the help of Dennie Briggs. A sociologist by training, Briggs had previously researched the reintegrative behaviors of former Japanese American internees, had studied racism among navy recruits, and had observed group morale and interpersonal relationships among nuclear submariners during long periods of confinement. The work of Wilmer and Briggs for the Veterans Administration received much attention among professionals and the public. In 1956, the president of the American Psychiatric Association, Dr. Francis J. Braceland, wrote that "Wilmer should be endowed and sent throughout the nation as a teacher and as a catalyst." Wilmer's work became the basis for the television drama *People Need People,* which was introduced in 1961 by none other than war hero and fleet admiral Chester Nimitz on ABC's *Alcoa Premiere,* hosted by Fred Astaire.[25]

---

[23] Ibid., 130; Eaton, *Stone Walls Not a Prison Make* (cit. n. 13), 172, 191; Eugene Wells, "Increased Correctional Effectiveness Unit" and "Therapeutic Community Living Program," *Correctional Review,* March-June 1965, 21–4, 20; Robert M. Harrison and Paul F. C. Mueller, *Clue-Hunting About Group Counseling and Parole Outcome,* Research Report no. 11, California, Department of Corrections Research Division (Sacramento, 1964); Glaser, *Preparing Convicts for Law-Abiding Lives* (cit. n. 1), 78.

[24] Maxwell Jones, *The Therapeutic Community: A New Treatment Method in Psychiatry* (New York, 1953), ix, xv, xvii.

[25] Harry Wilmer, *Social Psychiatry in Action: A Therapeutic Community* (Springfield, Ill., 1958); Dennie Briggs, "In the Navy: Therapeutic Community Experiment at the U.S. Naval Hospital, Oakland, California," Occasional Papers, ser. 1, no. 2, San Francisco, 2000, http://www.pettarchiv.org.uk/pubs-dbriggs-navyhtml.htm; Briggs, "Social Characteristics Associated with Ethnocentric Scores of Naval Recruits" (paper presented at Annual Meeting of the Pacific Sociological Society, Los Angeles, June 1952); Briggs, "Social Adaptation among Japanese American Youth: A Comparative Study,"

By 1958, Briggs, Grant, Wilmer, and Maxwell Jones (then a visiting professor at Stanford University Medical Center) constituted the core of a small transinstitutional network of experts that harbored increasing doubts about the benign nature of professional authority in the therapeutic setting. They shared the concerns of critics such as Ervin Goffman about the deadening, decivilizing, and emasculating effects of the state institution. Doubtful about the benign nature of professional authority, they sought to increase staff and inmate participation and shatter the divide between care and custody. Skeptical of mass society, they advocated community. Instead of co-opting prisoners into conformity through counseling, the therapeutic community tried to provoke crises and resolution; and instead of self-control only, prisoners were to achieve self-expression.[26]

This was hardly a suitable strategy for institutional order, but for the architects of the therapeutic community, well-managed instability promised political benefits. More than a specific clinical practice, the therapeutic community was to remain a constant experiment, something Grant later called the "institutionalization of the Hawthorne effect." Indeed, was that not the essence of a vibrant liberal democracy? "America after all," wrote Briggs, "from its beginnings, had been characterized as an experimenting society." With such confidence in the political nature of their enterprise, it is not surprising that the goal of these new correctional researchers continued to be the full realization of the democratic personality.[27]

The concerns about professional authority in Grant's circle reflected a broader trend among mental health professionals in the late 1950s and early 1960s, who opposed the warehousing that seemed to have become the hallmark of welfare institutions based on a medical model. Their ideas also found a receptive audience on a political level, namely in the leadership of the CDC and in the state legislature. There, both fiscal conservatives and welfare state liberals looked for alternatives to the mass institutions that seemed to have bred two key sources of unrest: idleness and homosexuality.[28] The central role of labor recommended the therapeutic community concept to

---

Sociology and Social Research 38(5) (1954): 293; Harry A. Wilmer, *Social Psychiatry in Action,* ix; Henry Greenberg, "People Need People," in *ABC: Alcoa Premiere,* 10 July 1961, quoted from http://www.pettarchiv.org.uk/pubs-dbriggs-navyhtml.htm.

[26] The belief among social critics in postwar America that "open-minded, flexible, autonomous, and creative thinking" constituted the essence of a democratic personality is the subject of a recent dissertation: Jamie Cohen-Cole, "Thinking about Thinking in Cold War America" (PhD diss., Princeton Univ., Princeton, N.J., 2003), iii.

[27] Briggs, *In Prison* (cit. n. 20), 12, Douglas J. Grant, *Changing Times and Our Institutions: Participants, Not Recipients* (Sacramento, Calif., 1965), 15–6.

[28] See, e.g., the essay of Clinton T. Duffy, former warden of San Quentin, "The Prison Problem Nobody Talks about," *Los Angeles Times,* 21 Oct. 1962, TW 8; California Senate Special Committee on Governmental Administration, *Study on Building Needs of State Correctional Institutions, Political, Journal of the Senate,* 1955, appendix vol. 1, 23, 32–3. California Fact-Finding Committee on Governmental Administration, *Expanded Use of Prison Inmates in the Conservation Program, Journal of the Senate,* March 1961, appendix vol. 1, 25–6. Report of the California Senate Committee on Correctional Facilities, *Journal of the Senate,* March 1961, appendix vol. 1, 82. James Chriss, *Counseling and the Therapeutic State* (New York, 1999), 21; and Erving Goffman, *Asylums: Essays on the Social Situation of Mental Patients and Other Inmates* (New York, 1961). The anti-institutional community movement in the mental health profession has been addressed by James H. Capshew, *Psychologists on the March: Science, Practice, and Professional Identity in America, 1929–1969* (Cambridge, 1999). It has also been noted by criminal justice sociologists. See David Garland, *The Culture of Control: Crime and Social Order in Contemporary Society* (Chicago, 2001), 123; Stanley Cohen, *Visions of Social Control* (Oxford, 1985); and D. R. Karp, *Community Justice: An Emerging Field* (New York, 1998).

conservative and liberals alike. Additionally, the state's budget analyst appreciated the cost-efficient emphasis on the treatment capabilities of prisoners and regular staff—what Leslie Wilkins and Hermann Mannheim had called the "lad on lad effect." Last but not least, the new statistical tool of the base expectancy score limited the political risk that came with a necessary reduction of custody levels.[29]

INTENSIVE TREATMENT: THE PILOT ROCK AND PINE HALL COMMUNITIES

California's first correctional therapeutic community experiment, the so-called Intensive Treatment Project Phase II (ITP II) started out, in February 1960, in one of Chino prison's forest labor camps in the San Bernardino Mountains called Pilot Rock.[30] Project director Dennie Briggs recruited inmates from volunteers at Southern California's Chino prison nearby. The "typical" candidate for the experiment strongly resembled the "rebel without a cause" that had been the subject of the nationwide concern with juvenile delinquency since the war. With maturity levels ranging from 3 to 4, they were mostly white young first termers of working class background, check forging being the most common offense but rape, murder, assault, and robbery being equally represented. Two out of three had been in the military, and many of those had received dishonorable discharges for disciplinary reasons. Aware that the Pilot Rock population was slightly more aggressive than the average inmate, Briggs nonetheless hoped for the spawning of "an atmosphere in which a group of rebellious young men could evolve a social system . . . [and] in the process of so doing, grow."[31]

As in Maxwell Jones's project design, labor was supposed to serve as "an anchor for social life," to use Jonathan Simon's description. However, at Pilot Rock, the labor demands of the Division of Forestry soon turned into a major obstacle for the therapeutic community.[32] Foresters cared little about interpersonal maturity levels and bypassed Briggs to place the most productive and skilled workers from the main prison. Neither were forestry foremen interested in developing a community with their captive labor force. In fact, resistance was common among the free foremen in a variety of jobs. "The horizons of busy work foremen extend only to the completion of their assignments to finish a construction job," a counseling supervisor noted with some

[29] Glaser, *Preparing Convicts for Law-Abiding Lives* (cit. n. 1), 74; Eaton, *Stone Walls Not a Prison Make* (cit. n. 13), 172; Wilkins and Mannheim had pointed toward using the inmate culture for positive gains. Briggs, Whiteley, and Turner, *Dealing with Deviants* (cit. n. 11), 104.

[30] Briggs, Whiteley, and Turner, *Dealing with Deviants* (cit. n. 11), 101; Glaser, *Preparing Convicts for Law-Abiding Lives* (cit. n. 1), 75; Dennie Briggs to Milton Burdman, 29 April 1960, F3717:401 Corrections, Conservation Camp Services, CSA.

[31] ICE—Statistics 1962–1965, 1 July 1965, 110–1, F3717:635 Corrections, Correctional Program Services, CSA. Over 60 percent of the ITP population was younger than twenty-nine, compared with less than 30 percent in regular camps. In contrast to 45 percent of the regular population, 72 percent of Pilot Rock prisoners were in prison for the first time. Over 80 percent were classified as Caucasian, compared with 55 percent in other camps. Significantly fewer prisoners in Pilot Rock had drug or robbery convictions than elsewhere, but the proportion of check forgers was almost double the camp average. There are no data on military service records for the general camp population. Dennie Briggs, "A Social-Therapeutic Community in a Correctional Institution: Some Implications for Education," *Popular Government* (April 1964): 19–24, on 21.

[32] Floyd Chamlee, "Administrative Considerations in the Correctional Community," in *The Correctional Community: An Introduction and Guide,* ed. Norman Fenton, Ernest G. Reimer, and Harry A. Wilmer (Berkeley, Calif., 1967), 29–51, at 32–3; Fred Fromm, *The Intensive Treatment Program Phase II: A Condensation of Working Papers,* Department of Corrections Research Division, Sacramento, Dec. 1966, 10, F3717:351 Corrections, Administration, Reports and Studies, Intensive Treatment Program 1966, CSA. The quotation is from Simon, *Poor Discipline* (cit. n. 1), 75.

frustration, "to meet plumbing crises, or clean the administration building." While Briggs and his peers worked to get prisoners to open up to them and each other, forestry personnel were trained "not to become free with inmates, not to listen to their problems, and to be overly suspicious of any requests."[33]

After an eight-hour day in the forest, the 100 Pilot Rock inmates gathered for their daily plenary community meetings. Sessions generally started with the reading of a logbook, which noted residents' "significant behavior . . . which the entire community ought to know about," so that prisoners could "practice for themselves and as an organized social group, many of the democratic processes which most of us subscribe to at least in principle." Most prisoners seemed to dislike the large meetings, however, preferring to practice their "democratic rights" in the more familiar setting of their work gang rather than in a circle of 100 chairs.[34] Despite the foremen's lack of interest in the formation of a community, prisoners working out on the grade reported closeness within the crew that was not matched in other work projects. In fact, the less than "therapeutic" disposition of their foremen seemed to suit prisoners just fine. Six out of ten prisoners in a subsequent study spoke favorably of their project foremen.[35]

The intense daily community meetings, however, exhausted many participants. An inmate confessed, "They work us all day and squeeze our head all night, and that is more than I think is good for me." A visitor from department headquarters reported: "It was just too much therapy. . . . It got to be more than they could take." Indeed, some could not take it. Within a year, 11 of the 100 community members had escaped. It also became increasingly difficult to recruit new prisoners from the main prison, who "weighed the advantages of treatment against the camp's proximity for visits," Corrections researcher Fred Fromm noted. Of course, group meetings were stressful by design. Pushing "acting-out" personalities to the brink of personality change required what Maxwell Jones had called "painful communication." It certainly felt like that to the participants. The organizers pushed group communication, conflict resolution, and democratic decision making to a point that could make members despair. The ideal method of towel distribution, for example, could occupy the community for a week. New arrivals found the following eerie welcoming note on their pillows: "You are now part of the group. And it is a part of you." The group, the note continued, "is here to draw out any problems you have in your personality traits, not always coming up with the answers but kicking around the problem, leaving you to think about it."[36]

Concerned about the conflicting purposes of productive forest labor and therapeutic

[33] Fromm, *Intensive Treatment Program Phase II* (cit. n. 32), 11; Briggs interview (cit. n. 15); Howard Ohmhart, "Institutional Preparation for the Correctional Community," in *The Correctional Community: An Introduction and Guide,* ed. Harry Aron Wilmer, Norman Fenton, and Ernest G. Reimer (Berkeley, Calif., 1967), 13–28, on 27; Meeting of Pilot Rock Advisory Committee, May 1960, 3–4, Conservation Camp Services 1955–61, F3717:417 Department of Corrections Records, CSA.

[34] Briggs, "A Social-Therapeutic Community in a Correctional Institution" (cit. n. 31), 20.

[35] Richard B. Heim, *Perceptions and Reactions of Prison Inmates to Two Therapeutic Communities* (Sacramento, 1964), 38–9.

[36] Maxwell Jones, *Beyond the Therapeutic Community: Social Learning and Social Psychiatry* (New Haven, Conn., 1968), x; Camp Pilot Rock Proposal, 3 Nov. 1959, 2, 5, Conservation Camp Services 1955–61, F3717:417 Department of Corrections Records, CSA; Chamlee, "Administrative Considerations" (cit. n. 32), 38–42; Untitled document, 26 April 1962, Administration, Suggestions & Complaints 1955–1962, F3717:366 Department of Corrections Records, CSA; Fromm, *The Intensive Treatment Program Phase II* (cit. n. 32), 12, 41; CSA F3717:351 Corrections, Administration, Reports and Studies, Intensive Treatment Program 1966; Grant, Grant, and Briggs, "Personality Integration" (cit. n. 15), 3. The pillow note is quoted in Jones, *Beyond the Therapeutic Community,* x.

communications and a political fallout from the high escape rate, the Department of Corrections moved the project to the Chino prison grounds in May 1961. Occupying a separate wing of the prison, the newly designated Pine Hall community went to work in the laundry, where, the prison administration hoped, the community could put an end to a twenty-year history of corruption and theft. But community members faced hostility from con bosses, foremen, and regular prisoners who relied on the extra income from black-market deals in laundry supplies and services. Moreover, the communal approach to both work and living arrangements separated the project participants from the code of male individualism shared by the majority of inmates and guards and put the experiment under heavy pressure. In the yard, Pine Hall residents faced harassment and were shunned as traitors. Guards tried to recruit community members as spies. Both regular inmates and correctional officers suspected that the community project was a cover-up for homosexual activities and sexual promiscuity.[37]

It was against such opposition that the new workers from Pine Hall tried to improve the work project and shorten their working hours, however. Some began to conduct their own time and motion studies and quality control. The group even hired an industrial psychologist as a consultant and produced a blueprint for a more efficient workshop. Their suggestion to offer services currently only available through bribes, however, gained them no favors with the foremen or the inmates, and their reform efforts stalled. Witnessing the intransigence of the foremen, even guards assigned to the therapeutic community sided with the prisoners. "Mr. A. [the laundry superintendent] said the best way to handle it was to send most of them to the fields to work," reported one guard in his logbook. "I believe the best way would be to send Mr. A. to the fields and let the men work it out as they have planned."[38]

Guards were generally reluctant to identify with the community for good reason. Playing sports or socializing with inmates was strictly forbidden. Yet over the course of a year or so, project leader Dennie Briggs was able to convince a few officers to trade their uniforms for civilian clothing, abandon the practice of keeping a logbook on all inmate movements, and even leave the night watch to community members. To dissolve the oppositional inmate culture, Briggs was willing to break the conventions of guards and the etiquette of his own profession. He urged his fellow counselors on the project to withhold their professional opinion and leave all decisions to the group. At work, he even took an inmate's orders in the laundry. Few counselors, however, were willing to give up their roles of authority. Asked to join the laundry crew, one counselor replied, "I went to college so I could wear white shirts, not wash them."[39]

Frustrated with the deadlock situation at the laundry, community members eventually went on strike, and Briggs took the community off the job. Now, three years into the project, the community had to develop its own work program in a democratic process. The prisoners established a personnel committee that assigned jobs to appli-

---

[37] Briggs interview (cit. n. 15). Fromm, *The Intensive Treatment Program Phase II* (cit. n. 32), 45; Glaser, *The Effectiveness of a Prison and Parole System* (cit. n. 15), observed the low status of laundry and kitchen work in almost all California prisons. Davidson, *Chicano Prisoners* (cit. n. 11), addresses the variety of rackets that existed in the prison underground economy and its subversion of the official prison economy of maintenance, vocational training, and industries. James W. L. Park, "Visiting Hazard at Camp Don Lugo," 17 Feb. 1961, F3717:401 Corrections, Conservation Camp Services, 1959–1961, CSA.

[38] Quoted from the Correctional Officers' log. See Briggs, Whiteley, and Turner, *Dealing with Deviants* (cit. n. 11), 121–6, on 126.

[39] Ibid.; Briggs interview (cit. n. 15); Briggs, *In Prison* (cit. n. 20).

cants. Most of the work consisted of maintenance and gardening, but in time, the committee established inmate positions as therapists, watch standers, staff trainers for similar projects in other prisons, and researchers. By now, the project was almost completely isolated from the main prison, and community members enjoyed a whole range of unusual freedoms, from owning the keys to their cells to voting on the admission of new members to guarding themselves at night.[40] The therapeutic community established a collective authority well beyond that of its coordinator Briggs. It frequently confronted individual members about their lack of progress and at times even exercised its own punishment, sending members into isolation for a week or two. The group had become in charge of its own imprisonment. With not a single escape during the last year of the program, they were good at it, too.[41]

Briggs and his colleagues described the professional and scientific accomplishment of reforming a community of prisoners explicitly in political terms. The improvement in human relations skills anticipated by Grant and Jones had not only inner-directed psychological consequences but social ones as well. "Communication, communalism, and democratization" had been the group's top three objectives, providing, in the words of staff researcher Fred Fromm, "continual opportunities for learning the meaning of freedom and its responsibilities and obligations." Briggs praised "the degree of sophistication which the men have achieved in understanding, structuring, and operating their own community, utilizing many democratic processes; more so since nearly all could be seen as 'failures' to live by these processes prior to incarceration." The therapeutic community, it seemed, had indeed brought the practice of democratic citizenship to the total institution.[42]

C-UNIT: A POLITICAL COMMUNITY

Pilot Rock and Pine Hall were not the only therapeutic community experiments the CDC Division of Research initiated. In fact, the establishment of the Research Division ensured that the therapeutic community approach would multiply into a variety of new projects.[43] In 1960, the CDC diverted funds earmarked for a new 1,200-bed prison into a statewide expansion of a new therapeutic community project. In 1962, the department transformed the former naval hospital at Corona, Southern California, into the California Rehabilitation Center (CRC) for drug offenders; the center drew heavily on the ideas developed at Pilot Rock and Pine Hall.[44] However, such a clinical

---

[40] Briggs, *In Prison* (cit. n. 20), 116–20, 130.

[41] Ibid.; Fromm, *The Intensive Treatment Program Phase II* (cit. n. 32), xv–xvi, 41.

[42] Ibid., 18–20; Briggs, "A Social-Therapeutic Community in a Correctional Institution" (cit. n. 31), 21. "Strange indeed," Briggs mused, "to see the experiment grow and mature in a prison, which traditionally, has been and is the antithesis of democracy." Briggs was quite conscious of the relationship between therapeutic treatment and the practice of citizenship. He wrote: "It is interesting that these projects have evolved in prisons with men and women who legally have no 'civil rights'. . . . They learn the meaning of rights and how to use them effectively for their own as well as the community's good." Briggs, "A Social-Therapeutic Community in a Correctional Institution," 21.

[43] Eaton, *Stone Walls Not a Prison Make* (cit. n. 13), 176.

[44] Untitled document, 30 March 1962, Parole Programming, 1961–1965, F3717:320 Corrections Administration California Rehabilitation Center, CSA; Simon, *Poor Discipline* (cit. n. 1), 91. In 1968, Alfred N. Himelson and Blanche M. Thoma noted the total failure of the drug rehabilitation program that began at Chino in 1960. See Himelson and Thoma, *Narcotic Treatment Control Program: Phase Three,* Research Report no. 25 Department of Corrections Research Division, Sacramento, Calif., June 1968, ix, where they note: "Men directly released from prison did significantly better on parole than those sent to the Narcotic Treatment Control Unit at CIM prior to their releases on parole."

conception of the aforementioned rehabilitative practices should not distract from the fact that therapeutic communities were imbued with essentially political meanings. In fact, in 1960 another community project at the Deuel Vocational Institution near Tracy focused specifically on the construction of a political community of prisoners without the scientific measurement of personality change. Rather than an engineered therapeutic environment with few parallels in the outside world, C-Unit, as it was called, tried to resemble life in a free community, focusing on practical issues. "New experiences, self-determined and fully-lived . . . in a democracy sharpen the individual's creativity so that he does not become dulled by conformity," explained C-Unit director Elliott Studt. Once again, the production of democratic relationships within an authoritarian institution was a major concern.[45]

A product of wartime research in group dynamics, this community project did not conceive of prisoners as "patients" but postulated their "normality, competence, and worth" as virtual citizens of an experimental polity. "If offenders are to be dealt with as human beings," explained Phillip Selznick, which was the central message of the project, "it must be assumed that they are basically like everyone else; only their circumstances are special. Every administrative device that negates these principles, and any therapy that ignores it, must be questioned, and if possible, set aside." Not the family, the factory, the military, the hospital, or the school provided the institutional model for C-Unit on which other narratives of rehabilitative prisons had relied historically. Instead, "the community in the real world [served] as the social institution within which the basic moral code [had to] develop [among the] disparate and potentially conflicting interests." A political organization of prisoners more akin to future community action programs than to group counseling, C-Unit pushed not painful communication but problem solving along the principles of a social-democratic society. For example, after getting the approval of a vast majority of the prisoners in the unit, an inmate committee decided to set up the community's own welfare program for the poor.[46]

The inmate committee's survey in C-Unit also discovered fissures along racial lines in the prison polity. The use of the record player, it turned out, was a major bone of contention. "The Negroes liked blues played very loudly," a report stated, while "Mexicans wanted guitar music," with whites being divided into several different factions. In response to these racial divisions, the community, with the guidance of counseling staff, put together a race-relations working group that produced a C-Unit Emergency Plan for extreme racial incidents that exempted them from the institution-wide lockdowns. "Inmates of C-Unit," Studt wrote, "had learned that to be one community it was necessary not to eliminate subgroups, but to provide for differences in needs and interests." Prison communities could, Studt believed, teach the social values of pluralism and diversity rather than impose the "parody of social order that was conformity."[47]

[45] Glaser, *Preparing Convicts for Law-Abiding Lives* (cit. n. 1), 90. The first phase of PICO had consisted of intensive individual psychoanalytic casework interviewing and changed radically in 1960 under the direction of the new Research Division. *Manual of Operations,* PICO Phase III, F3717:362 Corrections Administration, Research 1960–1965, CSA; Inmate Staff Community Project Report, 1 July 1962 through 31 Oct. 1962, 1–2, in F3717:362 Corrections Administration, Research 1960–1965, CSA.

[46] Elliott Studt, Sheldon L. Messinger, and Thomas P. Wilson, *C-Unit: Search for Community in Prison* (New York, 1968), viii, xv, 7, 81.

[47] Ibid., 83–4, 86.

EMPOWERED PRISONERS: PUTTING THE GENIE BACK IN THE BOTTLE

Limits to a pluralist community existed both inside and outside the institution, of course. The emergence of the Nation of Islam in California's prisons in the 1950s illustrated precisely these limitations of therapeutic corrections and their embrace of the group—or community—as a socializing and rehabilitative agent. Moreover, the growing membership of the Nation of Islam among California's black prisoners highlighted the fact that research in correctional rehabilitation had consistently assumed its subjects to be white and the communities to which they returned to be white. The prison administration had labeled the Nation of Islam a "cult" rather than a culture and thus not a legitimate prisoner organization. Muslims in the C-Unit put Studt's claims to the test. "How can we recognize sub-cultures and give them dignified status and identity," Studt asked, "without upsetting the institutional society in which they live?" Furthermore, what good would assimilation do in prison if life on the outside remained culturally divided and racially segregated? "Are we asking the minorities to adopt the white middle class standards and values," Studt asked, rhetorically, "only to have these same inmates revert to their old standards and environmental settings on parole, where the code of the jungle prevails and where it is still hard to get jobs?" Without an integrated civil society, the assimilation of black prisoners would only rob them of survival skills particular to their own world, the urban "jungle."[48]

Ironically, it had been the CDC's very embrace of the group concept in therapeutic corrections that had given imprisoned Muslims good reason to organize. Starting in the 1950s, the Department of Corrections had begun to authorize the formation of "inmate activity clubs" to help prisoners develop a perspective for life after prison and practice civil society on a small scale. These inmate clubs ranged from purely recreational hobby groups to meetings on business and vocational training to debate clubs and religious gatherings. No apparent difference existed between these clubs and the Nation of Islam. Yet over and over again, the department rejected petitions of the Muslims, on the grounds that the Nation was a political, rather than a religious, organization preaching "racial hatred." Rebuffed, Muslim prisoners responded with nonviolent resistance and labor stoppages. Corrections officials lacked specific evidence that Muslim prisoners wanted to congregate for anything other than prayer. But the Nation's separatist politics on the outside rendered it, according to the CDC, an unfit organization for rehabilitating prisoners to a national community reluctant to address the politics of racial inequality.[49]

Of course, censorship and prohibitions were not the only responses available to prison authorities, and Corrections authorities tried to alleviate the tensions with educational "human relations institutes" for staff. However, the growing political consciousness of black prisoners was hard to stuff back into its therapeutic framework. Reviewing the historical decline of liberal corrections—and thus his own legacy, Richard McGee, head of Corrections from 1944 to 1966, wrote in 1981 that "[r]acial

---

[48] PICO III Report, 1 Nov. 1962 through 31 March 1962, 2, F3717:362 Corrections Administration, Research 1960–1965, CSA.

[49] Inmate Activity Clubs, 1960–1963, F3717:674 Corrections, Correctional Program Services, Central Files, CSA. James P. Alexander, "Information Bulletin: Muslim Cult in Prisons," 3 April 1961, F3717:336 Corrections Administration, Meetings, Misc., 1961, CSA. Milton Burdman to Director McGee, 16 Aug. 1960, Projects 7 Programs, Incidents in Folsom, 1949–1960, F3717:588 Corrections, Correctional Program Services, CSA. A.B. 61/40 Islamic Literature, 4 April 1961, Administrative Bulletin, 1961, F3717:1387 Corrections, Administration, CSA.

troubles began creating major disciplinary and management problems . . . only in the 1950s and 1960s [with] the emergence of activist groups, Left and Right." Convinced that prisoners' political engagement had been purely the product of "outside agitators," McGee got caught off guard by the changes in California's prisons. He failed to see that the therapeutic practice of restoring an inmate's civic identity in group and community settings had, in fact, legitimized their political understanding of themselves and thus made plausible their claims for rights of citizenship.

Prisoners—both black and white—had indeed recognized that the therapeutic role of group life was political in nature as well and that the nurture of a democratic personality in an institutional authoritarian setting also fostered claims for civic participation that was likely to create tensions. Whether in the civic community experiment in C-Unit, in therapeutic communities at Pilot Rock and Pine Hall, or in the inmate activity clubs, the community was not just therapeutic but political as well.

To undermine the prisoners' ability to import their own meanings into the correctional community, the CDC later defined "voluntary self-help groups" more narrowly. "These groups should have as their prime objective the educational, social, and vocational self-improvement which leads to re-socialization." Trying to separate personal growth and rehabilitation from issues of citizenship, the definition further prohibited prisoners' "involvement in political, administrative, national, or legal issues."[50] Black prisoners first and foremost made it clear, however, that education, vocational training, social betterment, and an improved sense of self indeed meant challenging the prison administration, addressing political issues, considering national trends and developments, and fighting legal battles.

The fact that therapeutic corrections could not contain a growing social conflict over race and civil rights, but instead seemed to exacerbate it, proved crucial for the growing strength of a law-and-order opposition. Beginning in October 1964, student protests at Berkeley provided advocates of law and order with a new focal point. As in so many other respects in California history, however, it would be the riots in Watts that proved a decisive turning point.[51] On August 11, 1965, the Los Angeles neighborhood erupted in violence, which led to the death of thirty-five people. African Americans in the impoverished, segregated neighborhood took to the streets, prompted by yet another instance of police brutality, expressing their anger and frustration over discrimination in housing and employment. This was not, however, the first such riot; it was one in a nationwide series of urban disorders lasting from 1964 to 1969. Nor was this the first indication of black frustration and the resultant white backlash. Just a year earlier, the state's referendum on a fair housing law had made it clear where the racial rift divided public opinion. But it was Watts that gave conservative critics of California's efforts in rehabilitation the confidence to systematically brand supporters of liberal reforms as "soft" on murderers and drug dealers.[52]

Just days before the riots, the conservative *Oakland Tribune,* owned by the power-

---

[50] "Promoting Equal Opportunity," Admin. Bulletin 69/21, 27 May 1969, F3717, CSA; Alexander, "Information Bulletin: Muslim Cult in Prisons" (cit. n. 49).

[51] Richard A. McGee, *Prisons and Politics* (Lexington, Mass., 1981), 72.

[52] Ruth Wilson Gilmore, "Globalisation and US Prison Growth: From Military Keynesianism to Post-Keynesian Militarism," in *Race and Class* 40(2–3) (1998–1999): 171–88; Richard A. Berk, Harold Brackman, and Selma Lesser, *A Measure of Justice: An Empirical Study of Changes in the California Penal Code, 1955–1971* (New York, 1977), 53.

ful Republican Knowland family, had run a three-day series on California's Department of Corrections, which it had praised as the world's "foremost laboratory for the development of new methods of [converting] criminals into useful citizens." But with Watts in flames, criminals turned from potential citizens into public enemies, if not animals. Los Angeles police chief William Parker explained the uprising of African American Angelinos in the unadorned racist language for which he became known. "One person throws a rock and then, like monkeys in a zoo, others started throwing rocks." His first rule of criminal justice: "The only way we can stop this is to arrest, arrest, arrest." Los Angeles mayor Sam Yorty agreed that tough law and order was the only way to quash the violence in South Central LA.

Yorty subsequently used Watts as a platform, during the state's 1966 Democratic primaries, from which to launch his attacks against the liberal two-term governor Pat Brown, under whose watch the Department of Corrections had developed its rehabilitative research and experiments. The governor had coddled prisoners, the mayor charged, had done nothing to halt the proliferation of drugs, and had idly stood by while communist agitators fomented riots on campus. Worst of all, he had failed to crack down on black rioters in Watts.[53]

This was the opportunity for a successful Republican challenge to many of the tacit assumptions behind liberal criminal justice. After the California electorate buried Pat Brown's formerly stellar reputation under a landslide majority of one million votes for his opponent in November 1966, law and order took center stage. "The time has come for us to decide," Governor Ronald Reagan declared in his 1967 inaugural address, "whether . . . we can afford everything and anything we think of simply because we think of it," or whether the state's social programs "were just goodies dreamed up for our supposed betterment." In other words, the time for social experimentation was over. True to his promise, Reagan cut the budget of the Correctional Research Division, prompting its director, Douglas Grant, and Richard McGee, director of Corrections for more than twenty years, to leave public service.

This did not mean a disengagement from the war on crime—quite the opposite. The Reagan administration's new paramilitary California Specialized Training Institute (CSTI) offered courses in riot and crowd control useful in any venue, from university campuses to prisons to urban ghettos. The purpose of the California Crime Technological Research Foundation (CCTRF), founded in 1967, was to "analyze the etiology of civil disorders," improve law enforcement, surveillance, and investigation, and produce "formula models . . . predictive of individual or group criminality." Many of these technologies came, once again, from military research. The predictive formulas for the identification of those most deserving of punishment and incapacitation, however, were precisely the statistical tools Douglas Grant had introduced to calculate the

---

[53] "State Unlocks Prisons to Rehabilitate Criminals," *Oakland Tribune,* 8 Aug. 1965, 1–3. Pride in the institution's scientific and professional leadership pervaded much of the press coverage on California corrections in the first half of the 1960s. "The entire nation has only 50 prison psychiatrists," the *Tribune* reported in its Aug. 8 article, adding proudly that "California has half of them." "Convicts Break with Tradition," *Oakland Tribune,* 9 Aug. 1965, 1–2; Matthew Dallek, "Liberalism Overthrown," *American Heritage* 47(6) (1996): 39–52; Robert J. Minton Jr., ed., *Inside: Prison American Style* (New York, 1967), 75; Walter Lear Gordon, "Political, Ideological, and Institutional Aspects of Comprehensive Criminal Law Reform in California: 1960–1975" (PhD diss., Univ. of California, Los Angeles, 1981).

risk of more open therapeutic communities.[54] Designed to shore up support for scientific experimentation, quantitative evaluation had become a tool for undermining the claims of correctional experts by the end of the 1960s.

Whereas the Nation of Islam had revealed the racial limits to the experimentation with democratic pluralism within the institution, the riots in Watts had tipped the political balance on the outside from liberal support for research on the possibilities of rehabilitation to conservative demands for correctional research of new techniques of incarceration. Not the disoriented veterans and abandoned youths of postwar California were the target group of this new approach, but rather free-speech demonstrators, Vietnam War protesters, and—most important—angry urban black males. This political revision of the institutional subject produced a gradual reorientation of correctional practice from research on methodologies of reintegration to strategies of containment and exclusion. This, of course, was the case elsewhere in the American welfare state. Of this, modern therapeutic corrections was a central part, providing one of the most disturbing continuities between the carceral imagination of postwar liberalism and the plain carceral force of the law-and-order state.

[54] Berk, Brackman, and Lesser, *A Measure of Justice* (cit. n. 52), 56. The decisive signal that correctional welfare and therapeutic corrections were in decline, remembered John Irwin, "was the abandonment of many programs in progress or in the planning stages after Governor Reagan took office in 1967." Irwin, *The Felon* (cit. n. 8), 53; A. James Reichly, "Ronald Reagan Faces Life," *Fortune,* July 1967, 98–157; *California Crime Technological Research Foundation* (Sacramento, 1967); *California Specialized Training Institute* (Sacramento, 1968).

"New Soviet Man" Inside Machine:
Human Engineering, Spacecraft Design, and the Construction of Communism

By Slava Gerovitch[*]

ABSTRACT

Soviet propaganda often used the Soviet space program as a symbol of a much larger and more ambitious political/engineering project—the construction of communism. Both projects involved the construction of a new self, and the cosmonaut was often regarded as a model for the "new Soviet man." The Soviet cosmonauts publicly represented a communist ideal, an active human agency of sociopolitical and economic change. At the same time, space engineers and psychologists viewed human operators as integral parts of a complex technological system and assigned the cosmonauts a very limited role in spacecraft control. This article examines how the cosmonaut self became the subject of "human engineering," explores the tension between the public image of the cosmonauts and their professional identity, and draws parallels between the iconic roles of the cosmonaut and the astronaut in the cold war context.

INTRODUCTION

On April 12, 1961, Yurii Gagarin's historic spaceflight shook the world, sending enthusiastic crowds of Soviet citizens onto the streets to celebrate. Just a few months later, the Twenty-Second Congress adopted a new Communist Party program, which set the goal of building the foundations of communism in the Soviet Union by 1980. This all-out drive toward communism had two crucial components: the construction of a material and technical basis of communism, and the development of the "new Soviet man"—"a harmonic combination of rich spirituality, moral purity, and physical perfection."[1] Who better than Gagarin to embody this new ideological construct? The

[*] Science, Technology and Society Program, E51-185, Massachusetts Institute of Technology, 77 Massachusetts Avenue, Cambridge, MA 02139; slava@mit.edu.

Earlier versions of this paper were presented in October 2004 at the University of Georgia conference "Intelligentsia: Russian and Soviet Science on the World Stage, 1860–1960," in Athens, Georgia, and in April 2005 at a seminar at Eindhoven University of Technology, in the Netherlands. I wish to thank the participants of these forums and two anonymous referees of *Osiris* for their very useful comments. I am especially grateful to David Mindell and Asif Siddiqi for their invaluable insights into the history of the American and Soviet space programs. The staff of the Russian State Archive of Scientific and Technical Documentation in Moscow and the staff of the National Air and Space Museum in Washington, D.C., were most helpful in locating relevant documents. Research for this article was supported by the National Science Foundation under Grant No. SES-0549177. Unless otherwise noted, translations are my own.

[1] *Materialy XXII s'ezda KPSS* (Moscow, 1962), 411.

Soviet media machine quickly generated a propaganda cliché: "the Soviet cosmonaut is not merely a victor of outer space, not merely a hero of science and technology, but first and foremost he is a real, living, flesh-and-blood *new man,* who demonstrates in action all the invaluable qualities of the Soviet character, which Lenin's Party has been cultivating for decades."[2]

In the first half of the 1960s, the Soviet space program boasted one success after another—the first man's flight, the first group flight, the first woman's flight, the first multicrew mission, and the first space walk.[3] Ordinary people became genuinely fascinated with the Soviet triumphs in space. "Gagarin's achievement was our greatest pride," recalled one member of the "Sputnik generation."[4] According to the 1963 poll of the readers of a popular youth-oriented Soviet newspaper, Gagarin's flight was by far the greatest human achievement of the century, and Sputnik the greatest technological feat.[5]

Soviet propaganda vividly portrayed cosmonaut heroes bravely flying their spacecraft into the unknown, but the cosmonauts, in fact, were assigned a very limited role on board a spacecraft. Soviet spaceships were fully automated. Although systems of manual control were installed, their functions and use were severely limited. Gagarin's *Vostok* had only two manual control functions: attitude correction and firing the retrorocket for reentry—and those could be used only in case of emergency.[6] The designer of *Vostok*'s manual control system jokingly summed up Gagarin's instructions in four words: "Do not touch anything!" The Soviet engineers' vision of a manned flight was that of a cosmonaut flying *on board* a spacecraft, rather than flying a spacecraft.[7]

On later models of spacecraft, the cosmonauts gradually gained more control functions, but they still served mostly as backup for failed automatics; the standard mode of control remained automatic. Soviet cosmonauts were "designed" as part of a larger technological system; their height and weight were strictly regulated, and their actions were thoroughly programmed. Soviet space politics, one might say, was inscribed on the cosmonauts' bodies and minds, as they had to fit, both physically and mentally, into their spaceships.

The cosmonauts strongly opposed this trend, which they labeled "the domination of automata."[8] With their professional background as pilots, they felt that greater human control of spacecraft would increase the reliability and effectiveness of space missions. Some cosmonauts regarded the domination of automata in the Soviet space program as the manifestation of a general ideological attitude toward the individual

    [2] Evgenii Riabchikov, "Volia k pobede," *Aviatsiia i kosmonavtika,* no. 4 (1962): 10–19, on 19 (emphasis added).

    [3] The most comprehensive history of the Soviet space program is Asif A. Siddiqi's thoroughly researched *Challenge to Apollo: The Soviet Union and the Space Race, 1945–1974* (Washington, D.C., 2000), which includes an excellent bibliographic essay.

    [4] Donald J. Raleigh, trans. and ed., *Russia's Sputnik Generation: Soviet Baby Boomers Talk about Their Lives* (Bloomington, Ind., 2006), 133.

    [5] Boris A. Grushin, *Chetyre zhizni Rossii v zerkale oprosov obshchestvennogo mneniia,* vol. 1, *Zhizn' 1-ia: Epokha Khrushcheva* (Moscow, 2001), 403.

    [6] Valentina Ponomareva, "Osobennosti razvitiia pilotiruemoi kosmonavtiki na nachal'nom etape," in *Iz istorii raketno-kosmicheskoi nauki i tekhniki,* no. 3, ed. V.S. Avduevskii et al. (Moscow, 1999), 132–67.

    [7] Boris E. Chertok, *Fili—Podlipki—Tiuratam,* vol. 2 of *Rakety i liudi,* 3rd ed. (Moscow, 2002), 428.

    [8] Georgii Beregovoi, as quoted in Valentina Ponomareva, "Nachalo vtorogo etapa razvitiia pilotiruemoi kosmonavtiki (1965–1970 gg.)," in *Issledovaniia po istorii i teorii razvitiia aviatsionnoi i raketno-kosmicheskoi tekhniki,* nos. 8–10, ed. Boris Raushenbakh (Moscow, 2001), 150–74, on 166.

as an insignificant cog in the wheel.[9] They viewed the strict regulation of their activities as part of a general pattern of social control in the Soviet state.[10] In my view, exploring this tension between the cosmonauts' public identity as icons of communism and their conflicted professional identity may throw light on some fundamental contradictions in the Soviet discourse on the communist self in the Khrushchev era.

Soviet historians have long focused their attention on the attempts to reform the human self, and they bring up the new Soviet man as being essential to the Soviet project. The "totalitarian model" of Soviet society traditionally considered "the cog in a wheel" as a central metaphor for the new Soviet man.[11] This metaphor embodied the notion of the passive individual subsumed under the collective and implied the machinelike operation of the party and state apparatus controlling social life.

Recently, scholars began to question the passive nature of the "totalitarian self" and to explore the historical evolution of the Soviet notions of the self. Vladimir Papernyi has suggested that two opposing cultural patterns coexisted in Soviet society, dominating in different periods: the first, which privileged the mechanism and collectivism, dominated in the 1920s; the second, which focused on the human and individualism, prevailed in the 1930s–1950s.[12] Igal Halfin and Jochen Hellbeck have argued that the Stalinist subject was not merely a passive recipient of official ideology. In their views, young Soviet people internalized communist values and made active attempts to reform themselves, striving for the alluring ideal of the new Soviet man.[13] Sheila Fitzpatrick has found more mundane reasons for individuals' attempts to construct new identities for themselves. The Soviet state, she argues, discriminated on the basis of class, and resourceful individuals often resorted to self-fashioning, impersonation, and outright imposture to claim their "proletarian" origins and revolutionary identity.[14]

The transition from the Stalin era to Khrushchev's political "thaw" led to a marked shift in the prevailing conception of the self. Historians differ, however, on the exact direction of that shift. Elena Zubkova has described the Stalin era as an age of collectivism, followed by the "turn to the individual" in the Khrushchev years.[15] Oleg Kharkhordin, by contrast, has suggested a historical trajectory from the collectivism of the 1920s to the individualism of the 1930s–1950s to the new collectivism of the 1960s. He provocatively argues that there was more room for individual freedom under Stalin than under Khrushchev. Whereas Stalinist terror was punitive and haphazard, Khrushchev's policies were aimed at a pervasive rational system of preventive mutual surveillance.[16]

[9] Valentina Ponomareva, *Zhenskoe litso kosmosa* (Moscow, 2002), 207.

[10] Ponomareva, "Nachalo vtorogo etapa" (cit. n. 8), 170.

[11] Mikhail Heller, *Cogs in the Soviet Wheel: The Formation of Soviet Man,* trans. David Floyd (London, 1988).

[12] Vladimir Papernyi, *Architecture in the Age of Stalin: Culture Two,* trans. John Hill and Roann Barris (New York, 2002).

[13] See Igal Halfin, *From Darkness to Light: Class, Consciousness, and Salvation in Revolutionary Russia* (Pittsburgh, 2000); Halfin, *Terror in My Soul: Communist Autobiographies on Trial* (Cambridge, Mass., 2003); and Jochen Hellbeck, *Revolution on My Mind: Writing a Diary under Stalin* (Cambridge, Mass., 2006).

[14] Sheila Fitzpatrick, *Tear off the Masks! Identity and Imposture in Twentieth-Century Russia* (Princeton, N.J., 2005).

[15] Elena Zubkova, "Turning to the Individual: The Paths from Above and from Below," in *Russia after the War: Hopes, Illusions, and Disappointments, 1945–1957,* trans. Hugh Ragsdale (Armonk, N.Y., 1998).

[16] Oleg Kharkhordin, *The Collective and the Individual in Russia: A Study of Practices* (Berkeley, Calif., 1999); see, especially, 299–300.

If the Soviet policy on the new Soviet man is still baffling historians, it must have looked even more confusing to contemporaries. It is precisely the ambiguity of the new Soviet man as an ideological construct that will be addressed in this paper. Instead of viewing this ambiguity as a result of policy inconsistencies, I will interpret it as a product of fundamental ideological tensions in the Soviet discourse on the self.

Soviet propaganda often used large technological projects, such as the space program, as symbols of the construction of socialism and communism. I will examine the notion of the new Soviet man through its iconic representations—from the heroic aviator in the Stalin period to the cosmonaut in the Khrushchev era. In these cases, the self was viewed as an active agency and, at the same time, defined as part of a technological system. The first quality implied autonomy; the second, discipline and subordination. I will argue that this tension gave rise to the paradox of "disciplined initiative," which plagued both the cosmonaut self and the new Soviet man.

Based on recently declassified archival documents related to the Soviet space program, private papers of leading space engineers and officials, and interviews with spacecraft designers and cosmonauts, I will examine the application of principles of "human engineering" to the training of Soviet cosmonauts, the formation of their professional identity, and the clash between their professional and public identities. In conclusion, I will draw parallels between the iconic roles of the cosmonaut and the astronaut in the cold war context.

"A FLAMING MOTOR FOR A HEART": NEW SOVIET MAN IN THE SKY

In his pioneering study of Soviet technology under Lenin and Stalin, the historian Kendall Bailes noted that in the 1930s, famous Soviet aviators became "prime exhibits of the 'new Soviet men' whom the authorities wished to create."[17] In April 1934, Mikhail Vodop'ianov and Nikolai Kamanin and five other pilots, all of whom had distinguished themselves during the Arctic rescue of the crew of the stranded icebreaker *Cheliuskin,* became the first Soviet citizens to be awarded the newly established title of Hero of the Soviet Union. As the historian Jay Bergman aptly put it, air heroes became "ideological prototypes, precursors of the people who would inhabit the future, from whose achievements . . . the Soviet people could develop a sense of what living under communism would be like."[18]

In November 1933, Stalin put forward a new slogan, calling on Soviet aviators to fly farther, faster, and higher than anyone else, and the Soviet Union jumped into the international race for air records. By 1938, the Soviets claimed to have achieved sixty-two world records, including, as requested by Stalin, the longest, fastest, and highest flights.[19] Aviation became one of the most spectacular "display technologies," showing off the Soviet technological prowess and implying the ideological superiority of the Soviet regime.[20]

As Bailes keenly observed, the regime skillfully exploited the public enthusiasm

---

[17] Kendall Bailes, *Technology and Society under Lenin and Stalin: Origins of the Soviet Technical Intelligentsia, 1917–1941* (Princeton, N.J., 1978), 391.

[18] Jay Bergman, "Valerii Chkalov: Soviet Pilot as New Soviet Man," *Journal of Contemporary History* 33 (1998): 135–52, on 139.

[19] Bailes, *Technology and Society under Lenin and Stalin* (cit. n. 17), 386.

[20] Paul R. Josephson, "'Projects of the Century' in Soviet History: Large-Scale Technologies from Lenin to Gorbachev," *Technology and Culture* 36 (July 1995): 519–59.

for aviation to counterbalance the sobering effect of the Great Purges of 1936–1938.[21] As hundreds of thousands perished in prisons and labor camps, Stalin used the celebratory occasions of record flights to stress his personal concern for human life. "Your lives are dearer than any machine," he frequently told aviators, urging them not to take unjustified risks.[22] Yet that was precisely what the aviators had to do in order to set records, so valuable on the propaganda front. In January 1934, the crew of the *Osoaviakhim* stratosphere balloon, dedicating their feat to the Seventeenth Party Congress, set a new world record in height. In doing so, however, they pushed the balloon beyond its technological limits and died in the ensuing crash. During the funeral, Stalin personally carried the ashes through Red Square.[23]

The political project of creating the public image of the aviator as a new Soviet man took precedence over the practical demands for the development of modern military aircraft. Concerned largely with the propaganda aspect of aviation as a "display technology," Soviet leaders neglected much-needed technological reforms in the aviation industry. Instead of designing swift, maneuverable aircraft with sophisticated electronic equipment, the Soviets produced heavy, slow, long-distance models, which were good for setting world records but useless in bombing or air combat.[24]

Stalin's famous toast to "the 'little cogs' of a grand state mechanism" at a June 1945 reception celebrating victory in the Second World War encapsulated a popular cultural image of the individual under Stalin's rule: a necessary but ultimately subservient and replaceable part.[25] The more loudly Stalin proclaimed that human life was "dearer than any machine," the more plainly his actual policies forced individuals to obey the relentless rhythm of the state machine.

The popular culture of the 1930s was filled with man-machine metaphors that reinforced the regime's ideological message. For instance, in the 1930s, in the well-known 1920s song "Aviation March," the word "reason" in "Reason gave us steel wings for arms, and a flaming motor for a heart" was replaced with "Stalin." Visual imagery in public spaces reinforced the metaphorical merger of humans and airplanes as well. For example, the ceiling of the Mayakovskaya subway station, completed in 1938, was decorated with mosaics depicting athletic men and women soaring in the sky like airplanes, "as if these people themselves were a technical achievement of the new Soviet epoch."[26] Along with aviation, the spectacular Moscow subway system itself became, in the words of a contemporary, "a majestic school in the formation of the new man."[27]

The widely propagated image of the new Soviet man was filled with inner tensions and ambiguities. The new man was both a distinct individual and a "little cog"; he strove for personal achievement and wanted to be a good member of the collective; he

[21] Bailes, *Technology and Society under Lenin and Stalin* (cit. n. 17), 381.

[22] Ibid., 387.

[23] Iaroslav Golovanov, *Korolev: Fakty i mify* (Moscow, 1994), 198.

[24] Bailes, *Technology and Society under Lenin and Stalin* (cit. n. 17), 390; Bergman, "Valerii Chkalov" (cit. n. 18), 151.

[25] Iosif Stalin, "Vystuplenie na prieme v Kremle v chest' uchastnikov Parada Pobedy" (1945), in *Sochineniia* 15 (Moscow, 1997): 232.

[26] Michael O'Mahony, "Zapiski iz podzemki: moskovskoe metro i fizkul'tura v 30-e gody XX veka," *Neprikosnovennyi zapas (NZ)* 23 (2002); http://www.nz-online.ru/index.phtml/index.phtml?aid=25011375 (accessed 1 June 2005).

[27] Quoted in Andrew Jenks, "A Metro on the Mount: The Underground as a Church of Soviet Civilization," *Technol. Cult.* 41 (Oct. 2000): 697–724, on 697.

was to be a master of technology, yet he merged with technology as its intrinsic part. Stalin publicly encouraged air heroes to choose their own courses of action during flights, ignoring, if necessary, advice from the ground, and at the same time he instructed them not to take any risks. Paradoxically, "while individual initiative, even disobedience, were qualities that Stalin considered admirable and highly desirable in the new Soviet man, they were also things that, in Stalin's view, would be strictly limited in the communist society he envisioned," a society that would be "rigidly hierarchical" and "informed by an ethos of deference and obedience."[28] The ideological constructs of the new Soviet man and of the bright communist future did not quite match up. This did not particularly upset professional ideologues, however: those constructs were to be believed, rather than rationally examined.

The contradictory nature of Stalin's new man stemmed from the fundamental ambiguity of Stalinist official discourse.[29] Soviet ideology constantly oscillated between belief in the power of technology and trust in active human agency. Stalin's 1931 slogan, "in the reconstruction period, technology decides everything,"[30] was replaced in 1935 by its exact opposite, "cadres decide everything."[31] Despite the clear signal from the top marking a significant ideological shift, the public discourse had inertia of its own, and the two slogans—"technology decides everything" and "cadres decide everything"—coexisted in popular writings and speeches for quite a while, creating much confusion about the correct party priorities with respect to people and machines. Whereas the old slogan presented technology as a measure of progress, the new one placed an equally high value on human skill and personal sacrifice.

In the 1960s the cosmonaut quickly supplanted the aviator as a top model for the Soviet self. The role of the new Soviet man in a complex technological system, however, remained ambiguous: Will he become the master of technology or its servant?

FROM STALIN'S "FALCONS" TO KOROLEV'S "LITTLE EAGLES"

In the 1930s, the Soviet media habitually referred to the aviation heroes as "Stalin's falcons," implying their "extra-human, and even superhuman, characteristics and abilities."[32] Sergei Korolev, the chief designer of Soviet spacecraft, echoed this cultural image, calling the first cosmonauts "my little eagles." He expected the cosmonauts to be ready for self-sacrifice, just like the famous aviators of the 1930s. At a meeting of the Military Industrial Commission two weeks before Gagarin's flight, Korolev admitted the considerable risks of the mission but cited the courage of the *Osoaviakhim* stratosphere balloon crew: "They died but held a record for the Soviet Union for 22 years."[33]

Which personal qualities were required of the Soviet cosmonaut became a matter of serious debate. In January 1959, top scientists, physicians, and spacecraft design-

[28] Bergman, "Valerii Chkalov" (cit. n. 18), 143, 149.

[29] Slava Gerovitch, *From Newspeak to Cyberspeak: A History of Soviet Cybernetics* (Cambridge, Mass., 2002), chap. 1, "The Cold War in Code Words: The Newspeak of Soviet Science."

[30] Iosif Stalin, "O zadachakh khoziaistvennikov" (1931), in *Sochineniia* 13 (Moscow, 1951): 29–42, on 41.

[31] Iosif Stalin, "Rech' v Kremlevskom dvortse na vypuske akademikov Krasnoi armii" (1935), in *Sochineniia* 14 (Moscow, 1997): 58–63, on 61.

[32] Bergman, "Valerii Chkalov" (cit. n. 18), 138.

[33] Boris Chertok, 29 March 1961, notebook no. 41, Chertok Papers, National Air and Space Museum, Smithsonian Institution, Washington, D.C.

ers gathered at the Soviet Academy of Sciences to discuss the criteria for cosmonaut selection. The physical requirements were clear: because of the small size of the *Vostok* spacecraft, the candidates had to be no taller than 1.75 meters (5'7") and no heavier than 72 kilograms (158 pounds). Opinions divided over the question of future candidates' professional background. Some participants thought submarine sailors, missile forces officers, and even race car drivers should be considered. Korolev argued, however, that fighter jet pilots were best prepared for space missions: "A fighter pilot has the universal skills that we need. He flies in the stratosphere on a one-seat airplane. He is a pilot, a navigator, and a radio operator in one. It is also important that he is a regular military man and therefore possesses such necessary qualities for a future cosmonaut as assiduousness, self-discipline, and unwavering determination to reach the set goal."[34]

Selection was made from among fighter pilots age twenty-five to thirty in perfect health; no requirements were set for their piloting skills. As a result, most of the twenty selected candidates had relatively little flying experience—230 hours in Gagarin's case; the Mercury astronauts, by contrast, had to have a minimum of 1,500 hours. Nineteen out of the twenty cosmonauts were fighter pilots with no training in engineering; the Mercury seven were skilled test pilots with strong engineering backgrounds. Soviet spacecraft designers believed that the high degree of automation of spacecraft control allowed them to run the entire mission in the automatic or semiautomatic mode, thus making high piloting and engineering skills unnecessary. Korolev explained: "As has been repeatedly demonstrated in our automated flights and those with animals on board, our technology is such that we do not require, as the American Mercury project does, that our early cosmonauts be skilled engineers."[35]

The task of cosmonaut training was assigned to the air force, which in 1960 established the Cosmonaut Training Center in an isolated area eighteen miles northeast of Moscow, now widely known as Star City. Lieutenant General Nikolai Kamanin was appointed the deputy chief of the air force's General Staff in charge of cosmonaut selection and training. It was the same Kamanin who received the highest Soviet honor, Hero of the Soviet Union, for his role in the 1934 *Cheliuskin* rescue mission. One of the most famous aviators of the 1930s, a public icon of Stalin's regime, Kamanin had strong convictions and a commanding personality. He did not hesitate to confront an equally authoritative Korolev and the powerful leadership of the air force and the Ministry of Defense whenever they did not go along with his uncompromising views on space policy.

Kamanin's vision of the role of the cosmonauts in the space program forcefully clashed with Korolev's position. While Korolev extolled the virtues of automation and proudly asserted that on his spacecraft even "rabbits could fly,"[36] Kamanin insisted that the cosmonauts be assigned a greater role in spacecraft control.

In preparation for Gagarin's historic launch, Korolev suggested that Gagarin should limit his actions during the flight to visual inspection of onboard equipment and should not touch any controls. Korolev's cautious approach may have been prompted by the responsibility placed on him by the political authorities. At a meeting of the

[34] Iaroslav Golovanov, *Nash Gagarin* (Moscow, 1978), 50–1.
[35] Quoted in Siddiqi, *Challenge to Apollo* (cit. n. 3), 244.
[36] Nikolai Kamanin, *Skrytyi kosmos,* vol. 3, *1967–1968* (Moscow, 1999), 335 (diary entry of 12 Dec. 1968).

Presidium of the Party Central Committee on April 3, 1961, just a few days before Gagarin's launch, Nikita Khrushchev himself raised a question about the cosmonaut's working capacity and psychological stability in orbit. Korolev had to give his personal assurances to the Soviet premier.[37] Not relying entirely on the disciplining force of the cosmonaut's written instructions, spacecraft designers took some technological measures to prevent any accidental damage by the cosmonaut should he lose his psychological stability. They blocked the manual orientation system for reentry with a digital lock. There was some debate about whether to give the combination to the cosmonaut or to transmit it over the radio in case of emergency. Eventually, they decided to put the combination in a sealed envelope and to place it on board so the cosmonaut could open it in an emergency.[38]

Supported by flight physicians, Kamanin proposed giving Gagarin a broader set of functions, such as checking equipment before launch, writing down his observations and instrument readings in the onboard journal, and reporting those over the radio.[39] As doctors explained, keeping the cosmonaut busy would help deflect his attention from possible negative emotions during g-loads and weightlessness.[40] Kamanin prevailed, and Gagarin performed his monitoring functions very well, while the flight itself was conducted in the automatic mode.

Kamanin carefully supervised the official reports written by the cosmonauts after their flights. Based on his suggestions, the cosmonauts Andrian Nikolaev and Pavel Popovich, who tested the possibility of carrying out various military tasks during their *Vostok 3* and *Vostok 4* flights, reported that the human was "capable of performing in space all the military tasks analogous to aviation tasks (reconnaissance, intercept, strike)." Kamanin then used their reports to substantiate his view that "man can maintain good working capacity in a prolonged spaceflight. The 'central character' in space is man, not an automaton."[41]

Kamanin envisioned the cosmonaut as a quintessential pilot of a space vehicle, in full control of his craft and of his mission. Korolev, by contrast, viewed the cosmonaut as part of a complex technological system—a part that had to obey the logic of system operations as faithfully as any other part. Despite their conflicting visions of the overall cosmonaut role, the two men often agreed on cosmonaut training, though they emphasized different aspects. Korolev stressed the cosmonaut's ability to fit into the machine, to carry out precisely programmed actions.[42] Kamanin, for his part, demanded strict military discipline and political loyalty. While spacecraft designers standardized cosmonauts' bodies, Air Force officials regularized their thoughts. Together Korolev and Kamanin attempted to engineer the Soviet cosmonaut, a living embodiment of the new Soviet man. They were aided in this project by specialists in human engineering.

---

[37] Nikolai Kamanin, *Skrytyi kosmos,* vol. 1, *1960–1963* (Moscow, 1995), 23 (diary entry of 2 March 1961), 43 (diary entry of 4 April 1961).

[38] As it turned out, two people independently told Yurii Gagarin the combination before the launch so that he would not waste time in a real emergency. See Chertok, *Fili—Podlipki—Tiuratam* (cit. n. 7), 428–9.

[39] Siddiqi, *Challenge to Apollo* (cit. n. 3), 264.

[40] Kamanin, *Skrytyi kosmos* (cit. n. 37), 1:23 (diary entry of 2 March 1961).

[41] Ibid., 174 (diary entry of 13 Sept. 1962), 149 (diary entry of 16 Aug. 1962).

[42] Siddiqi, *Challenge to Apollo* (cit. n. 3), 244.

HUMAN ENGINEERING AND THE DESIGN OF A COSMONAUT

Human engineering emerged in the Soviet Union in the early 1960s under the name engineering psychology. This field developed under the wide umbrella of cybernetics and also became known as cybernetic psychology.[43] The Council on Cybernetics of the Soviet Academy of Sciences set up a psychology section, which included a committee on human engineering that coordinated nationwide research in this field. Soviet specialists in engineering psychology defined their discipline as a "study of humans as part of a control system" and included in their area of interest such fields as applied psychology, experimental psychology, biomechanics, psychoacoustics, ergonomics, operations research, and the study of human-machine systems.[44] Adopting the cybernetic conceptual framework, they viewed both humans and machines as cybernetic systems governed by the same feedback mechanism. Blurring the boundary between human and machine, cybernetics legitimized the idea of designing, or human engineering, the self.

The Council on Cybernetics coordinated research on human perception, information processing, and the impact of emotional states on control functions at several universities and research institutes, including the Air Force Institute of Aviation and Space Medicine. The institute set up a department of spacecraft simulators, which was responsible for the adaptation of onboard equipment to cosmonauts' psychological and physiological characteristics and for the development of specifications for ground simulators.[45] By early 1967, the institute had conducted several hundred flight experiments and more than 1,000 tests on simulators to find an optimal division of function between human and machine.[46] Moscow University and Leningrad University also conducted a number of studies focused on the human operator on board a spacecraft. They examined various statistical characteristics, work efficiency, interaction among human operators, and selection and special training of personnel for working with various types of control systems.[47]

As "cybernetic psychologists," Soviet specialists in human engineering conceptualized the spacecraft control system as a "cybernetic 'human-machine' system."[48] They defined the cosmonaut as a "living link"[49] in this system, and analyzed this living link in cybernetic terms, borrowed from control theory and information theory—the same terms as applied to the other links in this system. They discussed how efficiently a human operator could perform the functions of a logical switchboard, an

[43] On Soviet cybernetics, see Gerovitch, *From Newspeak to Cyberspeak* (cit. n. 29).

[44] E. I Boiko et al., "Kibernetika i problemy psikhologii," in *Kibernetiku—na sluzhbu kommunizmu,* ed. A. I. Berg (Moscow, 1967), 5:314–50, on 316.

[45] V. I. Iazdovskii, *Na tropakh Vselennoi* (Moscow, 1996), chap. 1.

[46] Georgii T. Beregovoi et al., *Eksperimental'no-psikhologicheskie issledovaniia v aviatsii i kosmonavtike* (Moscow, 1978), 64–7.

[47] Records of the Psychology Section of the Council on Cybernetics, 1962, f. 1807, op. 1, d. 24, ll. 27–9, Archive of the Russian Academy of Sciences, Moscow.

[48] Viktor G. Denisov, "Nekotorye aspekty problemy sochetaniia cheloveka i mashiny v slozhnykh sistemakh upravleniia," in *Problemy kosmicheskoi biologii,* ed. N. M. Sisakian and V. I. Iazdovskii, vol. 2 (Moscow, 1962), 54–67, on 54.

[49] V. G. Denisov, A. P. Kuz'minov, and V. I. Iazdovskii, "Osnovnye problemy inzhenernoi psikhologii kosmicheskogo poleta," in *Problemy kosmicheskoi biologii,* ed. N. M. Sisakian and V. I. Iazdovskii, vol. 3 (Moscow, 1964), 66–79, on 77.

amplifier, an integrator, a differentiator, and a computer.[50] They described the "static and dynamic characteristics" of a human operator in terms of delay time, perception speed, reaction speed, bandwidth, and so on.[51] The "human channel capacity," for example, was estimated at 0.8 bit per second. Based on this estimate, human engineering specialists concluded that, if forced to make a decision within ten seconds, a human could take into account no more than two or three factors.[52]

The cybernetic framework effectively set a standard for evaluating human performance in machine terms. Based on quantitative evaluations, human engineering specialists argued that the human was better than the machine in intelligence, reasoning, and overall flexibility (receiving and processing diverse types of information, learning, and performing diverse tasks). The machine, however, was vastly superior in receiving and processing large amounts of information, performing precise operations, multitasking, work capacity, computation, and discarding unnecessary information.[53] Purely human qualities were seen as a mixed blessing: "The machine does not feel boredom, irritation, hesitation in decision making, apathy, fear, or lack of self-confidence. Neither does the machine possess élan, responsibility, the ability to take risks, or imagination."[54]

The psychologists concluded that the human could be either the strongest or the weakest link in the system, depending on how the functions were divided between human and machine.[55] They formulated the principle of an "active operator" and developed basic guidelines for the joint human/machine, or semiautomatic, control. Researchers recommended, for example, trusting rendezvous and repair operations to the cosmonaut and routine equipment operation to the machine.[56] If these considerations were taken into account, they argued, a human operator could increase the reliability, and in some cases reduce the weight and bulk, of onboard equipment.[57]

Although these conclusions seemed to support a greater role for the cosmonaut on board, the cybernetic framework underlying this approach fundamentally assigned the human operator a secondary role. Ultimately, the function of the human operator was to enhance the operations of machines, not the other way around.

OPERATOR TRAINING: TOWARD A PERFECT AUTOMATON

The spacecraft designers tended to be more skeptical about the human abilities in space than were the psychologists. Most engineers viewed the cosmonaut on board as a weak link, a source of potential errors. For example, Konstantin Feoktistov, the lead-

---

[50] P. K. Isakov, V. A. Popov, and M. M. Sil'vestrov, "Problemy nadezhnosti cheloveka v sistemakh upravleniia kosmicheskim korablem," in *Problemy kosmicheskoi biologii,* ed. N. M. Sisakian, vol. 7 (Moscow, 1967), 5–11, on 6.

[51] Denisov, "Nekotorye aspekty" (cit. n. 48), 55.

[52] Aleksandr I. Men'shov, *Kosmicheskaia ergonomika* (Leningrad, 1971), 14.

[53] V. N. Kubasov, V. A. Taran, and S. N. Maksimov, *Professional'naia podgotovka kosmonavtov* (Moscow, 1985), 6; Men'shov, *Kosmicheskaia ergonomika* (cit. n. 52), 11.

[54] Kubasov, Taran, and Maksimov, *Professional'naia podgotovka kosmonavtov* (cit. n. 53), 6.

[55] Men'shov, *Kosmicheskaia ergonomika* (cit. n. 52), 10.

[56] Denisov, Kuz'minov, and Iazdovskii, "Osnovnye problemy" (cit. n. 49), 67; Men'shov, *Kosmicheskaia ergonomika* (cit. n. 52), 220.

[57] Denisov, Kuz'minov, and Iazdovskii, "Osnovnye problemy" (cit. n. 49), 66–7; Isakov, Popov, and Sil'vestrov, "Problemy nadezhnosti cheloveka" (cit. n. 50), 5; Men'shov, *Kosmicheskaia ergonomika* (cit. n. 52), 237.

ing integration designer of the *Vostok* spacecraft, openly told the cosmonauts that "in principle, all work will be done by automatic systems in order to avoid any accidental human errors."[58] He put forward the principle that "every operation that can be automated on board a spaceship should be automated."[59]

The perception of human operators as unreliable was not entirely due to slow human reaction or limited memory capacity. Engineers discovered that quantitative characteristics of human activity in flight often differed from the characteristics measured during ground-training sessions. Thus the main problem was not that the human was not capable; the main problem was that the human was not fully predictable. Engineers therefore recommended that the manual control regime be used only in emergencies.[60] As one candidate cosmonaut put it, "They trusted hardware and did not trust the human being."[61]

Cosmonaut training was geared toward reducing this fundamental human unpredictability and turning the cosmonaut into a perfect machine. Korolev's Experimental Design Bureau No. 1 set up a special department to design cosmonaut activity so that it conformed to the logic of onboard automatics. Spacecraft designers viewed cosmonaut activity as auxiliary to the spacecraft's automatic control system, and therefore avoided the word "pilot" and preferred the term "spacecraft guidance operator."[62]

Spacecraft designers took to heart advice given by Igor' Poletaev, a leading Soviet cybernetics expert. He argued that the way to avoid human error was to train the human to operate like a machine: "The less his various human abilities are displayed, the more his work resembles the work of an automaton, the less [the human operator] debates and digresses, the better he carries out his task."[63] Yurii Gagarin recalled how the cosmonauts were "getting used to every button and every tumbler switch, learning all the movements necessary during the flight, making them automatic."[64] The *Vostok 5* pilot, Valerii Bykovskii, was praised in his official evaluation for "the high stability of automation of skill."[65] A cosmonaut training manual explicitly stated that "the main method of training is repetition."[66] The cosmonaut Vladimir Shatalov had to perform 800 dockings on a ground simulator before he was allowed to carry out the first manual docking of *Soyuz 4* and *Soyuz 5* in January 1969.[67] Later on, the requirement for crews training for rendezvous missions was reduced to 150 simulated dockings.[68]

The planning of cosmonaut activity in orbit was detailed and thorough. The timing and length of every action was predetermined on the ground. The control system engineer and cosmonaut Aleksei Eliseev designed a step-by-step procedure (a *cyclogram*) for a transfer from one spacecraft to another by spacewalk, which he himself carried out during the *Soyuz 4/Soyuz 5* mission. Eliseev specified all the actions and

[58] Vladimir Komarov, 1961, workbook no. 39, Gagarin Memorial Museum Archive, Gagarin, Smolensk, Russia (hereafter cited as GMMA); http://web.mit.edu/slava/space/documents.htm (accessed 28 Aug. 2006).

[59] Viktor D. Pekelis, *Cybernetic Medley,* trans. Oleg Sapunov (Moscow, 1986), 287.

[60] Kubasov, Taran, and Maksimov, *Professional'naia podgotovka kosmonavtov* (cit. n. 53), 190.

[61] Ponomareva, *Zhenskoe litso kosmosa* (cit. n. 9), 207.

[62] Kubasov, Taran, and Maksimov, *Professional'naia podgotovka kosmonavtov* (cit. n. 53), 278.

[63] Igor' A. Poletaev, *Signal: O nekotorykh poniatiiakh kibernetiki* (Moscow, 1958), 281.

[64] Yurii Gagarin, *Doroga v kosmos* (Moscow, 1961), 137.

[65] Quoted in A. N. Babiichuk, *Chelovek, nebo, kosmos* (Moscow, 1979), 209.

[66] Kubasov, Taran, and Maksimov, *Professional'naia podgotovka kosmonavtov* (cit. n. 53), 138.

[67] Vladimir A. Shatalov, *Trudnye dorogi kosmosa,* 2nd ed. (Moscow, 1981), 129.

[68] Kubasov, Taran, and Maksimov, *Professional'naia podgotovka kosmonavtov* (cit. n. 53), 138.

code words for every crew member. The procedure was recorded on a four-meter-long scroll of paper.[69]

For every deviation from the established procedure during the flight, cosmonauts received a citation. An error could be as small as flipping the wrong switch, even if it did not affect the operation of any systems. On average, two-person crews accumulated fifty to sixty citations during a several-month mission. This amounted to only one or two transgressions per person per week.[70] The cosmonauts truly achieved automaticity in their actions.

The cosmonauts occasionally complained about the "excessive algorithmization" of their activities, which, they claimed, turned them into automatons and stripped them of the possibility of planning their own actions.[71] If a cosmonaut finished a certain task before the specified time, he or she was not allowed to start the next task earlier than was specified in the cyclogram. This often led to idling, loss of valuable observation time, and waste of limited resources. During their seven-month-long stay on the Salyut-7 station in 1982, the cosmonauts Anatolii Berezovoi and Valentin Lebedev often chose to perform the most interesting experiments on their days off because on those days they could work at their own pace, without waiting for instructions from the ground.[72]

PSYCHOLOGICAL TRAINING: TOWARD TOTAL SELF-CONTROL

The psychologists who participated in cosmonaut training came largely from the field of aviation psychology, and they conceptualized cosmonaut activity in essentially the same terms as piloting. They stressed that the activities of the cosmonaut and the pilot had the following characteristics in common: (1) "continuity of work"—constant participation in controlling the most critical phases of flight, even if an autopilot is available; (2) "a mandatory or compulsory order of operations"—no change in the order of operations is allowed; the prescribed length of every operation must be followed; (3) "time deficit"—limits on flight operations, reception and processing of information from the ground and from onboard equipment; and (4) "mediated sensory inputs"—hearing is mediated by the radio, vision by optical equipment, and so on.[73]

Space psychologists had limited influence within the space program, and their advice was taken very selectively. Spacecraft designers embraced the idea of "a mandatory or compulsory order of operations," for it fitted well with their insistence on the automaticity of operations. They were skeptical, however, about the proposed parallels between piloting and cosmonaut work. The leading control system designer Boris Chertok wrote: "We, engineers who designed the control system, believed that controlling a spacecraft is much easier that controlling an aircraft. All processes are extended in time; there is always time to think things over. The craft will not suddenly break into a downward spin; . . . the laws of celestial mechanics will not let the spacecraft leave its orbit."[74] Spacecraft designers not only denied the significance of the

---

[69] Aleksei Eliseev, *Zhizn'—kaplia v more* (Moscow, 1998), 91.

[70] Kubasov, Taran, and Maksimov, *Professional'naia podgotovka kosmonavtov* (cit. n. 53), 235.

[71] Beregovoi et al., *Eksperimental'no-psikhologicheskie issledovaniia* (cit. n. 46), 31.

[72] Valentin V. Lebedev, *Moe izmerenie* (Moscow, 1994), 246–7 (diary entry of 3 Sept. 1982).

[73] F. D. Gorbov and F. P. Kosmolinskii, "Ot psikhologii aviatsionnoi do psikhologii kosmicheskoi," *Voprosy psikhologii,* no. 6 (1967): 46–58, on 49.

[74] Boris E. Chertok, *Goriachie dni kholodnoi voiny,* vol. 3 of *Rakety i liudi* (cit. n. 7), 237.

time deficit factor but also neglected the principle of continuity of work. They preferred to keep the crew in "cold reserve," passively monitoring the operations of an automatic control system. Only if the automatic system failed was the crew expected to resort to manual control. The cosmonauts complained that being in cold reserve effectively kept them out of the control loop. Without regular participation in control operations, the crew would find it exceedingly difficult to switch from passive observation to active control in case of emergency.[75]

Space psychologists described the model cosmonaut as "a human being with great self-discipline, with a high degree of self-control, capable of thinking clearly and acting decisively in uncertain situations."[76] To prepare them psychologically for the dangers of spaceflight, the trainees were flown on high-performance airplanes and helicopters, performed parachute jumping, and escaped from a submarine through the torpedo compartment. Such life-threatening exercises were meant to re-create the level of emotional tension characteristic of spaceflight.[77] After 1963, parachute jumping was no longer a requisite skill for the cosmonauts, yet, despite occasional traumas, such as broken legs, it was retained in the training program as a means of "shaping the psychological structure of the cosmonaut as practitioner in a dangerous profession."[78] Cosmonaut training was based on the principle "the safety of spaceflight through the dangers of training."[79]

Particular attention in cosmonaut training was given to psychological stability in the presence of various disturbing factors. The model cosmonaut, space psychologists argued, "must be able to pick out relevant signals, even in the presence of interfering speech."[80] Cosmonauts were trained to control spacecraft motion and monitor eight onboard systems at the same time while being distracted by constant questioning about control panel readings.[81] Gagarin was selected for the first piloted mission based on his mastery of such skills. According to his official evaluation, he "showed high precision in performing various experimental psychological tasks, high resistance to interference from sudden and strong irritants," and "the ability to control himself in various unexpected situations."[82]

Cosmonauts not only had to perform their tasks flawlessly but also, just as important, retain perfect composure under stress. During simulated docking tests, the examiners closely watched the trainees' faces. As one cosmonaut recalled, "It was imperative not merely to carry out the procedure, but to do it calmly, confidently, without visible strain."[83]

[75] Ponomareva, "Nachalo vtorogo etapa" (cit. n. 8), 170. For specific examples of how the Soviet approach to automation of control affected the course of various space missions, see Slava Gerovitch, "Human-Machine Issues in the Soviet Space Program," in *Critical Issues in the History of Spaceflight,* ed. Steven J. Dick and Roger D. Launius (Washington, D.C., 2006), 107–40.

[76] Gorbov and Kosmolinskii, "Ot psikhologii aviatsionnoi do psikhologii kosmicheskoi" (cit. n. 73), 50.

[77] Georgii T. Beregovoi et al., "Ob otsenke effektivnosti raboty cheloveka v usloviiakh kosmicheskogo poleta," *Voprosy psikhologii,* no. 4 (1974): 3–9, on 7.

[78] V. A. Dovzhenko et al., "Spetsial'naia parashiutnaia podgotovka kosmonavtov," in *Materialy XXXVII chtenii K. E. Tsiolkovskogo* (Kaluga, 2002); http://www.museum.ru/gmik/readings.htm (accessed 30 June 2005).

[79] R. B. Bogdashevskii et al., "Psikhologicheskaia podgotovka i bezopasnost' kosmicheskogo poleta," in *Materialy XXXVII chtenii K. E. Tsiolkovskogo* (Kaluga, 2003); http://www.museum.ru/gmik/readings.htm (accessed 30 June 2005).

[80] Gorbov and Kosmolinskii, "Ot psikhologii aviatsionnoi do psikhologii kosmicheskoi" (cit. n. 73), 50.

[81] Isakov, Popov, and Sil'vestrov, "Problemy nadezhnosti cheloveka" (cit. n. 50), 10.

[82] Golovanov, *Nash Gagarin* (cit. n. 34), 137.

[83] V. N. Kubasov, *Prikosnovenie kosmosa* (Moscow, 1984), 125.

The trainers tried to boost the cosmonauts' "self-control and self-regulation of action in extreme circumstances" by requiring a continuous verbal report in the course of a complex parachute jump.[84] The cosmonauts were asked to report their every action, velocity, distance to their partners, and so on.[85] By comparing the acoustic characteristics of the reportage with regular speech patterns recorded on the ground, psychologists made conclusions about the degree of stress and self-control exhibited by the trainee. To avoid being disqualified for lack of self-control, the cosmonaut candidates had to manage their vocabularies and intonations very carefully, trying to minimize the emotional element in their speech—all while performing a difficult parachute jump.

Speech control proved a very useful skill in actual spaceflight. During prolonged flights, psychologists thoroughly analyzed communication sessions to determine the degree of psychological stability of the crew and their ability to continue the flight. As Valentin Lebedev, who spent more than seven months on board the Salyut-7 station, confessed, "One must keep himself in check all the time; one must control every word."[86] He likened the need to control speech to "prolonged abstinence from intercourse: it is painful, but it has to be endured."[87]

To train cosmonauts to deal with the emotional tension of a life-threatening situation, space psychologists not only placed cosmonaut candidates in dangerous conditions but also used hypnotic experiments. Under hypnosis, experimental subjects were given the instruction, "Your life will depend on how you perform your job. Any error may lead to a catastrophe." The psychologists were hoping that this instruction would hold power even after the subject was awakened from the hypnotic state and instill greater care in the subject's subsequent actions.[88]

Space psychologists suggested that the cosmonaut must be able to endure the feeling of disconnection from the Earth, solitude, limited sensory input, and noise-ridden communications.[89] To prepare for such eventualities, the cosmonauts were confined individually to a "silence chamber" for ten to fifteen days. During the entire period, the experimental subject remained alone, isolated from any outside light, sound, or other sensory input, and limited to four one-way communication sessions a day, during which the subject sent reports but received no reply. The subject's physiological parameters were constantly monitored. The high degree of "emotional stability" displayed by Gagarin during his silence chamber test may have contributed to his selection as the first cosmonaut.[90]

Space psychologists further insisted that the cosmonaut must feel equally comfortable in an unlimited ("empty") space and in a narrowly confined space. The candidate cosmonauts were subjected to short-term zero gravity during parabolic trajectory flights on a specially equipped airplane; psychologists also successfully experimented with suggesting a state of weightlessness under hypnosis. Studies of forced limitation on body movements were also conducted by restraining subjects' limbs with multiple

[84] Dovzhenko et al., "Spetsial'naia parashiutnaia podgotovka" (cit. n. 78).

[85] Irina Solov'eva, interview with author, 9 June 2004, Zvezdnyi Gorodok (Star City).

[86] Lebedev, *Moe izmerenie* (cit. n. 72), 281 (diary entry of 19 Sept. 1982).

[87] Ibid., 272 (diary entry of 15 Sept. 1982).

[88] Georgii T. Beregovoi and Andrei I. Iakovlev, *Modelirovanie system poluavtomaticheskogo upravleniia kosmicheskikh korablei* (Moscow, 1986), 59.

[89] Gorbov and Kosmolinskii, "Ot psikhologii aviatsionnoi do psikhologii kosmicheskoi" (cit. n. 73), 49.

[90] I. B. Ushakov, V. S. Bednenko, and E. V. Lapaeva, eds., *Istoriia otechestvennoi kosmicheskoi meditsiny* (Voronezh, Russia, 2001), chap. 16.

belts or plaster casts. In contrast to the ecstatic feelings expressed during weightlessness sessions, movement restriction tests led to a "severe psychological condition."[91] Yet one of the most spectacular Soviet space feats—the 1964 launch of a three-man crew on board the *Voskhod* spacecraft—was achieved by exploiting the cosmonauts' ability to operate in a narrow space. Instead of designing a larger spacecraft, Korolev's engineers fitted three cosmonaut seats side by side in the space previously occupied by only one cosmonaut on *Vostok*. As a result, the three cosmonauts had five times less space and air per capita than had the *Vostok* cosmonauts.[92]

In a certain way, the extraordinary trials that the cosmonauts went through during their training and actual flights presented in a concentrated form some familiar experiences of Soviet citizens. As social life in the Soviet Union was highly regulated, the cosmonauts' activity was also subjected to "a mandatory or compulsory order of operations." As Soviet citizens scrambled to find grains of information in the propaganda-filled official discourse, the cosmonauts trained to "pick out relevant signals" in the presence of noise. As Soviet citizens were virtually isolated from the outside world, the cosmonauts endured isolation tests in a silence chamber. As the secret police and an army of informers constantly watched ordinary citizens, ready to persecute them for any sign of political disloyalty, physicians constantly monitored the cosmonauts' physiological and psychological parameters, ready to disqualify anyone who showed a deviation from the norm. Most important, as Soviet citizens had to constantly watch themselves not to allow any slip, the cosmonauts had to exercise ultimate self-control, carefully choosing every action and every word. Like ordinary Soviet citizens, the cosmonauts had to follow the rules. As one candidate cosmonaut has observed, "The social behavior of the Soviet man is strictly regulated; similarly, for the cosmonauts instructions and guidelines of various sorts play a very significant role."[93]

A model cosmonaut was truly an exemplary Soviet citizen. Yet the cosmonauts' resentment toward the excessive algorithmization of their activity and their efforts to preserve their professional identities as pilots made them less than perfect candidates for the public embodiment of the new Soviet man. To turn the cosmonauts into walking emblems of the communist self, their political overseers had to reshape their public personas, just as the engineers and the psychologists had remodeled the cosmonauts' bodies and minds.

SHAPING THE COSMONAUTS' PUBLIC IDENTITY

If before the flight the cosmonauts' training was largely technical, their activity after the flight was to a large extent political. For many months after completing their space missions, Gagarin, Gherman Titov, and the other first cosmonauts toured the world, serving as "agitators for communism." Their visits had a particular political importance in the third world, where their public appearances were carefully planned to support pro-Soviet politicians. During a trip with his wife to India, Gagarin privately complained to Kamanin that their schedule was overloaded: "Too much politics, and nothing for ourselves; we did not even see any elephants."[94] In the course of one day of his visit to Ceylon, for example, Gagarin traveled more than 300 miles, visited nine

[91] Gorbov and Kosmolinskii, "Ot psikhologii aviatsionnoi do psikhologii kosmicheskoi" (cit. n. 73), 51.
[92] Siddiqi, *Challenge to Apollo* (cit. n. 3), 416.
[93] Ponomareva, "Nachalo vtorogo etapa" (cit. n. 8), 170.
[94] Kamanin, *Skrytyi kosmos* (cit. n. 37), 1:76 (diary entry of 7 Dec. 1961).

towns, and gave more than fifteen speeches. During his numerous foreign trips, he endured a total of nearly 150 days of such political marathons. No wonder he considered worldwide fame his heaviest burden.[95] During 1961–1970, the cosmonauts made 200 trips abroad.[96] The first woman in space, Valentina Tereshkova, alone made 42 foreign trips; she was able to escape the political speech circuit only when she was several months into her pregnancy.[97]

Groomed by the Soviet political leadership to serve as ideological icons of communism, cosmonauts had to appear personally at all major public forums inside the country. Kamanin, who oversaw the cosmonauts' schedules, received dozens of requests daily. In 1961–1970, the cosmonauts attended more than 6,000 public events in the Soviet Union.[98] Kamanin carefully scripted their public appearances, rehearsed their speeches, and corrected their "mistakes." He took upon himself not only the formal supervision of the cosmonauts' selection and training but also their moral bearing. Kamanin did not spare any effort to make the cosmonauts conform to their public images as exemplary Soviet citizens, scolding them for marital troubles and withholding their promotions in rank for drunken driving incidents. His own role as a famous aviator and a public icon in the 1930s served as a model for his efforts to shape the cosmonaut self.

Under Kamanin's supervision, the Cosmonaut Training Center introduced a program of enculturation to broaden the fighter pilots' intellectual horizons. The cosmonauts went on group trips to museums, art galleries, and historic sights, visited the Bolshoi and other theaters, and attended concerts by performers from Czechoslovakia, Cuba, and the United States. They listened to lectures about ancient Greece and Rome, the Renaissance men, Peter the Great, and famous Russian painters and opera singers.[99] Political education was made part of the formal curriculum. The first group of six candidate cosmonauts, including Gagarin, received forty-six hours of instruction in Marxism-Leninism, 8 percent of their total training time.[100] Any overt sign of political dissent was quickly suppressed. After one trainee refused to give a ritual speech at a party meeting and told a senior party official, "I will not speak to a Party of swindlers and sycophants!" he was immediately expelled from the cosmonaut corps.[101]

The attempts to make the cosmonauts into exemplary communists proceeded with difficulty. Cosmonauts privately exchanged political jokes, such as the double entendre slogan "Officers of the Missile Forces, Our Target Is Communism!" Even some of their supervisors laughed at ideological clichés. One cosmonaut recalled that the deputy director of the center in charge of political education "understood everything, believed that the cosmonauts would not give him away, and did not make pretenses with us. . . . When asked 'How are things?' he invariably replied, 'Our country is on the rise.' If we mockingly asked 'And how is the Party?' he replied with an equal measure of irony, 'The Party teaches us that heated gases expand.'"[102]

In 1961, Gagarin and Titov were elected deputies of the Twenty-Second Congress

[95] Golovanov, *Nash Gagarin* (cit. n. 34), 207, 211, and 183.
[96] Nikolai Kamanin, *Skrytyi kosmos,* vol. 4, *1969–1978* (Moscow, 2001), 252 (diary entry of 23 Jan. 1971).
[97] Kamanin, *Skrytyi kosmos* (cit. n. 37), 1:399 (diary entry of 21 Dec. 1963).
[98] Kamanin, *Skrytyi kosmos* (cit. n. 96), 4:252 (diary entry of 23 Jan. 1971).
[99] Golovanov, *Nash Gagarin* (cit. n. 34), 56.
[100] "Akt o rezul'tatakh ekzamenov," 18 Jan. 1961, GMMA.
[101] Siddiqi, *Challenge to Apollo* (cit. n. 3), 817.
[102] Eliseev, *Zhizn'—kaplia v more* (cit. n. 69), 120, 93.

of the Communist Party. The congress would adopt a new party program, which set a triple goal of creating a material and technical basis of communism, forming the new communist social relations, and bringing up the new Soviet man. The "Moral Code of the Builder of Communism," to be approved by the congress, included the ethical imperatives of honesty, sincerity, moral purity, modesty, and conscientious work. Gagarin and Titov were supposed to sit in the presidium of the congress and to show-case the tangible achievements of the regime both in high technology and in the up-bringing of the new man. The plans went awry, however, when a few days before the congress, Gagarin broke a facial bone jumping out of the window following a wom-anizing incident. Gagarin missed the opening of the congress, and he and Titov were dropped from the presidium list. Khrushchev was furious when he learned about the behavior of Gagarin, next to whom he had stood on top of Lenin's mausoleum during the May Day celebrations just a few months before.[103] Gagarin's transgressions were soon forgiven, however, and he was sent on a propaganda mission abroad, escorted by his wife and by Kamanin.

As the popularity of the cosmonauts grew, it became more and more difficult for Ka-manin to control their behavior. They were well known around the world as the public face of the space program. Because of the shroud of secrecy that surrounded Soviet rocketry, the leading designers of spacecraft, such as Sergei Korolev, remained anony-mous, and the public tended to view human space flights as the cosmonauts' personal achievements. Not all the cosmonauts could carry the burden of celebrity with honor. Kamanin bitterly complained in his private diary that "the cosmonauts overestimate the significance of their personal accomplishments and take at face value everything that is written, said, and shown about every human spaceflight in the media."[104]

Both foreign and domestic audiences viewed the cosmonauts as an emblem of the Soviet regime. For many people around the world, these young, enthusiastic, and technically proficient people were a living embodiment of the Soviet communist dream. "For all of us, Yurii [Gagarin] personified the whole generation of Soviet people, whose childhood was singed by the war," recalled one cosmonaut.[105] In the 1960s, more than thirty feature movies were made and hundreds of books and news-paper articles written about the cosmonauts, all extolling their glorious achievements.

More controversial representations of human space exploration were discour-aged.[106] No information about equipment failures or crew mistakes during space mis-sions was publicly released. As a result, Kamanin admitted, "people get the impres-sion of 'extraordinary ease' and almost complete safety of prolonged space flights. In fact, such flights are very difficult and dangerous for the cosmonauts, not only physi-cally but also psychologically."[107]

The cosmonauts found it difficult to reconcile their professional selves with the ideal public images assigned to them. The role of a public hero whose feats suppos-edly did not involve any danger was uncomfortable for the cosmonauts originally trained as fighter pilots. Most of them preferred training for new space flights to public

[103] Kamanin, *Skrytyi kosmos* (cit. n. 37), 1:59–60 (diary entries of 4–17 Oct. 1961).
[104] Kamanin, *Skrytyi kosmos* (cit. n. 96), 4:116–7 (diary entry of 11 Jan. 1970).
[105] Golovanov, *Nash Gagarin* (cit. n. 34), 281.
[106] Kamanin refused to serve as consultant for Andrei Tarkovsky's movie *Solaris* because, as he ex-plained, such fiction "belittles human dignity and denigrates the prospects of civilization." Kamanin, *Skrytyi kosmos* (cit. n. 96), 4:152 (diary entry of 18 April 1970).
[107] Ibid., 4:182 (diary entry of 6 June 1970).

appearances. Valentina Tereshkova long resisted Kamanin's attempts to turn her into a professional politician and even entered the Air Force Engineering Academy, hoping to retain her qualifications for another spaceflight. Kamanin was convinced, however, that "Tereshkova as the head of a Soviet women's organization and of international women's organizations would do for our country and for our Party a thousand times more than she can do in space."[108] Eventually he prevailed, and Tereshkova left the cosmonaut corps and served as the head of the Soviet Women's Committee for more than twenty years.

The Soviet propaganda portrayal of heroism without risk was not the only inner contradiction that plagued the cosmonaut self. The organization of spacecraft control had profound implications for the cosmonauts' professional identities. Both the cosmonauts and the flight controllers struggled with the question of how to find a proper balance between personal courage, ingenuity, and creativity and the need to follow a strict sequence of flight operations.

PARADOX OF DISCIPLINED INITIATIVE

Korolev's design bureau was responsible not only for the design and construction but also for the operation of piloted spacecraft during the flight. Spacecraft designers therefore tended to view the cosmonauts as their subordinates. One of Korolev's leading engineers, who later headed the bureau, explained that the managers expected the cosmonauts "to carry out their prescribed tasks just like other employed specialists."[109]

From the very beginning, Soviet spacecraft designers adopted the principle that they have followed to this day: all critical systems have three independent control channels: automatic, remote (from the ground), and manual.[110] Control during the three main stages of the flight—reaching the orbit, orbital flight, and reentry—is automatic; instructions to switch programs between the stages are given either from the ground or manually by the cosmonaut. The cosmonaut, however, has to obtain permission from the ground for any critical action. The norms of cosmonaut activity therefore include not only a technical protocol of interaction with onboard equipment but also a social protocol of subordination to their superiors on the ground. A cosmonaut training manual clearly stipulates that "all the most important decisions are made by Mission Control."[111]

Spacecraft designers believed that comprehensive automation and the strict following of instructions by the crew would best guarantee flight safety, but the cosmonauts pointed out that it was often necessary to break the rules in an emergency. Although the engineers tended to regard any departure from the standard procedure as a human error, it was precisely this ability to deviate from the standard path that made human presence on board so valuable in an emergency situation.

During a space mission, cosmonauts often found themselves in situations unforeseen by mission planners on the ground, situations to which the original instructions did not apply. The crew then faced a dilemma: to follow the rules and fail the mission, or to take risks and break the rules. Such an emergency occurred, for example, during

[108] Kamanin, *Skrytyi kosmos* (cit. n. 37), 1:332 (diary entry of 1 Aug. 1963).
[109] Iu. P. Semenov, "Slovo monopolistu," *Aviatsiia i kosmonavtika,* no. 6 (1991): 40-41, on 41.
[110] Vladimir S. Syromiatnikov, *100 rasskazov o stykovke i o drugikh prikliucheniiakh v kosmose i na Zemle,* vol. 1, *20 let nazad* (Moscow, 2003), 145.
[111] Kubasov, Taran, and Maksimov, *Professional'naia podgotovka kosmonavtov* (cit. n. 53), 190.

the *Voskhod 2* flight in March 1965. After completing his historic space walk, the cosmonaut Alexei Leonov realized that his spacesuit had ballooned (his arms and legs were not even touching the inside of the suit), making it impossible for him to reenter the airlock. He was supposed to report all emergencies to the ground and wait for instructions. He later recalled: "At first I thought of reporting what I planned to do to Mission Control, but I decided against it. I did not want to create nervousness on the ground. And anyway, I was the only one who could bring the situation under control."[112] He may have calculated that various bureaucratic procedures and possible reluctance of some managers to take responsibility could critically delay vital decisions, and it would be unwise for him to spend his limited oxygen supply waiting for them. Leonov turned a switch on his spacesuit, drastically reducing the internal air pressure, which allowed him to regain control of his movements. Yet he was still unable to enter the airlock feet first, which was required in order to squeeze into the landing module. Once Leonov had broken one rule, he decided that he could not make the situation worse by breaking another, so he climbed into the airlock head first, in violation of established procedure. He then performed an incredible acrobatic feat by turning around inside a narrow airlock.

The *Voskhod 2* crew—Leonov and Pavel Beliaev, both military pilots—had been trained to follow the rules and to obey orders from the ground. After more than 150 training sessions on a space walk simulator, Leonov was said to have brought his skills "to the point of automatic performance."[113] Yet in a real emergency, Leonov had to perform actions for which he was not trained, to violate explicit rules concerning entry into the airlock, and to make decisions without consulting Mission Control. He thus ensured the success of his mission by *not* acting like a perfect machine.

Control system designers realized that there was a tension between centralized control and the need to maintain what they called "relative autonomy of subsystems and even individual elements,"[114] one such element being the cosmonaut. One of the walls in Korolev's design bureau was adorned with a 1910 memo of the prerevolutionary Russian Navy Engineering Committee: "No manual can enumerate all the responsibilities of an official, account for all individual cases, and provide full instructions ahead of time. For this reason, gentlemen engineers must show initiative and, guided by their specialized knowledge and consideration for the common good, must apply every effort to justify their vocation."[115] "This recommendation," Korolev's deputy Boris Chertok argued in 1972, "holds true today both for the engineers who control space systems and for the cosmonauts who control spacecraft."[116]

While encouraging initiative, mission planners also made it very difficult for space crews to deviate from their instructions. During their mission on the Salyut-7 station in 1982, the cosmonauts Anatolii Berezovoi and Valentin Lebedev showed remarkable ingenuity in fixing malfunctioning equipment and conducting scientific experiments

[112] David R. Scott and Alexei A. Leonov, *Two Sides of the Moon: Our Story of the Cold War Space Race* (London, 2004), 109.

[113] N. N. Gurovskii et al., "Trenazhery dlia podgotovki kosmonavtov k professional'noi deiatel'nosti po upravleniiu korablem i ego sistemami," in *Problemy kosmicheskoi biologii,* ed. N. M. Sisakian, vol. 4 (Moscow, 1965), 3–9, on 6; Siddiqi, *Challenge to Apollo* (cit. n. 3), 451.

[114] B. Evseev [Boris Chertok], "Chelovek ili avtomat?" in *Shagi k zvezdam,* ed. M. Vasil'ev [Vasilii Mishin] (Moscow, 1972), 281–7, on 286.

[115] Ibid., 284–5; Kubasov, *Prikosnovenie kosmosa* (cit. n. 83), 123; Lebedev, *Moe izmerenie* (cit. n. 72), 258 (diary entry of 8 Sept. 1982).

[116] Evseev [Chertok], "Chelovek ili avtomat?" (cit. n. 114), 284–6.

that would have otherwise been canceled. Yet they received advice from the ground "to do less improvisation": their performance was evaluated not by the amount of research successfully done on board, but by the number of minor errors, which they occasionally made by trying their innovations. "Here is a paradox," wrote Lebedev in his diary. "If we had not improvised . . . and just followed orders and instructions, the end result would have been worse, but we would not have had any citations."[117] The engineer cosmonaut Valerii Kubasov made up a list of ten "cosmonaut's commandments," two of which perfectly illustrated the ambivalence of mission planners about cosmonauts' initiative: "always try to consult with Mission Control, but also take your own initiative"; "initiative is good, but always try to stick to the rules, otherwise you will be considered undisciplined, and your grades will be lowered."[118]

The cosmonaut self thus fractures into two barely compatible parts: an active, autonomous agent and a disciplined subordinate. Valentina Ponomareva, a member of the first women's cosmonaut group, has captured this contradiction in her vision of the model cosmonaut:

> The requirements [for being a cosmonaut] are very strict. They include readiness to take risks, the sense of highest responsibility, the ability to carry out complex tasks in harsh conditions, high dependability of the operator's work, advanced intellectual abilities, and physical fortitude. . . . In addition, the cosmonaut must possess such qualities as curiosity and the ability to break rules. . . . Regulations work well only when everything goes as planned. . . . The ability to act in extraordinary situations is a special quality. In order to do that, one has to have inner freedom . . . the ability to make nontrivial decisions and to take nonstandard actions. In an extreme situation, the very life of the cosmonaut depends on these qualities.[119]

Despite her high qualifications as an engineer and a pilot and her excellent test marks, Ponomareva was not selected for the first woman's flight, and she never got a chance to fly. In his private diary, Kamanin admitted that two female candidates, Ponomareva and Irina Solov'eva, were better prepared for the mission than Tereshkova. Yet they "would never be able to compete with her in the skill of influencing the crowd, in the ability to attract warm sympathies of people, or in the readiness to speak well before any audience. These qualities of Tereshkova determined her selection as the first female cosmonaut."[120] The first female cosmonaut's public persona proved more important than her professional skills.

The need for the cosmonaut to be both obedient and creative, to follow the rules and to break them, one might call a paradox of "disciplined initiative." The historian Sonja Schmid, in her study of Soviet nuclear power station operators, observed a similar contradiction in the way the operators were viewed by nuclear reactor designers: both as a "weak link" and a "reliable cog in the wheel."[121] Both spacecraft designers and nuclear engineers viewed the human operator as part of the technology, which must always function according to the rules, yet at the same time they expected the operators to show human qualities such as initiative and inventiveness.

[117] Lebedev, *Moe izmerenie* (cit. n. 72), 258 (diary entry of 8 Sept. 1982).
[118] Kubasov, *Prikosnovenie kosmosa* (cit. n. 83), 123.
[119] Ponomareva, *Zhenskoe litso kosmosa* (cit. n. 9), 285.
[120] Kamanin, *Skrytyi kosmos* (cit. n. 37), 1:391 (diary entry of 28 Nov. 1963).
[121] Sonja Schmid, "Reliable Cogs in the Nuclear Wheel: Assigning Risk, Expertise and Responsibility to Nuclear Power Plant Operators in the Soviet Union" (paper presented at The Forty-Fifth Meeting of the Society for the History of Technology, Amsterdam, the Netherlands, 7–10 Oct. 2004).

The leaders of the Soviet space program constantly vacillated between belief in the power of technology and trust in human skill and creativity. Echoing the duality of Stalin's old slogans, Dmitrii Ustinov, the secretary of the Party Central Committee in charge of the military-industrial complex, told top space managers in February 1971: "One should not jump to the extremes—either the human decides everything or the machine does. . . . The human must not enter into competition with the machine in pressing keys on a keyboard but must engage in research, in discovery, where his creative faculties and brain abilities are needed most." He acknowledged that it was difficult to take advantage of human creativity on fully automated spaceships: "We have not been using these [creative] capabilities in space."[122]

One could suggest that this paradox reflected a fundamental contradiction in the Soviet approach to the role of the human in large technological systems and perhaps more broadly to social control and government. According to the "Moral Code of the Builder of Communism," a model Soviet citizen was expected to be an active member of society and to take "an uncompromising attitude" toward any injustice or insincerity. At the same time, an exemplary citizen was supposed to have "a strong sense of social duty."[123] As the historian Polly Jones has noted, two opposite trends paradoxically combined in the Khrushchev era: the new emphasis on individual identity, personal well-being, and private freedoms was held in check by the policy of mass mobilization to participate in public events and collective action.[124] Although Stalinism was followed by a political thaw, the Soviet ideological discourse preserved its signature trait—fundamental ambivalence. The new man had to be both an active agent of change and a disciplined member of the collective, dutifully fulfilling orders.

CONCLUSION: NEW SOVIET MAN MEETS AMERICAN HERO

The communist ideal of the 1960s was imagined as a "harmonic merger" of a technological utopia, the construction of a material and technical basis of communism, and a humanist utopia, the creation of the spiritually fulfilled new Soviet man. The tension between the two parts of this project—technological and human—can be traced throughout Soviet history. Early Bolshevik ideas of the "machinization of man" paradoxically combined traditional images of machinery as an exploitative force and futurist visions of a creative merger of workers and machines.[125] A similar field of ideological tension was maintained in the 1930s by Stalin's dual slogans, "In the reconstruction period, technology decides everything!" and "Cadres decide everything!" The aviator hero, who personified Stalin's new Soviet man, also had a split self: both a distinct individual and a little cog, a master of technology and a part of the machine.

In the space age, the old tensions resurfaced in the debates over automation of spacecraft control. The division of function between human and machine on board determined the cosmonauts' degree of autonomy in the control of their missions and, more broadly, both reflected and shaped the role of the cosmonaut corps within the

---

[122] Boris E. Chertok, *Lunnaia gonka,* vol. 4 of *Rakety i liudi* (cit. n. 7), 249.

[123] *Materialy XXII s'ezda* (cit. n. 1), 411.

[124] Polly Jones, introduction to *The Dilemmas of De-Stalinization: Negotiating Cultural and Social Change in the Khrushchev Era,* ed. Polly Jones (London, 2006), 1–18, on 9.

[125] See Richard Stites, *Revolutionary Dreams: Utopian Vision and Experimental Life in the Russian Revolution* (New York, 1989), chap. 7, "Man the Machine."

space program. The cosmonaut identity itself was constructed as part of spacecraft control system design.

The attempts to appropriate the cosmonaut as an exemplar of the new Soviet man revealed that the chosen model was far from perfect. The cosmonauts resisted their transformation into propaganda icons, just as they resisted their full integration into a technological system. Perhaps they appealed to ordinary Soviet people precisely because they were not perfect embodiments of ideology but living beings with their own thoughts and doubts.

While the Soviets designed the cosmonaut as a prototype for the new Soviet man, the Americans turned their astronauts into public icons as well. As the historian Roger Launius observed, "Both NASA officials and the astronauts themselves carefully molded and controlled their public images every bit as successfully as those of movie idols or rock music stars." Combining youth, vigor, playfulness, and virile masculinity, the astronaut image represented the American ideal, the quintessential American hero. The astronauts served as "surrogates for the society that they represented."[126]

Soviet space engineers and cosmonauts often regarded the U.S. space program as the paragon of a human-centered approach to spacecraft design. One of Korolev's deputies, for example, remarked: "Americans rely on the human being, while we are installing heavy trunks of triple-redundancy automatics."[127] Yet the Soviet perception of the American emphasis on manual control was to a large extent based on a myth. In fact, the astronauts did not manually fly their spacecraft to the Moon and back. As the historian David Mindell has shown, a tight coupling of the crew and the onboard computer was required for effective control of Apollo operations. The astronaut served as "a systems manager, coordinating a variety of controls as much as directly controlling himself." Working in close contact with flight controllers on the ground, the astronauts carried out such crucial operations as spacecraft docking and lunar landing via the computer. Dealing with a computer alarm in the final moments of the *Apollo 11* lunar landing, Neil Armstrong performed the landing manually, and NASA "narrated the landing as the victory of a skilled human operator over fallible automation—a result that highlighted the heroic goals of the program."[128] In fact, as MIT engineers later pointed out, the crew had failed to turn off a switch, which led to computer overload and produced an alarm signal.

Just like the cosmonauts, the astronauts were working within a complex technological system, and their actions were strictly regulated and controlled from the ground. Both American and Soviet engineers chose to rely on automation, even though the means of automation in the American case (the computer) proved more complex and versatile. The cybernetic vision of human-machine merger gave rise to the notion of "cyborg," first formulated by U.S. space psychologists and also contemplated by Soviet physicians.[129] Although they did not resort to cyborglike modifi-

---

[126] Roger D. Launius, "Heroes in a Vacuum: The Apollo Astronaut as a Cultural Icon," AIAA Paper 2005-702 (paper presented at the Forty-Third Aerospace Sciences Meeting and Exhibit of the American Institute of Aeronautics and Astronautics, Reno, Nevada, 10–13 Jan. 2005); http://klabs.org/history/roger/launius_2005.pdf (accessed 30 June 2005), 1, 10.

[127] Sergei Okhapkin, as quoted in Chertok, *Goriachie dni kholodnoi voiny* (cit. n. 74), 257.

[128] David Mindell, "Human and Machine in the History of Spaceflight," in Dick and Launius, *Critical Issues in the History of Spaceflight* (cit. n. 75), 141–62, on 153 and 158.

[129] See Manfred E. Clynes and Nathan S. Kline, "Cyborgs and Space" (1960), in *The Cyborg Handbook,* ed. Chris Hables Gray (New York, 1995), 29–33; N. Sisakian, "Biologiia i osvoenie kosmosa," *Aviatsiia i kosmonavtika,* no. 2 (1962): 24–30.

cations of the human body, both American and Soviet specialists in "human engineering" took an active part in reshaping the space explorer's self. Firm discipline and the ability to function as part of control machinery were equally important for the cosmonauts and the astronauts. In different political contexts, however, the same professional qualities were reinterpreted to build two opposite ideological constructs—the American "right stuff" and the new Soviet man.

Ideological declarations of the cold war rivals differed, but the figures they chose to represent those declarations proved remarkably similar. Both sides viewed the space race as a proxy for the cold war, and both sides chose to personify the technological competition with a human space explorer. "From a larger perspective, our designers are probably right in their intention to create fully automated piloted spaceships," grudgingly admitted Kamanin in his private diary. "Perhaps in the future, when communism triumphs over the entire planet, people will fly into space on such ships. But in our time, one must not forget about the severe struggle between two opposing ideologies."[130] In the U.S. and in the Soviet Union, the main reasons for building piloted ships were political, rather than technological or scientific. Instead of showcasing the difference of ideologies, the appropriation of the cosmonauts and astronauts as public icons illustrated the similar dependence of the two superpowers on the cold war mindset.

[130] Kamanin, *Skrytyi kosmos* (cit. n. 36), 3:348 (diary entry of 28 Dec. 1968).

Supervising Spoiled Selfhood:

Inquiry and Interpretation in the History of Modern American Child Adoption

*By Ellen Herman**

ABSTRACT

This article uses child adoption to explore the science and the politics of selfhood in modern U.S. history, especially forms of spoiled selfhood associated with difference and deviance. In particular, it traces the debate about whether and how parents should inform children of their adoptive status. Why was "telling" both imperative and troublesome? What theories about normal and abnormal identity and belonging were implicated in the practice?

The case of the adopted self suggests why child development and family life became the subjects of ambitious legal intervention, professional management, and research in the human sciences during the twentieth century. Normalization was a favored governmental activity that also exposed a central tension between liberalism, on the one hand, and conceptions of nature at odds with it, on the other.

INTRODUCTION

Child adoption illustrates a novel phenomenon in modern U.S. history: the spread of social operations devoted to negotiating the distance between stigmatized and normal identities. More than four decades ago, sociologist Erving Goffman pointed out that managing conflicts between subjective experiences of difference and social expectations of conformity "spoils . . . social identity; it has the effect of cutting [the individual] off from society and from himself so that he stands as a discredited person facing an unaccepting world."[1] In cases in which deviance from the norm could be kept secret—homosexuality and drug addiction, for instance—questions about disclosure, passing, and covering were chronic: "To display or not to display; to tell or not to

* Department of History, University of Oregon, Eugene, OR 97403; eherman@uoregon.edu.

Many thanks to Greg Eghigian, Andreas Killen, and Christine Leuenberger, who organized the symposium titled "The Self as Scientific and Political Project in the Twentieth Century: The Human Sciences between Utopia and Reform," sponsored by the National Science Foundation and Pennsylvania State University in October 2003. Archival sources for this article include: Child Welfare League of America Papers, Social Welfare History Archives, University of Minnesota, Minneapolis, Minn. (hereafter cited as CWLA Papers); Dorothy Hutchinson Papers, Rare Book and Manuscript Room, Columbia University Library, New York (hereafter cited as Hutchinson Papers); U.S. Children's Bureau Papers, National Archives II, College Park, Md., (hereafater cited as USCB Papers); Viola Wertheim Bernard Papers, Archives and Special Collections, Augustus C. Long Health Sciences Library, Columbia University, New York (hereafter cited as VWB Papers).

[1] Erving Goffman, *Stigma: Notes on the Management of Spoiled Identity* (Englewood Cliffs, N.J., 1963), 19.

tell; to let on or not to let on; to lie or not to lie; and in each case, to whom, how, when, and where."[2]

Decisions such as these transformed the self into a permanent object of surveillance by internal and external agencies alike—a "project" saturated in purposeful effort. Self-presentation, Goffman argued, was a drama rooted in "the arts of impression management," and the most assiduous performers frequently had numerous biographies.[3] Possessing more than one narrative to tell about the self was convenient for moving between social situations in which one was known and others in which one was unknown or merely known about. The moral careers of stigmatized subjects, Goffman concluded, illustrated how individuals and social institutions choreographed multiple, malleable, frequently conflicting appearances to camouflage signs of difference. They exemplified a basic challenge facing any social order: enlisting active support from those who were marginalized and therefore stood to gain more from dissent than from consent. In addition, the heavy emotional toll that stigma exacted suggested how powerfully cultural norms, themselves subjects of ongoing historical revision, shaped selfhood. That so many nonconforming people aspired to be or become persons who met the standard expectations of their time and place was not only a paradox. It was also a major problem for social theory.[4]

Goffman confined his discussion to adults, never mentioning adoptees or adoption, but his observations are relevant. Throughout the twentieth century, dominant ideologies of kinship in the United States stressed blood as the measure of all that was normal, natural, and real in family life, as in "blood is thicker than water."[5] Even during the second half of the nineteenth century, when states enacted adoption laws, many nonrelative adoptions did not attempt to mimic blood-based ties or make them appear "as if begotten." Adoptions often had as much to do with labor as with love and involved apprentice- and indenture-like arrangements whose benefits were practical and material: promises of housing, clothing, food, and skills. When relatives adopted—grandparents, aunts, and uncles, for instance—the logic of blood was reinforced. Well into the Progressive Era, adoption persisted as a way to preserve and re-create families as well as make them from scratch.[6] Relative adoptions were always numerous, and they have increased substantially since 1970, as divorce and remarriage have made stepparent adoptions more common.

Adoptions that made families up without benefit of blood, by forging new and permanent bonds of affection—the type of kinship most people describe as "adoption"— were pioneered by infertile couples.[7] The theory that "love makes a family," so that adoptive ties were equivalent to genetic connections, was still unusual at the dawn of the twentieth century, when children's value was in transition from productivity to

[2] Ibid., 42.

[3] Erving Goffman, *The Presentation of Self in Everyday Life* (New York, 1959).

[4] More recently, James Scott has built on Goffman's dramaturgical foundation to explain why open defiance of domination has been historically rare while indirect and daily forms of resistance have been common. Scott's analysis rests heavily on examples from caste, slave, and totalitarian societies, and its application to Western liberal democracies remains a matter of debate. James C. Scott, *Domination and the Arts of Resistance: Hidden Transcripts* (New Haven, Conn., 1980).

[5] Judith S. Modell, *Kinship with Strangers: Adoption and Interpretations of Kinship in American Culture* (Berkeley, Calif., 1994); David M. Schneider, *American Kinship: A Cultural Account,* 2nd ed. (Chicago, 1980).

[6] Chris Guthrie and Joanna L. Grossman, "Adoption in the Progressive Era: Preserving, Creating, and Re-Creating Families," *American Journal of Legal History* 43 (July 1999): 235–53.

[7] Barbara Melosh, *Strangers and Kin: The American Way of Adoption* (Cambridge, Mass., 2002).

pricelessness. On the one hand, adoption's modernization was linked to novel ideas about the innocence of childhood and its susceptibility to physical injury and emotional abuse, themselves closely tied to the emergence of a self-conscious middle class in an increasingly urbanized and bureaucratized industrial capitalist society.[8] On the other hand, this modernization built on expectations, in play since the Civil War, that the federal government was obligated to maintain public health and welfare, with special attention to the weak. Who could be weaker than dependent infants and children in need of parents?

Transitions from traditional to modern ideologies of childhood and liberal government mandated two new ideas: a conception of the child as exceptionally vulnerable in tandem with a conception of the state as exceptionally forceful. These cultural transformations were uneven and partial. Although many Americans welcomed "the century of the child" (a slogan made famous by Ellen Key's 1900 international best seller by that name), others remained convinced that countless children and parents were eugenically unfit and undeserving of aid, that household economies required children to work, and that centralized power over child and family life interfered unjustly with parental authority. All were stubborn obstacles to governing childhood rigorously from the center.

Orphans and children whose parents could not or would not care for them epitomized the helplessness that undergirded public obligations to children. Even heartrending images of babies in dire need, however, did not answer the question of how modern adoption could be governed without distorting the delicate balance between public power and private freedom. During the Progressive Era, turning strangers into kin through legal means alone was considered risky and unnatural, a process to be avoided whenever possible. Even the pioneers of professional child welfare never suggested that adoption should be anything but rare. Turning strangers into kin might help a small number of children deprived of parents by terrible tragedy or extreme depravity, but it was far less desirable than growing up in one's "own" family. For parentless children, nonrelative adoption was a last resort when relatives could not or would not assume responsibility. For childless couples, it was a last resort after the preferred method of biogenetic reproduction failed.

Over the course of the twentieth century, adoption retained its second-class status even as it came to symbolize broader trends in American kinship. "Love makes a family" eventually extended far beyond adoption to define virtually all forms of kinship as exercises in deliberate and planned nurture, yet the view of adoption as an inauthentic substitute for real family has endured. Since 1970, genetic research has surged and the foster care population has ballooned with children whose "special needs" mark them as the unwanted offspring of an underclass mired in crime, drug addiction, and generations of poverty and hopelessness. In an era in which new reproductive technologies have reconfigured relationships between genes and gestation in ways that accentuate biology while subjecting it to previously unimaginable levels of human intervention, the "new biologism" has kept adoption decidedly "second best."[9]

---

[8] Anne Higonnet, *Pictures of Innocence: The History and Crisis of Ideal Childhood* (New York, 1998); Viviana A. Zelizer, *Pricing the Priceless Child: The Changing Social Value of Children* (Princeton, N.J., 1985); Robert H. Wiebe, *The Search for Order, 1877–1920* (New York, 1967).

[9] Arlene Skolnick, "Solomon's Children: The New Biologism, Psychological Parenthood, Attachment Theory, and the Best Interests Standard," in *All Our Families: New Policies for a New Century,* ed. Mary Ann Mason, Arlene Skolnick, and Stephen D. Sugarman (New York, 1998), 236–55. See also

In vitro fertilization, embryo transfer, and sperm sorting have intensified the quest for "own" children and reinforced the allure of biological connection even as they raise startling new questions about the meaning of family and nature.

What little evidence we have about public opinion on adoption suggests that images of fortunate children and parents coexist with serious doubts about outcomes, an ambivalent mixture of idealization and skepticism. A majority of Americans in the late 1990s reported favorable attitudes about adoption, and few disagreed that it met urgent social needs. Equally prevalent, however, were convictions that adoptees were prone to academic, emotional, behavior, substance abuse, and medical problems. Americans also believed that adoptees were less likely to achieve happiness, self-confidence, and adjustment.[10] Adoption has become more familiar over time: personal contacts and exposure to adoption stories in the news and entertainment media have proliferated. At the same time, adoption is perceived as a method of family formation rooted in deep loss. It leaves permanent scars.

All of adoption's participants live under the sign of stigmatized difference. That is why efforts to dignify and equalize adoption in law, language, and literary representation have been so urgent. Birth parents gave children away; if that was not disgraceful enough, these children were often marked with other kinds of shame as well. Mental defect or illness, promiscuity, criminality, and grinding poverty meant that adoption simultaneously involved children considered of dubious quality or "risky" while invariably moving them up the socioeconomic and cultural ladder. By midcentury, adoption was virtually synonymous with infertility, also a stigmatized condition, and couples who raised children born to others were expected to do so only after a lengthy period of treating and resolving this terrible blow to their reproductive and psychological integrity. Children were tainted by hereditary defects and illegitimate birth, by the enduring suspicion that they were genetic lemons who might spread their "bad seed," and by sometimes chaotic early childhoods spent in orphanages or numerous foster homes. Even if they had been placed as infants, they were stigmatized by the simple fact that they were adopted.

THERAPEUTIC GOVERNMENT AND THE PROBLEM OF DIFFERENCE

"Kinship by design" is what I call the regulatory, scientific, and therapeutic reform of modern adoption in the United States. It is an example of the larger campaign to reduce stigma by erasing the difference between spoiled and normal identity. This case of liberal social engineering illustrates how child welfare reformers set out to govern selfhood, in the broadest sense, by applying legal, scientific, and clinical resources to family formation. Their goal? To normalize adoptive kinship. As many scholars have noted, normality has been a historically contingent category inseparable from its abnormal or subnormal counterparts.[11] This pattern is visible in constant efforts to hold

---

Jill Bialosky and Helen Schulman, eds., *Wanting a Child: Twenty-two Writers on Their Difficult but Mostly Successful Quests for Parenthood in a High-Tech Age* (New York, 1998); Richard Lewontin, *It Ain't Necessarily So: The Dream of the Human Genome and Other Illusions* (New York, 2000); Dorothy Roberts, *Killing the Black Body: Race, Reproduction, and the Meaning of Liberty* (New York, 1997).

[10] Princeton Survey Research Associates, *Benchmark Adoption Survey: Report on the Findings* (New York, 1997), 8–9; Dave Thomas Foundation for Adoption, "National Adoption Attitudes Survey," June 2002, 5–8, 20, chart 2. Skepticism increases when questions are asked about children adopted internationally and children adopted from the public foster care system.

[11] Georges Canguilhem, *The Normal and the Pathological* (New York, 1989).

adoption up to the mirror of nature and make it as "real" as the "real thing." Being adopted and being a "natural" or "own" child share a modern history in which inquiry and interpretation guide the transformative operations that people perform on themselves and others to eliminate or hide evidence of suspicious differences.

These operations have been concentrated on personalities, the appropriate unit of analysis and intervention during a modernizing period when a production-driven economy premised on external behavioral constraints gave way to a consumption-driven economy whose regulatory regime did not so much disappear as require strict internalization.[12] Individualism was firmly rooted in the early American republic, as Alexis de Tocqueville famously observed, but the rise of an encompassing consumer culture in the twentieth century reinforced the view that democracies blessed citizens with infinite possibilities about who to be and become. Making a project out of selfhood was not only desirable but also an inescapable fact of daily life in a market society awash in choices.

Self-operations have become widespread, lifelong preoccupations, but rarely are they more urgently executed or overseen than during childhood. Since the Enlightenment, human beings have been conceived as persons whose earliest years constitute a formative stage in the production of rationality, autonomy, discipline, and other prerequisites of democratic citizenship and market participation. By charting the course for individuality and anticipating the achievements and problems of adult life, childhood placed emphasis on learning social rules early—or else. Awareness that the developmental potential of young individuals determined the fate of nations made childhood at once a locus and an object of power and planning. Childhood was both barometer and forecast of social conditions. A historical process that unfolded over time, development literally connected the past to the present and the future.

During the early twentieth century, new genres of empirical inquiry nourished the campaign to mold character and conduct in the public interest. Research on patterns of growth, educational and occupational outcomes, and the origins of individual and group differences highlighted the simultaneous importance of childhood for science and social policy.[13] Studying children scientifically might finally reveal the nature of human nature while constructing a sturdy foundation for new technologies of being and behavior. Most states appointed special commissions to survey a wide range of child-related social problems between 1900 and 1920, from feeblemindedness to sanitation and the quality of milk, with the express purpose of recommending new and revised legislation.[14]

The Children's Commission of Pennsylvania was one of the first public bodies in the country to make an ambitious empirical study of legal adoption. It led to a number of legal reforms designed to reduce abuses by expanding knowledge and public oversight, including new requirements that judges be given more facts about the parties to adoption and interview them personally in court.[15] In 1912, the U.S. Congress

---

[12] Warren I. Susman, "'Personality' and the Making of Twentieth-Century Culture," in *Culture as History: The Transformation of American Society in the Twentieth Century* (New York, 1984), 271–85.

[13] Alice Boardman Smuts, *Science in the Service of Children, 1893–1935* (New Haven, Conn., 2006).

[14] U.S. Children's Bureau, *State Commissions for the Study and Revision of Child Welfare Laws*, Publication no. 131 (Washington, D.C., 1924).

[15] Neva R. Deardorff, "'The Welfare of the Said Child . . .'," *Survey Midmonthly* 53, 15 Jan. 1925, 457–60; Deardorff, "Scrutinizing Adoption," *Catholic Charities Review* 10 (Jan. 1926): 3–8.

established the Children's Bureau, a federal agency with jurisdiction over issues such as infant mortality and child labor. A generation of visionary female reformers and ambitious bureaucrats turned the bureau into a hub of creative research, advocacy, and social planning. They used childhood as a springboard to rethink vast areas of social policy and envision a welfare state that would offer all Americans—not only children—new kinds of social security.[16]

Development was an important border crossing between science and statecraft. According to Lawrence K. Frank, foundation power broker and patron of the new behavioral sciences, the outstanding trend of the modern era was "the growing belief in the possibility of directing and controlling social life through the care and nurture of children."[17] Decisive intervention in a wide range of social institutions followed logically so as to capture "the marvelous potentialities already within our grasp."[18] Engineering development was nothing if not an ambitious knowledge project, and efforts to gauge and guarantee children's well-being in schools, courts, clinics, and communities around the country gained momentum during and after the Depression. The conception of child welfare as a public resource was institutionalized by the New Deal, which held an expanded state role in economic and social life necessary for both negative and positive reasons: to safeguard the public from hazards that could not be predicted or controlled by individuals acting on their own, and to equip the population with the requisite skills needed to build a constructive future.

The expansion of schooling as well as the establishment of diverse public and private organizations devoted to child health, education, and recreation during the twentieth century offer tangible evidence of the social imperative to govern children's minds and bodies and transform childhood into a standardized—hence more readily governable—subject from birth through adolescence.[19] Parent education, child guidance, and family therapy movements, which emerged during the first two-thirds of the century, epitomized the therapeutic sensibility permeating professions from psychiatry to social work. Delinquents, dropouts, students with learning disabilities, and victims of abuse and neglect all needed individual help, to be sure, but they were also examples of emotional and material deprivation that could be ameliorated only through the political and policy process.[20] With their developmental progress, educational careers, leisure activities, aesthetic inclinations, and behavioral idiosyncrasies made ever more legible to state agencies and private businesses, American children were

---

[16] Kriste Lindenmeyer, *"A Right to Childhood": The U.S. Children's Bureau and Child Welfare, 1912–46* (Urbana, Ill., 1997).

[17] Lawrence K. Frank, "Childhood and Youth," in *Recent Social Trends in the United States,* ed. The President's Research Committee on Social Trends (New York, 1933), 798.

[18] Ibid.

[19] Carita Constable Huang, "Making Children Normal: Standardizing Children in the United States, 1885–1930" (PhD diss., Univ. of Pennsylvania, 2004); Ann Hulbert, *Raising America: Experts, Parents, and a Century of Advice about Children* (New York, 2003); Peter N. Stearns, *Anxious Parents: A History of Modern Childrearing in America* (New York, 2003); Alexandra Minna Stern and Howard Markel, *Formative Years: Children's Health in the United States, 1880–2000* (Ann Arbor, Mich., 2002).

[20] LeRoy Ashby, *Endangered Children: Dependency, Neglect, and Abuse in American History* (New York, 1997); Linda Gordon, *Heroes of Their Own Lives: The Politics and History of Family Violence* (New York, 1988); Julia Grant, *Raising Baby by the Book: The Education of American Mothers* (New Haven, Conn., 1998); Kathleen W. Jones, *Taming the Troublesome Child: American Families, Child Guidance, and the Limits of Psychiatric Authority* (Cambridge Mass., 1999); Deborah F. Weinstein, "The Pathological Family: A History of Family Therapy in Post-World War II America" (unpublished manuscript, 2004).

brought into sharp focus individually and collectively, advancing the reach of the welfare state and consumer culture and stirring fears that the disappearing boundary between family and market had caused parents to lose control of children.[21] Rationalizing child development helped to turn exceedingly complex, varied, and obscure private transactions into simplified exchanges subject to documentation, quantification, and other forms of public monitoring.[22] Thus did many aspects of childhood selves-in-formation become social problems.

All of these general trends are visible in the history of twentieth-century adoption. The rationale for intensifying investigation and supervision during the adoption process rested on three interrelated claims. First, government has a fundamental responsibility to ensure citizens' safety by protecting them from avoidable risks, such as being saddled with unfit children or cruel parents. Second, exclusive ideals—heterosexual marriage, biogenetic reproduction, patriarchal relations of economic dependence—define what legitimate families are, which makes departures from these ideals into pathological deviations requiring adjustment rather than benign variations. Third, scientific methods and therapeutic interventions will make social management more efficient. Rationalizing the adoption process, according to this way of thinking, would inevitably improve it. A plethora of new tests, studies, and legal rules were therefore marshaled in the name of progress and child protection. Adoption "standards" promised to "safeguard" the parties involved, compensating for problematic deviation from the biogenetic norm.[23]

Matching was the paradigm that set out to make adoptive families real by simulating the physical resemblance and intellectual similarities understood to be natural causes of love and obligation while also concealing the dangerous signs of difference. Matching required adults who acquired children born to others to be married heterosexuals who looked, felt, and behaved as if they could have conceived those children themselves. This approach to family formation gave rise to policies such as confidentiality and sealed records, first instituted by Minnesota in 1917. These policies prevented adoptive and natal relatives from discovering one another's identities or establishing contact; their architects envisioned adoption as a wholesale replacement of children's natal ties rather than a supplement to them. In state after state, authorities hid original birth records from public view and issued new birth certificates to make it appear as if adoptees had been born to their new parents. Here was a novel government practice that managed the threat of difference by denying the relevance, even the existence, of the social processes that made families up. At the same time, adoptive parents were supposed to inform children of their adoptive status. In private, where "telling" took place, the troublesome dimensions of adoption could be overseen without attracting public scandal, insult, or pity.

---

[21] Daniel Thomas Cook, *The Commodification of Childhood: The Children's Clothing Industry and the Rise of the Child Consumer* (Durham, N.C., 2004); Lisa Jacobson, *Raising Consumers: Children and the American Mass Market in the Early Twentieth Century* (New York, 2004); Christopher Lasch, *Haven in a Heartless World: The Family Besieged* (New York, 1977). For contemporary perspectives, see Susan Linn, *Consuming Kids: The Hostile Takeover of Childhood* (New York, 2004); Juliet Schor, *Born to Buy: The Commercialized Child and the New Consumer Culture* (New York, 2004).

[22] James C. Scott, *Seeing Like a State: How Certain Schemes to Improve the Human Condition Have Failed* (New Haven, Conn., 1998).

[23] Ellen Herman, "The Paradoxical Rationalization of Modern Adoption," *Journal of Social History* 36 (Winter 2002): 339–85; Herman, "Families Made by Science: Arnold Gesell and the Technologies of Modern Adoption," *Isis* 92 (Dec. 2001): 684–715.

Well established by midcentury, such policies and practices were the work of benevolent reformers determined to reduce the stigma that adoptees faced and secure for them the full benefits of authentic kinship. By making adoption confidential, professionals hoped to place obstacles between an uninformed and nosy public and the parties to adoption while still ensuring that those parties retained access to relevant information. By midcentury, however, adoption mediators had abandoned their historical role as agents of disclosure, and adoptees were cut off from information about their births and backgrounds.[24] Post-1945 adoptees were angry at the poisonous secrets surrounding them and, as adults, attacked the adoption closet that had frustrated reunions with natal relatives. Adoptees themselves played leading roles in adoption reform after midcentury, and they were just as certain as zealous professionals that they acted in children's best interests. Whereas advocates of matching, confidentiality, and sealed records hoped to shield children from stigma, advocates of openness and reunification claimed that pretense harmed children far more than evidence of difference. It was simply a lie to insist that children in adoptive families had only one mother, one father, and one family when they actually had two.

The social fictions that accompanied adoption modernization ultimately fooled no one. The most obvious characteristic about adoptive family making throughout the twentieth century was its distinctiveness, and the association between distinctiveness and danger was precisely why adoption needed special oversight. In addition to being different, adoption was unusual. In comparison with other legal procedures that transformed kinship—divorce, for example, which was still quite unusual in 1900—adoptions were exceedingly rare. One well-informed observer after the turn of the century reasoned that encountering attorneys and judges was "thoroughly formidable to families who have never in their lives had anything to do with 'the law.' They naturally shrink from it and from the possible publicity entailed."[25] Americans, however, gradually grew more comfortable with bringing family matters to court, and the tangible advantages of legal ties (social security benefits, for example) made the expense of adopting a child more reasonable as time passed. At adoption's high point, around 1970, about 175,000 children were adopted each year in the United States by strangers and relatives. The total number, as well as the adoption rate, then declined sharply until the late 1980s. Curiously, this numerical decline from 1970 has been accompanied by trends in the adoption world as well as those in the broader culture that have made adoption more visible than ever. Growing numbers of children have been adopted across national, cultural, and racial borders. These children look nothing at all like their parents, a fact that has attracted curiosity and sustained media attention to international and transracial adoptions.[26]

In recent years, total adoptions by U.S. citizens have hovered around 125,000, with

[24] E. Wayne Carp, *Family Matters: Secrecy and Disclosure in the History of Adoption* (Cambridge, Mass., 1998).

[25] Sophie van Senden Theis, *How Foster Children Turn Out*, New York State Charities Aid Association, Publication no. 165 (1924), 127. Theis studied 910 children placed by the New York State Charities Aid Association after 1898 who were eighteen as of Jan. 1, 1922. Only 269 were legally adopted.

[26] For a small sample of recent books, see Sara K. Dorow, *Transnational Adoption: A Cultural Economy of Race, Gender, and Kinship* (New York, 2006); Karin Evans, *The Lost Daughters of China: Abandoned Girls, Their Journey to America, and the Search for a Missing Past* (New York, 2000); Randall Kennedy, *Interracial Intimacies: Sex, Marriage, Identity, and Adoption* (New York, 2003); Barbara Katz Rothman, *Weaving a Family: Untangling Race and Adoption* (Boston, 2005); Sharon E. Rush, *Loving across the Color Line: A White Adoptive Mother Learns about Race* (Lanham, Md.,

adoptees accounting for approximately 2.5 percent of the population under eighteen years of age. Conservative estimates suggest that 2–4 percent of families have adopted and five million adoptees of all ages live in the United States. "Adopted son/daughter" was included as a census category for the first time in U.S. history in 2000, so we now have statistical confirmation of what people have long believed: adoptive families are atypical. Not only are they few in number. They are more racially diverse, better educated, and more affluent than are families in general.[27]

Adoption involves a tiny and unrepresentative minority of children and adults. Yet from the Bible to stories that compare adoption to immigration, exploration, and other cherished themes in U.S. history, people have seen adoption as a symbol of national, even universal, themes. This is important. In relation to selfhood, adoption had something in common with many other modern campaigns to refashion subjectivity and transform social relationships. It made a complex and labor-intensive social project out of qualities, such as naturalness and authenticity, that were supposed to be simple and automatic—not made through social action at all.

No aspects of identity and solidarity are ever simple and automatic. Despite the assumption that families are natural kinds, all families (including those made biogenetically) are "made," and the identity they impart is made as well. One might even claim that all kinship is "adoptive" to the extent that it is authorized by law and social convention, in contrast to kinship that exists entirely beyond formal categorization: all informal kinship, for example, or arrangements specifically proscribed by law, such as same-sex unions in most of the United States. People in adoptive families are necessarily aware that their kinship is a project, but it is this awareness, not unnaturalness, that sets adoption apart.[28]

It is paradoxical that so many adoption reforms, which subjected identity and belonging to systematic social design in the name of children's worth, denied that identity and belonging were made through social design and exalted a conception of nature indifferent to the very cultural and moral values that made adoption imaginable. The antithesis between the privileged authority of natural facts and that of purposeful

---

2000); Rita J. Simon and Rhonda M. Roorda, *In Their Own Voices: Transracial Adoptees Tell Their Stories* (New York, 2000); Jane Jeong Trenka, Julia Chinyere Oparah, and SunYung Shin, eds., *Outsiders Within: Racial Crossings and Adoption Politics* (Cambridge, Mass., 2006).

[27] Statistics on twentieth-century domestic adoptions are unreliable and must be used cautiously. A national reporting system existed only between 1945 and 1975, when the U.S. Children's Bureau and the National Center for Social Statistics collected data voluntarily supplied by states and territories. Useful sources of statistical data for the post-1945 period are Penelope L. Maza, "Adoption Trends: 1944–1975," Child Welfare Research Notes #9 (U.S. Children's Bureau, Aug. 1984), table 1, Folder: "Adoption–Research–Reprints of Articles," Box 65, CWLA Papers (SW55.1); and Kathy S. Stolley, "Statistics on Adoption in the United States," *The Future of Children* 3 (Spring 1993): 26–42. Contemporary statistics have been compiled by the Evan B. Donaldson Adoption Institute, online at http//:www.adoptioninstitute.org//research/adoptionfacts.php; the Child Welfare Information Gateway, online at http://www.childwelfare.gov/adoption/index.cfm; Anjani Chandra et al., "Adoption, Adoption Seeking, and Relinquishment for Adoption in the United States," *Advance Data from Vital and Health Statistics of the Centers for Disease Control and Prevention/National Center for Health Statistics,* no. 306, 11 May 1999; Victor E. Flango and Mary M. Caskey, "Adoptions, 2000–2001," *Adoption Quarterly* 8(4) (2005): 23–43; U.S. Census Bureau, "Census 2000 Special Reports, Adopted Children and Stepchildren: 2000" (Washington, D.C., Aug. 2003); U.S. Department of State, "Immigrant Visas Issued to Orphans Coming to the U.S.," online at http://www.travel.state.gov/family/adoption/stats/stats_451.html. Some of these sources and others are posted on The Adoption History Project, online at http://uoregon.edu/~adoption/topics/adoptionstatistics.htm.

[28] Sally Haslanger and Charlotte Witt, eds., *Adoption Matters: Philosophical and Feminist Essays* (Ithaca, N.Y., 2005); Rothman, *Weaving a Family* (cit. n. 26).

social decision making is both curious and characteristic in a liberal democratic culture such as that of the United States.[29] It suggests that even such cherished ideals as individualism, freedom, and choice have paled in comparison with beliefs that ascription and blood are thicker, more profound, and more durable bases for personhood and affiliation.

"TELLING": MAKING A PROJECT OUT OF ADOPTED SELFHOOD

"Telling" refers to informing a child of his or her adoptive status. Telling is both a momentous event and a conversation about adoption that continues over the course of childhood, even over a lifetime. In contrast, nonadopted children are not typically "told" anything about their birth status. What could be less worthy of mention than being raised by one's own parents? Growing up with natal relatives is a cultural default, so axiomatic that it does not appear to be an arrangement at all. Naturalized to the point of transparency, it appears simply to be.

Adoption has not come naturally. It is an effort-laden and labor-intensive project, which means it can be bungled or accomplished gracefully. That the project's prospects supposedly depend upon how adoptive parents tell is the reason why so many regard telling with trepidation. Telling offers an example of the difficult challenges posed by stigmatized difference and underlines the gap between the deliberately constructed identities of adoptees and the allegedly effortless selfhood of their non-adopted peers. The practice of telling and related theories about when, how, and why children should be told exemplify the strenuous campaigns of inquiry and interpretation that govern adoptees' problematic subjectivity. Theirs are spoiled selfhoods. They must be normalized and repaired on purpose, for the adoptees' own good.

Nothing in adoption made the management of dangerous difference—and its likely influence on emotional disorder and fragile selfhood—more obvious than the excruciatingly deliberate act of telling. Throughout the century, child welfare professionals maintained a firm consensus that children should be told, one of the few adoption-related issues about which they never changed their minds. Why? Because it was better for children to hear from loving parents than from unfeeling relatives, nosy neighbors, or cruel classmates. Telling had less to do with honesty than it did with therapeutic inoculation against pain and stigma.

This does not mean that all parents did tell. In cases of children adopted early in life, when children had no memories of other kin and no visible mismatching gave the fact of adoption away, not telling was always an option. Anecdotal evidence suggests that many parents never told their children or postponed telling for years. Why? Telling was a reminder of the stigmatized difference between adoptees and nonadopted children. Own children, of course, were never told anything because growing up with natal relatives was such a default arrangement that it did not appear to be an arrangement at all.

It is not surprising that many adoptive parents, reluctant to say anything that might hurt their children or diminish their own status as legitimate parents, considered telling "a dreaded job."[30] Not telling may have been "like . . . not saluting the American flag,"

---

[29] Lorraine Daston and Fernando Vidal, *The Moral Authority of Nature* (Chicago, 2004).

[30] Carl Doss and Helen Doss, *If You Adopt a Child: A Complete Handbook for Childless Couples* (New York, 1957), 28.

as adoptive parent Henrietta Whitmore put it, but she and her husband instantly regretted the anguish that telling caused their daughter, and she cautioned others not to make the same mistake.[31] The powerful intuition among so many adoptive parents—"that *you are our child*"—was worth more than the advice of "wise people" and "experienced, intelligent authorities," Whitmore concluded.[32] Many other parents surely agreed. Their decisions to pass met with expert disapproval, but they had a certain logic. To tell was to admit that adoptive kinship was different and inferior, so they refused to do so. In their view, not telling was not so much a lie as proof of the equality and authenticity of their family bonds. Hadn't they been encouraged to believe that their status was as real as real could be?

There is no way of knowing how many children grew up without knowledge that they were adopted, but the number may have been considerable—as many as half of all adoptees during the first several decades of the century.[33] "Many foster parents object so strenuously to telling their children that it is impossible to insist on it," lamented social workers Sophie Theis and Constance Goodrich in 1921.[34] In 1936, the director of the Child Welfare League of America, C. C. Carstens, observed that evasion of telling was widespread among parents; it constituted one of the major "pitfalls of adoption."[35] In the late 1930s, Edwina Cowan, a psychologist with the Wichita Child Welfare Research Laboratory, described the "program of forgetting" favored by so many adoptive parents as "one of the most certain causes of emotional strain and maladjustment" in adopted children.[36]

Some children, of course, were told during adolescence, others on the eve of marriage, and some even later. Many young draftees during the two world wars were surprised to discover their adoptions when the military's bureaucratic requirements, which included birth certificates, disclosed that the people who had raised them were not the same as the people who had brought them into the world.[37] After the 1930s, requests for social security and unemployment benefits also brought previously obscure adoptions to light. Adoptive parents were ultimately responsible for telling or not, but

---

[31] Henrietta Sloane Whitmore, "I Wish I Hadn't Told You," *McCalls,* Sept. 1959, 67.

[32] Ibid. (emphasis in original).

[33] See, e.g., Annie Hamilton Donnell, "The Adopted," *Harper's Monthly Magazine,* Nov. 1906, 927–33. In the 1920s, surveys in New York and California also showed that anywhere from half to two-thirds of children adopted before age five were kept in ignorance of their status. See Paul Popenoe, "The Foster Child," *Scientific Monthly* 29 (Sept. 1929): 244. A follow-up study of Minnesota adoptions between 1918 and 1928 found that 50 percent of children (all between the ages of five and fourteen) studied had been told. Because fear that the child's adoption would be revealed was a common reason that parents refused to participate in the study, we can safely assume that more than half of the adoptees in Minnesota at that time had not been told about their status. See Alice M. Leahy, "Nature-Nurture and Intelligence," *Genetic Psychology Monographs* 17 (Aug. 1935): 264. The first major outcome study, *How Foster Children Turn Out,* found that approximately 34 percent of adoptees had never been told. See Theis, *How Foster Children Turn Out* (cit. n. 25), 124, 231, table 60.

[34] The term "foster" denoted both temporary and permanent family placements until well after World War II. Sophie van Senden Theis and Constance Goodrich, *The Child in the Foster Home,* part 1, *The Placement and Supervision of Children in Free Foster Homes, A Study Based on the Work of the Child-Placing Agency of the New York State Charities Aid Association* (New York, 1921), 91.

[35] C. C. Carstens, "The Pitfalls of Adoption," *Child Welfare League of America Bulletin* 15 (Oct. 1936): 4.

[36] Edwina A. Cowan, "Some Emotional Problems Besetting the Lives of Foster Children," *Mental Hygiene* 22 (July 1938): 456.

[37] For one such story, see Eda Houwink, "An Adopted Child Seeks His Own Mother," *Child Welfare League of America Bulletin* 22 (April 1943): 1–4.

the growth of the New Deal state and its expanding functions in work, welfare, and war sometimes revealed truths that parents sought to keep secret.

Throughout the century, professionals maintained that secrecy, misinformation, and deception between parents and children had no place in adoption. Those same social workers, however, rarely offered clients full and honest information. When would-be adoptive parents were rejected, for example, professionals often reasoned that many were too lacking in insight to tolerate news about their own psychological defects. "Telling the whole truth may, instead of being a virtue, actually be an act of the crassest cruelty and may throw into disequilibrium a situation which has some stability."[38] Social workers also stopped short of full disclosure to adopters when children's backgrounds included such sensitive factors as racial mixture and illegitimacy. Adoption professionals were at their best when they grappled unflinchingly with the many uncertainties of family making, but uncertainty rarely, if ever, applied to the mandate that children be told.

Resistance to telling frustrated child welfare professionals, who redoubled their effort to make therapeutic interventions standard in adoption. Parents who did not tell, many social workers believed, felt insecure about their entitlement to children. However understandable, their doubts suggested they needed more professional interpretation and guidance. Before the 1950s, adult adoptees rarely expressed collective views about telling, but individuals certainly spoke out. "Do you really believe that it is possible to found secure happiness for any family on so basic an untruth?" asked one adoptee in sharp rebuke to the suggestion that it might be preferable never to tell. "We would all like our children not to suffer, to shield them from hurt, but . . . each of us must learn to accept and to adjust to the circumstances and experiences that life brings."[39]

Most adoptive parents agreed that their children needed to know about their adoptions, or at least they said so. Whether or not parents told their children, their worries made the subject a fixture in adoption literature. The first how-to book on adoption, published in 1936, included a chapter titled "What to Tell the Adopted Child."[40] The author urged parents to share their pride in being a real family and report knowing little or nothing about the child's natal background. "The child will accept this as he accepts many other things about which he has a normal curiosity that cannot be satisfied."[41]

Many parents suspected that telling would not be so easy and uncomplicated. During the 1930s, some agencies stopped pleading with adopters; instead, they refused to place children with parents who did not agree beforehand to tell. Agencies incorporated instructions about how-to-tell into the adoption process.[42] By midcentury, most agencies required adopters to pledge in writing that they would tell and provided guidance about how to answer children's questions about birth parents and the invariably

[38] Howard G. Aronson, "The Problem of Rejection of Adoptive Applicants," *Child Welfare* 39 (Oct. 1960): 23. See also Minutes of Seminar with Dr. Bernard, 28 Nov. 1950, Folder 4, Box 157, and Summary for Dr. Bernard's Seminar, 29 Nov. 1950, case of Norman and Anne S, Folder 5, Box 161, VWB Papers.

[39] Martha Miller (pseudonym) to Herbert R. Hayes, 14 Sept. 1959, Folder 6, Box 63, VWB Papers.

[40] Eleanor Garrigue Gallagher, *The Adopted Child* (New York, 1936), chap. 6.

[41] Ibid., 117.

[42] Ibid., 115–6; Iris Ruggles Macrae, "An Analysis of Adoption Practices at the New England Home for Little Wanderers" (master's thesis, Simmons College, School of Social Work, Boston, 1937), 94.

heartbreaking circumstances that resulted in surrender and adoption.[43] Telling strained the matching ideal by requiring extra effort from adoptive parents, effort that reminded them that the kinship they and their children shared could never match effortlessly. The danger that telling might undermine adoption and subject their children to ridicule surely explains why so many parents were apprehensive about telling. Professionals tried to calm the nerves of parents by assuring them that telling was most important within the circle of immediate family. Children, close friends, and relatives needed to know, but the whole world did not need to be told.[44] In fact, exiting the adoption closet too boldly could be as harmful as constructing it too tightly.

In addition to not telling at all, problems associated with tardy and mishandled telling received occasional notice early in the century. Ten young adults adopted around 1900 testified that adoptees were, on average, less happy and successful than natural children, but those who had been told early in life had fewer problems than those told later on.[45] One of them, Martha Vansant, concluded that "the successful ones among them today are those who in childhood were treated as rational beings who could safely be told that they were adopted, and not as mere emotional entities who had to be controlled by unthinking affection."[46]

By 1970, telling had been a major issue for a long time, and it came to symbolize the unavoidable trauma and loss of adoption. Not telling and telling were both associated with psychological trouble: the former amounted to lying, while the latter opened a Pandora's box of wrenching emotional dilemmas. "Being adopted means nothing unless it means that he [the adoptee] is different from other children, and so the issue for him is what this difference consists of," psychoanalyst Bruno Bettelheim wrote in 1970. "What is the solution?" he asked. "Maybe there is none. Maybe there are only problems that we must face without subterfuges. . . . There is no getting around telling him that his real mother just didn't appreciate him enough, which is a fact."[47]

Why Tell?

If telling reminded children of their original, painful rejection by birth parents, why make it mandatory? The answer had less to do with truth telling than with therapeutic action. Parents were wise to tell children about their adoptions early on, before they heard malicious comments in the world beyond the family. In 1930, *Children: The Magazine for Parents* urged telling because "tragedy may follow in the wake of concealment."[48] A 1933 *Ladies' Home Journal* article, "Shall I Tell My Child He is

---

[43] Louise Raymond, *Adoption . . . and After* (New York, 1955), 63.

[44] There were differences of opinion about how large the adoption closet should be. Some liberal advice givers suggested cards for "announcing the blessed event," gave tips on locating pediatricians whose practices included adoptees, and described annual adoption day celebrations that were designed to glorify adoption with birthdaylike rituals. Most advice givers were more cautious for two reasons. First, confidential information conveyed in a moment of excitement might not stay that way. Second, adoptees had a special need to fit in. Parents should think twice before telling teachers about the child's adoption, giving adoptees unusual family names, or throwing adoption day parties. For a discussion about how wide the telling circle should be, see Resume of Group Meetings with Adoptive Parents, 31 May 1955, Folder 10, Box 161, VWB Papers.

[45] Martha Vansant, "The Life of the Adopted Child," *American Mercury* 28 (Feb. 1933): 214–22.

[46] Ibid., 216.

[47] Bruno Bettelheim, "What Adoption Means to a Child," *Ladies' Home Journal,* Oct. 1970, 18, 21.

[48] "Tell the Child the Truth," excerpted in "Adoption," Folder 7-3-3-4, Box 406, USCB Papers.

Adopted?" warned, "Greater dangers . . . lie in wait for those who beg the question."[49] According to the professional consensus, the reason to tell was to erect an emotional shield against shock and stigma. "Tell them it's OK. There's nothing to worry about," declared one 12-year-old adoptee whose parents were apparently models of rectitude.[50] The point was not merely to boost children's self-esteem. If a child's parents faithfully exclaimed, "How glad we are we picked you out!" then he was likely to absorb the lesson that "his parents chose him while most people have to take what nature sends to them."[51] Selection made adoption special, even superior, and that would help adoptees prevail against misunderstanding and ridicule.

It is noteworthy that couples who used donor insemination during this era were not expected to share this fact with their children; they were able to sustain a fantasy of "real" (i.e., genetic) parenthood.[52] This underscores the pragmatic rationale for telling: the probability that adoptees would find out anyway. Helen and Carl Doss, whose own unorthodox adoption story, *The Family Nobody Wanted* (1954), became a best seller, noted that "artificial insemination" was an attractive alternative precisely because it eliminated the headache of telling. They also observed that a child who is "half theirs" may be better than "an adopted child who is a total stranger."[53]

Trust between parents and children was a secondary benefit of telling that, conveniently, might console children for the bewilderment that accompanied knowledge of their adoptions.[54] Intimacy was both the precondition for telling and its consequence, according to this way of thinking. Parents had nothing to fear by telling because truthful information could not ruin parent-child relationships that were loving and strong. This reassuring message belied the tricky balancing act that telling entailed. Telling forced parents to navigate precariously between constructively emphasizing and destructively overemphasizing adoption. They were supposed to acknowledge that adoption was different, at least to their children, but also to behave as if it were not. They were supposed to raise children who appreciated the momentousness of their adoptions but who did not ask probing questions about birth parents or seek to locate them.

Preoccupations with telling were inseparable from preoccupations with the power that parents—particularly mothers—wielded over children's development. Maternalism remained an intensely ambivalent subject throughout the century, thanks to a combination of Freudian psychology, insistent domesticity, dramatic changes in women's lives, and anxieties about the decline of American masculinity.[55] "You see," commented British object relations theorist D. W. Winnicott about telling, "you are dealing with the child's mind."[56] Most parents were all too aware of their responsibilities for conserving children's mental well-being. They took their jobs seriously and

[49] Marion L. Faegre, "Shall I Tell My Child He Is Adopted?" *Ladies' Home Journal,* June 1933, 32.

[50] Frances Lockridge and Sophie van Senden Theis, *Adopting a Child* (New York, 1947), 153.

[51] Lee M. Brooks and Evelyn C. Brooks, *Adventuring in Adoption* (Chapel Hill, N.C., 1939), 67.

[52] Margaret Marsh and Wanda Ronner, *The Empty Cradle: Infertility in America from Colonial Times to the Present* (Baltimore, 1996), 167; Jill Morawski, "Imaginings of Parenthood: Artificial Insemination, Experts, Gender Relations, and Paternity," in *Believed-In Imaginings: The Narrative Construction of Reality,* ed. Joseph de Rivera and Theodore R. Sarbin (Washington, D.C., 1998), 229–46.

[53] Doss and Doss, *If You Adopt a Child* (cit. n. 30), 46, 49.

[54] For an example of this argument, see "Problems in Adoption Must Be Faced Squarely," *Hygeia,* Aug. 1933, 757.

[55] Rebecca Jo Plant, "The Repeal of Mother Love: Momism and the Reconstruction of Motherhood in Philip Wylie's America" (PhD diss., Johns Hopkins University, Baltimore, 2002).

[56] D. W. Winnicott, "On Adoption (1955)," in *The Child and the Family: First Relationships* (London, 1957), 128.

credited telling with enormous import. Telling emphasized the uniqueness of adoptive kinship while also functioning as a narrative about the universal project of self-making in relation to others. Like coming out for gay men and lesbians, it told a story that was simultaneously personal and social. Every child had (or lacked) his or her specific story, but telling was common to all adoptive families. That made adoption a project suffused with choice, invention, and purpose, qualities antithetical to the inexorable force of a presocial nature.

It is one of the great paradoxes of modern adoption that the very acts of planning and consent that make adoption exemplary in a liberal culture have also positioned it at odds with realness. Because of the deliberation inherent in adoption, it is often represented as quintessentially American, expressing cherished democratic values. Yet adoption has never come naturally. The adoption case thus illustrates the challenges that have plagued many forms of voluntary solidarity and fueled suspicions that choice is a flimsy and superficial basis for belonging in comparison with blood.

How to Tell?

How exactly were parents supposed to tell? Typical advice epitomized the difficulty of acknowledging difference while containing the threat it posed to realness. Parents were urged to approach the explosive potential of adoption with an attitude of studied casualness. In 1929, Jessie Taft (an adoptive mother herself) compared telling to "sex instruction," another loaded topic. "It isn't something to be whispered about mysteriously, over-stressed emotionally, postponed to some more fitting time, or relegated to a formal occasion. It is a fact which should be breathed in naturally from the first where parents are secure in their love for each other and the child, and do not fear either past or future."[57] Benjamin Spock, one of the century's most trusted child-rearing experts, addressed telling in *The Common Sense Book of Baby and Child Care* (1946). He recommended that parents accept the fact of their child's adoption "as naturally as they accept the color of the child's hair" while simultaneously suggesting that adoption was anything but a neutral fact of this kind.[58] Adoption made children so fearful and insecure that "one threat uttered in a thoughtless or angry moment might be enough to destroy the child's confidence in them forever."[59] Adoption mattered enormously, but when telling was handled correctly, it would not seem that way.

The most popular advice capitalized on the trend toward parents' reading to their children daily.[60] Classic children's books about adoption, such as *The Chosen Baby* (1939) and *The Family That Grew* (1951), were examples from a burgeoning literature recommended by agency professionals and popular media.[61] Books literally

---

[57] Jessie Taft, "Concerning Adopted Children," *Child Study* 6 (Jan. 1929): 87.

[58] Benjamin Spock, "Adopting a Child," in *The Common Sense Book of Baby and Child Care* (New York, 1946), 506.

[59] Ibid., 507.

[60] Anne Scott MacLeod, "Children, Adults, and Reading at the Turn of the Century," in *American Childhood: Essays on Children's Literature of the Nineteenth and Twentieth Centuries* (Athens, Ga., 1994), 114–26.

[61] Valentine P. Wasson, *The Chosen Baby* (New York, 1939); Florence Rondell and Ruth Michaels, *The Family That Grew* (New York, 1951). For a magazine article promoting book reading as a telling method, see "How-What to Tell an Adopted Child," *Social Service Review* 21 (1947): 251–2. For evidence that agencies routinely encouraged book-reading, see Ruth F. Brenner, *A Follow-Up Study of Adoptive Families* (New York, 1951), chap. 5. A report from one Vermont agency that tracked how adoptive parents told the children placed with them found 100 percent reliance on book reading. See

made adoption go down as easily as a bedtime story and relieved parents of always having to tell in their own words. Parents who wished to move beyond preformulated texts were urged to individualize the ritual by filling in details specific to their families[62] or by customizing scrapbooks to include photographs of landmark events, such as the day the child came home and the day adoption was finalized.[63] Occasionally, the strategy of allowing adoptees to "adopt" something themselves—usually pets— was advocated as an aid to telling. One fictional adoptee took in two kittens—he named one Really and the other Truly—after a neighborhood child taunted him about his adoption and explicitly challenged his authenticity as a "really truly" member of his family.[64] As an activity with no guarantee of working out well (pets could misbehave, run away, or die), it had obvious disadvantages.

Detailed instructions about exact words, phrases, and sentences to use (and to avoid) in telling underlined the power and peril of adoption as a factor in children's emotional development. Sample scripts were common.[65] Parents were tutored to use the words "chosen" and "adopted" interchangeably and in a happy and relaxed tone of voice. Other suggestions included repetition of "my precious adopted daughter" and "my dear little adopted son."[66] Telling was supposed to happen early and often because children would assimilate their parents' feeling about adoption long before they could understand words or ask questions. Any hint of anger or frustration could defeat the purpose of telling by betraying adoption's negative associations. "*Under no circumstances, ever, should the child be reminded that he is adopted when the parent is feeling angry at him*," admonished Robert Knight of the Menninger Clinic. "The adoption should *never be mentioned* except as a pleasant matter."[67]

Constructive government of children's emotional development was inseparable from strict government of parents' emotional expression. Parents who passed the telling test advantaged their children while also proving they were real parents: mature, loving, able to prioritize their child's needs, and willing to acknowledge that adoption made their family different. Parents who failed, however, betrayed doubts about their own authenticity. The same experts who maintained that there was no formula for how to tell also maintained that there were right and wrong ways to feel about telling. "If you yourselves have fully accepted your child's adoption," one adviser wrote, "you will be able to make him accept it, fully and happily."[68] Inculcating the appropriate "feeling tone" and attitude in adults was a chief goal of the telling

---

Julia E. Hatch, "Telling Children about Adoption," *Child Welfare* 43 (July 1964): 365–6. Adoption poems also figured in this literature, though less prominently than stories. See Polly Lindquist and Prudence Lyle, "The Adoption Story," Folder 9, Box 1, Dorothy Hutchinson Papers.

[62] Florence Rondell and Ruth Michaels, *You and Your Child: A Guide for Adoptive Parents* (New York, 1951). This guide, designed to accompany *The Family That Grew* (cit. n. 61), included advice about how to read the book to children.

[63] Doss and Doss, *If You Adopt a Child* (cit. n. 30), 189; Raymond, *Adoption . . . and After* (cit. n. 43), 58. Such tools, eventually called life books, are frequently and erroneously presented as recent innovations, dating to the 1970s or 1980s. Kristina A. Backhaus, "Life Books: Tool for Working with Children in Placement," *Social Work* (Nov.-Dec. 1984): 551–4.

[64] Carolyn Haywood, *Here's a Penny* (New York, 1944), chaps. 1–2.

[65] Doss and Doss, *If You Adopt a Child* (cit. n. 30), 186–9; Spock, "Adopting a Child" (cit. n. 58), 503–7. For a contemporary version, see Linda Bothun, *Dialogues about Adoption: Conversations between Parents and Their Children* (Chevy Chase, Md., 1994).

[66] Gallagher, *The Adopted Child* (cit. n. 40), 116.

[67] Robert P. Knight, "Some Problems in Selecting and Rearing Adopted Children," *Bulletin of the Menninger Clinic* 5 (May 1941): 71 (emphasis in original).

[68] Raymond, *Adoption . . . and After* (cit. n. 43), 85.

enterprise.[69] Telling was equally crucial in forging parents' and children's identities. It made a project out of everyone involved.

When to Tell?

For decades, most adoption professionals agreed that children should be told about their adoptions early in life. Even infants, they insisted, would benefit from hearing adoption discussed lovingly. But in 1960, a report by psychiatrist Marshall Schechter renewed debate about when to tell.[70] Based on observations of sixteen troubled adoptees in his clinical practice, Schechter argued that adoptees were 100 times more likely than nonadoptees to develop serious emotional problems, some of them rooted in botched telling. He recommended that parents wait until children were school age, had negotiated the oedipal crisis, and were firmly identified with their adoptive kin. Psychoanalyst Lili Peller seconded this opinion. She argued that revelation early in life "did considerable harm" by drawing adoption into "the whirlpool of the child's sexual and sadistic fantasies."[71]

Telling early and often, according to these critics, sabotaged parents' best intentions to nurture their children's emotional welfare by reminding children that their belonging was less than real. That is why Schechter and Peller opted for the early school age or even the teenage years over early childhood as the preferred time of revelation. Few experts revised the telling timeline as radically as adolescent psychiatrist Joseph Ansfield, who suggested that children never be told. His position, he admitted, "goes against practically all stated beliefs, opinions and policy of the past and present." "My strong belief," he wrote in 1971, "is that adopted children should not be told that they are adopted. They should not be told for a very good reason, that is, the knowledge will hurt them."[72] His lonely dissent made the comprehensive telling consensus all the more obvious.

Theoretically, too much emphasis on adoptive status could be as noxious as too little or none. "Am I a 'dopted child?" one worried little girl inquired in a 1929 story. "You certainly *are,*" her mother replied, "but I wouldn't talk about it with the other children. It's not nice to boast."[73] The prospect that children might relish or even flaunt their adoptive status was precisely the opposite of what parents feared. Preoccupations with telling invariably centered on the damage that it might do, not the pride it would instill. That is why most advice before 1970 stressed the theme of "chosen children," a theme repudiated by reformers after 1970 for sugarcoating painful realities.[74] Although telling children they had been picked out on purpose conveyed the depth of parental desire, it also implied that relinquishment by birth parents had been equally deliberate.

---

[69] Meeting of Psychiatric Consultants on Telling the Child of His Adoption, 10 July 1962, Folder 6, Box 162, VWB Papers.

[70] Marshall D. Schechter, "Observations on Adopted Children," *Archives of General Psychiatry* 3 (July 1960): 21–32. For evidence that Schechter's findings prompted agencies to reevaluate their "telling" policies, see Louise Wise Services, Minutes of the Child Adoption Committee, 4 May 1966, 3–7, Folder 4, Box 155, VWB Papers.

[71] Lili Peller, "About 'Telling the Child' of His Adoption," *Bulletin of the Philadelphia Association for Psychoanalysis* 11 (Dec. 1961): 145.

[72] Joseph G. Ansfield, *The Adopted Child* (Springfield, Ill., 1971), 35–6.

[73] Agnes Sligh Turnbull, "The Great Adventure of Adopting a Baby," *American Magazine,* May 1929, 182 (emphasis in original).

[74] The post-1970 critique continued to embrace children's books and stories. See, e.g., Betty Jean Lifton, *Tell Me a Real Adoption Story* (New York, 1993).

What to Tell?

That children had been born to one set of parents and adopted by another was not the only fact to be told. Sensitive information about adoptees' natal backgrounds frequently included poverty, alcoholism, mental illness, criminality, sexual immorality, incest, and other sordid details. What, if anything, should be disclosed about these? This question tormented parents. Advice givers emphasized that emotional tone was more important than the substance of information conveyed. Whatever information they shared, adopters should strive to talk about birth parents and relatives calmly and easily. Curiosity about the people who had given them life was to be expected of adoptees. It was normal for them to ask questions, typically at the point when they were old enough to understand sex and reproduction.

The dilemma of what to say about birth parents symbolized the threat that unembellished information posed to the adoptee's personality, much as the fact of adoption did. Children needed to feel comfortable with whatever they were told about their "first" parents, advisers agreed, in order to sustain trust in their "second" and "real" parents.[75] In many cases, this involved highly censored communication, if not outright lies. Early studies that found high rates of telling also found that parents withheld background information.[76] One long-term follow-up of 100 New York agency adoptions from the 1930s found that the vast majority of parents (90 percent) had told their children, scrupulously following instructions to tell early, often, and with the appropriate words and tenor. No parents at all, however, had given their children full and candid details about the reasons for their adoption.[77] Illegitimacy, in particular, remained unmentionable through the 1960s. Many parents lied to their children, "killing off" birth parents rather than saying they had been unmarried.[78] To convey certain facts would publicize the moral failings of those parents and, by association, smear the children.

It is understandable that many parents withheld information they found embarrassing as well as menacing to their children. To prevent children from realizing too early that birth parents had been unable or unwilling to raise them, adopters were told to say that their children's original parents (particularly mothers) were good individuals who made selfless decisions for their children's sake. The contradictory pressures built into telling were irreconcilable on this point. The same culture that demonized women who refused to mother, or to be "good" mothers, expected adoptive parents and children to consider surrender the result of maternal devotion—rather than desperation, coercion, or occasionally, callous disregard. If some adults chose to become parents through adoption and dutifully reinforced the rhetoric of "chosen children," this

---

[75] How-to literature displayed great terminological sensitivity from the start. Referring to birth parents as "real" or "own" parents would not do, so several writers suggested the more neutral "first" and "second" or "a man and a lady." See Doss and Doss, *If You Adopt a Child* (cit. n. 30), 187; Raymond, *Adoption . . . and After* (cit. n. 43), 76; Rondell and Michaels, *You and Your Child* (cit. n. 62), 30.

[76] Georgina D. Hotchkiss, "Adoptive Parents Talk about Their Children: A Follow-Up Study of Twenty-Four Children Adopted through a Child Placing Agency" (master's thesis, Simmons College, School of Social Work, Boston, 1950), 42–48.

[77] Benson Jaffee and David Fanshel, *How They Fared in Adoption: A Follow-up Study* (New York, 1970), 129.

[78] See, e.g., Joan Lawrence, "The Truth Hurt Our Adopted Daughter," *Parents' Magazine,* Jan. 1963, 44–45, 105–106; and First Group Meeting, 19 April 1955, Folder 10, Box 161, VWB Papers. Before 1940, it was not unusual for agencies to urge parents to substitute death for illegitimacy when they answered questions about birth parents.

reflected the powerlessness of others: birth mothers too poor, frightened, and dependent to possess choice-making authority of their own.[79]

Telling children about birth parents mattered for therapeutic, rather than legal or moral, reasons. It was supposed to put children's minds at ease and comfort them with the thought that their birth parents were decent human beings. This was by no means an invitation to search for birth parents. Most advocates of telling prior to 1970 were also advocates of confidentiality and anonymity. There was no contradiction, in their view, between the routine practice of telling children and the equally routine practice of keeping secrets about birth parents' names and whereabouts. Critics of matching and champions of legal protections for children denied that adoptees were entitled to identifying information, arguing that "confidentiality is essential to a child's sinking his roots deep into his adoptive home."[80] After 1970, the Adoptees' Liberty Movement Association was galvanized by the compelling search narrative of its founder, Florence Fisher,[81] and thousands of adult adoptees and birth mothers pleaded with agencies for help in finding lost relatives. Adoption activism since then has increasingly targeted access to sealed birth records, demanded more openness, and championed reunions with natal kin.[82]

Before the records wars of the 1970s, which continue today, dominant views about telling emphasized that well-adjusted adoptees would not wish to find natal relatives. Jean Paton, for example, who founded the first adoptee search organization in the United States, Orphan Voyage (established in 1953), formulated a "search hypothesis" in which the impulse to seek out natal relatives was calibrated to the security of the adoptive home.[83] In other words, adoptees in normal, loving families had no reason to search and probably would not. That a renegade early search advocate conceded such a direct relationship between insecurity and searching suggests the extent of agreement about what mental health looked like in adoptees: "the best-adjusted adopted child is the one who almost never thinks about the facts of his natural birth."[84] Michael, a teenager in 1965, had "been preoccupied since 5 with the past and with the true identity of his parents," a situation interpreted by agency workers as evidence of "emotional disturbance" and need for help "away from his blind search for his mother to the more realistic approach of psychotherapy."[85] Children such as Michael, who dwelled on lost relatives, talked about them incessantly, or set out to locate them needed psychiatric help. They were living proof of adoption's problems. Divided loyalties doomed the aspiration to make adoption authentic.

This put adoptive parents in a terrible bind. They were expected to talk honestly to their children about their birth parents, but they were also expected not to upset the children's emotional equilibrium or support search fantasies. Helen and Carl Doss, whose own transracial and transnational adoptions flouted matching, implied that the fact of having two sets of parents was a more difficult difference to handle than others in adoption. They urged adoptive parents to *forget everything that would not be*

[79] Rickie Solinger, *Beggars and Choosers: How the Politics of Choice Shapes Adoption, Abortion, and Welfare in the United States* (New York, 2001).

[80] Justine Wise Polier, quoted in Curtis J. Sitomer, "What to Tell an Adopted Child," *Christian Science Monitor,* 25 July 1974, 7.

[81] Florence Fisher, *The Search for Anna Fisher* (New York, 1973).

[82] Carp, *Family Matters* (cit. n. 24), chaps. 5–6.

[83] Jean M. Paton, *The Adopted Break Silence* (Philadelphia, 1954), 114.

[84] Ernest Cady and Frances Cady, *How to Adopt a Child* (New York, 1956), 118.

[85] Summary of B Case, 16 July 1965, Folder 5, Box 162, VWB Papers.

helpful to your child."[86] Forgetting was the most poignantly revealing approach to the conflict between what parents knew about their children's natal backgrounds, the mandate to tell, and fears about what adoptees could bear. It diminished the danger of difference by disregarding knowledge of it.

"I don't know" was, ironically, the ideal answer to children's questions—and all the better if parents really did not. This helps to explain why a remarkable number of parents between 1940 and 1970 did not want background information about the children they adopted and actively declined it during the adoption process. The possibility that communicating prejudicial facts might cause anguish prompted agencies to revisit the question of "whether or not to inform the adoptive parents of history which might be a cause of constant concern to them."[87] One scenario in which such anxiety was probable involved mixed-race children (or children suspected of being mixed) who might be able to pass for white; another involved family histories of mental illness.[88] Agency practices varied widely on this. Some professionals maintained that agencies were obligated to convey whatever they knew. Others went so far as to suggest that adoptive parents be given no information at all. "Parents should be able to say truthfully that they do not know the reasons why the child was given up," reasoned one psychiatrist, and "this will enable the child to think that he comes from two loving parents who, by reason of some catastrophic event (the nature of which will be left to his own phantasy), could not keep him."[89]

One study of almost 500 Florida adoptions between 1944 and 1947 found that only 20 percent of adopters wished to know as much as possible about their children's birth parents and backgrounds; the other 80 percent wanted little or no information.[90] This was a contorted way to preserve children's self-confidence and the adopters' own credibility, to be sure, but if everything worked out, children would take whatever they were told in stride. "*If he knows in his mind he is adopted but feels in his heart that he belongs to his family,*" the adoptee would follow a course of normal mental and emotional development.[91] When telling was handled well—done for the right reasons, in the right ways, at the right times, and with the right information—children turned out to be real members of real families. Nothing more. Nothing less.

CONCLUSION

A theory of self-fashioning that privileged children's continuous emotional bonds with parents as the wellspring of healthy identity and solidarity was the premise underlying telling. For adoptees, these bonds were disrupted and remade through social

[86] Doss and Doss, *If You Adopt a Child* (cit. n. 30), 191 (emphasis in original).

[87] Minutes of the Free Synagogue Child Adoption Committee, 28 April 1943, p. 2, Folder 1, Box 155, VWB Papers.

[88] Louise Wise Services, Minutes of the Child Adoption Committee, 7 March 1956, p. 3, Folder 2, Box 155, VWB Papers; Florence Brown to Viola Bernard, 19 Feb. 1958, Folder 7, Box 157, VWB Papers; Gertrude Sandgrund to Viola Bernard, 16 Dec. 1963, Folder 1, Box 158, VWB Papers; Staff Meeting of Adoption Department, 20 Jan. 1964, Folder 1, Box 158, VWB Papers; Minutes of Adoption Dept. Staff Meeting, 30 Oct. 1961, Folder 11, Box 161, VWB Papers. This controversy has resurfaced in recent "wrongful adoption" suits. See Lisa Belkin, "What the Jumans Didn't Know about Michael," *New York Times Magazine,* 14 March 1999, 42–9.

[89] Seminar with Dr. Annamarie Weil, 9 May 1957, Summary of Discussion, p. 1, Folder 6, Box 157, VWB Papers.

[90] Helen L. Witmer et al., *Independent Adoptions: A Follow-Up Study* (New York, 1963), 93.

[91] Doss and Doss, *If You Adopt a Child* (cit. n. 30), 204 (emphasis in original).

practices of surrender and placement that shrouded their kin ties in mystery and shame. The discontinuity and multiplicity of adoptees' family ties were conceived as differences compounding the problem of genetic unlikeness. It was in response to these special risks that kinship by design—a series of rational interventions aiming to govern adoption and its outcomes by turning it into a knowledge project and object of legal and social regulation—emerged during the first two-thirds of the twentieth century in the United States.

Kinship by design, important in its own right, also exposed a shadow story embedded in the history of modern liberalism. The social policies of the American welfare state recognized and ratified the superiority of natural imperatives, on the one hand, while subjecting conventional understandings of human nature, identity, and belonging to review and revision, on the other. Kinship by design exposed the social architecture of kinship while simultaneously promoting designs, such as matching, that erased themselves and maintained the appearance of an inexorable, effortless, and authentic nature. This paradox put the historical balance between "the social" and "the natural" into motion and eventually subjected it to serious critique, now as familiar in science and technology studies as in adoption.[92] Adoption history in the modern United States illustrates the perils and possibilities of equating selfhood with choice, deliberation, and planning. Identities and solidarities that were socially achieved were idealized over identities and solidarities that were naturally ascribed even as they paled in comparison.

Telling has remained as central as ever in the adoption world, yet some researchers found that when and how children were told of their adoptions had little of the influence that previous generations of parents and professionals believed it did. In 1970, Benson Jaffee and David Fanshel were startled by their own data on the irrelevance of telling methods. Their discovery that how, when, and what children were told did not alter outcomes "stands in sharp conflict with what has long been a fundamental working assumption in the field of adoption placement."[93] If children's mental health did not depend on how emotionally explosive information was presented, had the differences associated with adoption been exaggerated or entirely misconstrued?

In comparison with other adoption upheavals that took place around 1970—such as the vociferous public debate about African American children being placed with white parents—little notice accompanied the controversy over telling. Most people have certainly continued to believe that telling is crucial to how children turn out. Making a project out of the selves affiliated with adoptive kinship was, after all, already a well-established part of adoption culture by the 1960s. Steeped in the science of attachment and loss, researchers, professionals, and family members alike were persuaded that security was unusually precarious in adoptive families, making guidance necessary. Weren't worries about telling proof enough of that?

Outside the adoption world, the civil rights revolution, nationalist mobilizations, massive new immigration, and the celebration of diversity and multiculturalism within the American nation paralleled and reinforced the attention to difference within it. H. David Kirk's *Shared Fate* (1964) was the first book to make adoption a major issue in the sociological literature on family and mental health, and it marked

---

[92] Marilyn Strathern, *After Nature: English Kinship in the Late Twentieth Century* (Cambridge, 1992).

[93] Jaffee and Fanshel, *How They Fared in Adoption* (cit. n. 77), 275.

the turning point between "rejection-of-difference" and "acknowledgment-of-difference" strategies within American adoption culture.[94] Kirk, a McGill University researcher and adoptive father who gathered data about 2,000 adoptive families in the United States and Canada during the 1950s, argued that owning up to the difference that difference made was the only viable option for adults and children whose "shared fate" was the experience of "role handicap" in family life. In the late 1960s and early 1970s, reform movements devoted to opening sealed records, promoting "open adoption," facilitating searches and reunions, and honoring transracial and transnational placements all converged on one theme: difference was the heartbeat of the new adoption revolution.[95]

Since adoption has existed as a formal arrangement, social regulation and scientific research have been justified on the basis of its distinctiveness. As the challenge of telling illustrates, difference became virtually synonymous with damage over the course of the twentieth century, spoiling adoptees' identities. Even as a new generation of reformers worked to normalize adoption after 1970, the association between adoption and trauma became more fixed in a culture in which the language of post-traumatic stress became ever more ordinary and encompassing. The proliferation of postadoption services and support groups during the past three decades is a case in point. It is ironic but true that greater honesty about adoption multiplied the possibilities for instructing participants and governing their family lives over the long term.[96] Adoption is no longer a process that produces fully legal, autonomous families and ends in court. It is a continuing journey with a life cycle all its own.[97] Selves in adoption are consequently emblematic projects, centered on difference as a permanent, problematic reality in need of inquiry and interpretation without end.

[94] H. David Kirk, *Shared Fate: A Theory of Adoption and Mental Health* (New York, 1964).

[95] Adam Pertman, *Adoption Nation: How the Adoption Revolution Is Transforming America* (New York, 2000).

[96] Nancy G. Janus, "Adoption Counseling as a Professional Specialty Area for Counselors," *Journal of Counseling & Development* 75 (March–April 1997): 266–75.

[97] Elinor B. Rosenberg, *The Adoption Life Cycle: The Children and Their Families through the Years* (New York, 1992); Jayne E. Schooler and Betsie L. Norris, *Journeys after Adoption: Understanding Lifelong Issues* (Westport, Conn., 2002).

Cultures of Categories:

Psychological Diagnoses as Institutional and Political Projects before and after the Transition from State Socialism in 1989 in East Germany

By Christine Leuenberger[*]

ABSTRACT

How can psychological categories be understood as historical, political, and cultural artifacts? How are such categories maintained by individuals, organizations, and governments? How do macrosocietal changes—such as the transition from state socialism in East Germany in 1989—correlate with changes in the social and organizational structures that maintain psychological categories? This essay focuses on how—pre-1989—the category of neurosis (as a mental disorder) became entwined with East Germany's grand socialist project of creating new socialist personalities, a new society, and a new science and on how diagnostic preferences were adapted, modified, and extended by local cultural and institutional practices. It also examines how post-1989 the category of neurosis became redefined in accord with a formerly West German psychotherapeutic paradigm and was eventually obliterated by the bureaucratic health care system of the new Germany. East German practitioners adopted new therapeutic guidelines and a new language to make sense of the "normal," "neurotic," and "pathological" self in terms of "individualizing forms of knowledge"[1] that tied in with efforts to remake East German citizens as liberal democratic subjects. At the same time, practitioners' clinical practice remained based upon face-to-face encounters in which formal guidelines and stipulations were often superseded by local, interactional, institutional, and cultural practices and contingencies.

INTRODUCTION

One of the seminal events of the twentieth century was the transition from state socialism in Eastern Europe and the subsequent increasingly global dissemination of capitalist forms of governance. On November 9, 1989, the Berlin wall fell. The wall

[*] Cornell University, Department of Science and Technology Studies, 301 Rockefeller Hall, Ithaca, NY 14853-7601; cal22@cornell.edu.

I am grateful to participants at the conference "The Self as Scientific and Political Project in the Twentieth Century: The Human Sciences between Utopia and Reform," Pennsylvania State University, October 2003, for useful comments. I also am indebted to numerous interviewees from the German psychological professions for their thoughtful discussions.

[1] Nikolas Rose, "Individualizing Psychology," in *Texts of Identity,* ed. John Shotter and Kenneth J. Gergen (London, 1992), 119–32; Rose, *Governing the Soul: The Shaping of the Private Self* (London, 1990).

had represented one of the great political, economic, and ideological divides of the twentieth century. Symbolizing the cold war, it had divided communist East Germany from capitalist West Germany for decades. The crumbling of the wall enabled the reunification of the two parts of Germany in 1991. The ensuing adoption of a market economy by former East Germany entailed a "paradigm shift" that left no aspect of its cultural production unchanged. As a new political, economic, and institutional order came into being, bodies of expert knowledge and practices, including the psychological sciences, were also reevaluated and brought in line with the requirements of a formerly West German health care system. This article focuses on how one set of specific psychological categories—neurotic disorders—was affected by these changes. A combination of in-depth interviews held with psychological practitioners and historical and archival sources are used to show how these diagnostic categories became implicated in the socialist and capitalist project of remaking concepts of self and society.[2]

Several treatises in the history of science and in sociology have illustrated that classificatory schemes and categorizations are neither natural nor universal.[3] Émile Durkheim revealed that categories, rather than representing natural kinds, are constituted by social and religious beliefs.[4] Scholars have also focused on such aspects as how categories arise and facilitate our understanding by becoming a "mental infrastructure"[5]; how they are maintained by social institutions[6]; and what social consequences they entail for the modern state.[7] How the psychological sciences are entwined with

[2] The data presented here is based upon a longitudinal sociological study of the East German psychotherapeutic community between 1990–2003. The study commenced in 1990, shortly after the fall of the Berlin wall. It consists of forty-six in-depth interviews (cited interview extracts are differentiated by letter code and year of interview). The interviews provide an oral history that is set against published sources that espouse and enforce socialist psychology. The historical memories thereby rendered are themselves "cultural documents" (Robert Perks and Alistair Thomson, eds., The Oral History Reader [London, 1998], 36) that speak as much to East Germany's socialist past as to the cultural and institutional circumstances that face psychotherapeutic practitioners at present. (See also Christine Leuenberger, "Socialist Psychotherapy and Its Dissidents," Journal of the History of the Behavioral Sciences 37 [2001]. 261–73.) The data corpus also includes numerous formal and informal discussions with East and West German practitioners and participant observation of psychotherapeutic and psychiatric practices (including thirty-six audio recordings of therapy sessions), as well as historical and archival research on East and West German psychology and psychotherapy before and after 1989.

[3] See George Lakoff, Women, Fire, and Dangerous Things: What Categories Reveal about the Mind (Chicago, 1987); Bruno Latour, Science in Action: How to Follow Scientists and Engineers through Society (Milton Keynes, UK, 1987); Mark S. Micale, "Charcot and the Idea of Hysteria in the Male: Gender, Mental Science, and Medical Diagnosis in Late Nineteenth-Century France," Medical History 34 (1990): 363–411; Micale, Approaching Hysteria: Disease and Its Interpretations (Princeton, N.J., 1994); Carl N. Degler, In Search of Human Nature (Oxford, 1991); Kurt Danziger, Naming the Mind: How Psychology Found Its Language (London, 1997).

[4] Émile Durkheim, The Elementary Forms of the Religious Life (New York, 1995).

[5] Geoffrey C. Bowker and Susan L. Star, Sorting Things Out: Classification and Its Consequences (Cambridge, Mass., 1999).

[6] David Bloor, "Durkheim and Mauss Revisited: Classification and the Sociology of Knowledge," Studies in the History and Philosophy of Science 13 (1982): 267–92; Mary Douglas, How Institutions Think (Syracuse, N.Y., 1986); Ian Hacking, "World Making by Kind Making: Child Abuse for Example," in How Classification Works: Nelson Goodman among the Social Sciences, ed. Mary Douglas and David L. Hull (Edinburgh, 1992), 180–238; Hacking, Rewriting the Soul: Multiple Personality and the Sciences of Memory (Princeton, N.J., 1995); Stuart A. Kirk and Herb Kutchins, The Selling of DSM: The Rhetoric of Science in Psychiatry (New York, 1992).

[7] Michel Foucault, The Order of Things: An Archeology of the Human Sciences (London, 1970); Foucault, "Governmentality," in The Foucault Effect: Studies in Govermentality, ed. Graham Burchill, Colin Gordon, and Peter Miller (Chicago, 1991).

the practice of classifying and categorizing people and their behavior has been taken up by philosophers,[8] feminist scholars,[9] sociologists,[10] and historians.[11]

The focus on East Germany's societal transformation provides the opportunity to trace empirically how individual, organizational, and governmental efforts before and after 1989 were implicated in creating, maintaining, and disseminating particular categories of mental disorders. The shift in concepts of "neuroses" exemplifies how the "macro" realm of political ideology, institutional parameters, and economic constraints combined with changes in the "micro" realm of everyday and professional knowledge and practices of categorization. Overall, the article shows how psychological categories became political and institutional projects under socialism and under the aegis of a market economy.

The first part of this article, "Socialist Psychology and Psychodiagnostics," deals with developments in the psychological sciences pre-1989. I show how categories of neuroses were embedded in East German psychological theory and redefined as part of a socialist vision. These socialist concepts of neuroses are then juxtaposed with interviewees' accounts that speak to how such psychological categories were maintained and contested by local and institutional practices. The second part, "From 'Socialist Personalities' to a Democratic Citizenry," focuses on the transition from state socialism in 1989 and its aftermath. I show how certain diagnostic categories that were in line with bureaucratic health care requirements became institutionalized. At the same time, we shall see how psychological practice yet again created informal cultural and institutional spaces for contesting and tinkering with formal psychological categories.

SOCIALIST PSYCHOLOGY AND PSYCHODIAGNOSTICS

After World War II, the Federal Republic of Germany (FRG) (Bundesrepublik Deutschland, BRD) and the German Democratic Republic (GDR) (Deutsche Demokratische Republik, DDR) were to be ideological experiments of two kinds. The FRG was to be transformed into a parliamentary democracy based on Anglo-American principles.[12] The GDR was to be turned into a socialist society based on the principles of justice and equality. These visions of a good society were implemented in different ways. The victorious allied powers "made strenuous efforts to 're-educate' the Germans, and . . . to transform the German 'personality' into a new mould."[13] West Germans were to be reeducated as democratic citizens.[14] The focus in this section is on how East Germans were to be turned into socialist personalities.

---

[8] Ian Hacking, "The Invention of Split Personalities," in *Human Nature and Natural Knowledge,* ed. Alan Donagan, Anthony N. Perovich Jr., and Michael V. Wedin (Boston, 1986), 63–85; Hacking, "Making Up People," in *Reconstructing Individualism,* ed. Thomas C. Heller, Morton Sosna, and David E. Wellbery (Palo Alto, Calif., 1986), 222–36; Hacking, *Mad Travellers* (London, 1998).

[9] Janet Wirth-Cauchon, *Women and Borderline Personality Disorder* (New Brunswick, N.J., 2001).

[10] Andrew Abbott, *The System of Professions* (Chicago, 1988); Kirk and Kutchins, *The Selling of DSM* (cit. n. 6); William W. Eaton, *The Sociology of Mental Disorders* (Westport, Conn., 2001); Michael Lynch, "Turning Up Signs in Neurobehavioral Diagnosis," *Symbolic Interaction* 7 (1984): 67–86.

[11] See John Carson, "Minding Matter/Mattering Mind: Knowledge and the Subject in Nineteenth-Century Psychology," *Studies in the History and Philosophy of the Biological and Biomedical Sciences* 30 (1999): 345–76; Georges Canguilhem, *The Normal and the Pathological* (Cambridge, Mass., 1991); Elisabeth Lunbeck, *The Psychiatric Persuasion* (Princeton, N.J., 1994).

[12] Mary Fulbrook, *The Two Germanies, 1945–1990* (London, 1992).

[13] Ibid., 63.

[14] Ibid.; Wolfgang Zapf, *Die Modernisierung moderner Gesellschaften* (Frankfurt, 1990).

By the 1950s, Marxist-Leninists had "glorified" science[15] and technology as tools for solving social and economic problems and establishing "scientific socialism."[16] In other words, socialism was to be built on and extend the principles of science. To implement its socialist vision, East Germany's ruling Socialist Unity Party (Sozialistische Einheitspartei Deutschlands, SED) sought greater control over higher education. Educational reforms included transforming scientific organizations (e.g., the Academy of Sciences) in accord with the institutional structures of the Soviet sciences, increasing academic links to the Soviet Union, and introducing and regulating "cadre policies" that ensured that jobs went to party members, sympathizers, and people with proletarian backgrounds.[17]

The health services and psychology were also harnessed to the socialist project. Shortly after WWII, the East German health services had been plagued with funding difficulties, inadequate facilities, and a shortage of qualified personnel. Consequently: "48% of all psychiatrists and neurologists in the Soviet-Occupied Zone had been members of the Nazi party."[18] Their practices were still guided by predominant physiological explanations inherent in German psychiatry since the late nineteenth century.[19] With the beginning of a new socialist era, Nazi-era clinicians were to be replaced (through cadre policies), and the psychological sciences were to be reformed and harnessed to the project of building "scientific socialism." Psychology was to enhance socialism in various ways. Transferring psychological services from the private to the public sector[20] was thought to improve the mental and physical health, living conditions, and educational standards of the population. A close link between applied and basic research would ensure psychology's scientific and social relevance to socialist society. Such reforms within the psychological sciences would also spur productivity, prosperity, and the economy[21] and thereby contribute to the "victory of socialism over capitalism."[22]

[15] See Fulbrook, *The Two Germanies, 1945–1990* (cit. n. 12), 17.

[16] Kristie Macrakis, "The Unity of Science vs. the Division of Germany: The Leopoldina," and Peter Noetzoldt, "From German Academy of Sciences to Socialist Research Academy," in *Science under Socialism: East Germany in Comparative Perspective*, ed. Kristie Macrakis and Dieter Hoffmann (Cambridge, Mass., 1999).

[17] Peter Nötzoldt points out that even though SED members were recruited to reform the academy, they did not necessarily "let the party discipline them, and some even fought among themselves." (Nötzoldt, "From German Academy of Sciences to Socialist Research Academy" [cit. n. 16], 148.) However, archival sources from the Ministry of Health reveal the scrupulous hiring and firing process that could ensue if members were not sufficiently enforcing party policies. The Ministry of Health Gruppe Kontrolle to Frau Staatssekretär Matern, 14 Dec. 1957, Betrifft: Folgen schlechter Parteiarbeit im Kreiskrankenhaus Apolda, Bundesarchiv DQ-1/2769, Abteilung Deutsches Reich und DDR sowie Stiftung der Parteien und Massenorganisationen der DDR, Bundesarchiv Berlin-Lichterfelde (hereafter cited as Abteilung Deutsches Reich und DDR, BAB-L).

[18] Greg Eghigian, "Was There a Communist Psychiatry? Politics and East German Psychiatric Care, 1945–1989," *Harvard Review of Psychiatry* 10 (2002): 365.

[19] Ibid., 365.

[20] Hans-Dieter Schmidt, "Psychology in the German Democratic Republic," *Annual Review of Psychology* 31 (1980): 195–209; Hans-Dieter Rösler, "Clinical Psychology in the Mental Health Service of a Socialist Community," *Journal of Psychiatric Nursing and Mental Health Services* 8 (1970): 36–7.

[21] Adolf Kossakowski, "Psychology in the German Democratic Republic," *American Psychologist* 35 (1980): 450–60; see also Harry Schröder, "Persönlichkeitspsychologie in der DDR," in *Psychologie in der DDR: Entwicklungen—Aufgaben—Perspektiven*, ed. Friedhart Klix, Adolf Kossakowski, and Walter Mäder (Berlin, 1980), 68–79; Boris F. Lomow, "Über Entwicklungsstand und Perspektiven der psychologischen Wissenschaft in der UdSSR," *Zeitschrift für Psychologie* 186 (1978): 1–24.

[22] Dr. Krüger, Ministerium für Gesundheitswesen (Hauptinspektion Gesundheitsschutz in den Betrieben) to Organ der Bezirksleitung der SED, Erfurt, letter, date unkown, Bundesarchiv Q-1 DQ1/4382, Abteilung Deutsches Reich und DDR, BAB-L.

As a result of academic exchanges and cooperation between East German and Soviet scientists and institutions, Soviet theories of human development became increasingly influential during the 1950s. In particular, Ivan Pavlov (1849–1936), a Russian physiologist and founder of the study of conditioned reflexes, had a transformative influence on the psychological sciences. His study of the nervous system and its adaptive features provided techniques for human behavior modification.[23] The underlying assumption of Pavlovian psychology, that human beings are malleable and transformable, meshed well with "scientific socialism." In 1952, the Ministry of Health inaugurated a "Pavlov-commission" to conduct workshops, conferences, and colloquiums with practitioners to "familiarize our physicians with Pavlov's ideas" to make them "a taken-for-granted part of" their knowledge base.[24] A "Pavlov-wave"[25] ensued, supporting biologically based treatment methods and behavioral therapy, whereby the social determinants (such as class background) were increasingly explicable by drawing upon Marxism-Leninism.[26] Consequently, human beings were understood as "biosocial units"[27] who were the product of the dialectics between environment, organism, and historical context.[28] Social determinants, however, were thought to primarily shape human nature. Therefore the assumption was that individual characteristics and social relationships could be reformed to transform people into "socialist personalities."[29]

The socialist notion of human nature as malleable and transformable paralleled developments in the social and human sciences in Western Europe and the United States at the time.[30] For instance, the Chicago School in the United States, similar to other conceptual traditions at the time, developed a notion of the self that was no longer anchored in biology, transcendental reason, or God but was amenable to social and cul-

[23] Jeffrey A. Gray, *Ivan Pavlov* (New York, 1980).

[24] Dr. Hans-Günther Giessmann, Über die Anwendung der Lehre Pawlows in der DDR. Presseartikel 1956–57, Bundesarchiv DQ-1/6636, 2–4, Abteilung Deutsches Reich und DDR, BAB-L.

[25] Kurt Höck, *Psychotherapie in der DDR: eine Dokumentation zum 30. Jahrestag der Republik* (Berlin, 1979), 14.

[26] Hans-Dieter Schmidt, "Einige Bemerkungen zum Problem der biologischen Grundlagen der Persönlichkeit," *Zeitschrift für Psychologie* 185 (1977): 214–24; William Woodward and S. C. Clark, "The Reflection of Soviet Psychology in East German Psychological Practice," in *Post-Soviet Perspectives on Russian Psychology,* ed. Vera Koltsova et al. (Westport, Conn., 1996), 236–50.

[27] Kossakowski, "Psychology in the German Democratic Republic" (cit. n. 21); Schmidt, "Problem der biologischen Grundlagen der Persönlickeit" (cit n. 26).

[28] Stefan Busse, "Gab es eine DDR-Psychologie?" *Psychologie und Geschichte* 5 (1993): 40–62; Busse, "'Von der Sowjetwissenschaft lernen': Pawlow—der Stein des Anstosses," *Psychologie und Geschichte* 8 (2000): 200–29.

[29] The "socialist personality" was to be the carrier of ideology, the state, culture, and social morals. Personality was seen as constituted by sociality (including social and occupational needs and requirements). It represented a "social quality" that extended beyond "character" (that was taken to be a reductionist definition of personality). Individual needs were subordinated to the goals and aims of the social unit. Consequently, "psychology of personality" was disavowed in favor of social psychology. See Busse, "Gab es eine DDR-Psychologie?" (cit. n. 28), 50; Schmidt "Psychology in the German Democratic Republic" (cit. n. 20); John Erhard, "Socialist Needs—Socialist Personality Development and the Working Class," *Society and Leisure* 4 (1974): 49–78; Ekkehard Sauermann, "Probleme der Marxistisch-Leninistischen Persönlichkeitstheorie," *Wissenschaftliche Zeitschrift Martin-Luther-Universität Halle-Wittenberg: Gesellschafts- und Sprachwissenschaftliche Reihe* 5 (1974): 7–23; Hans-Günther Eschke, "Zur Entwicklung der sozialistischen Persönlichkeit beim Aufbau der entwickelten sozialistischen Gesellschaft," *Wissenschaftliche Zeitschrift: Gesellschafts- und Sprachwissenschaftliche Reihe* 23(4) (1974): 441–9; Adolf Kossakowski, "Gesellschaftliche Anforderung und Weiterentwicklung der Psychologie in der DDR," *Zeitschrift für Psychologie* 184 (1976): 1–16.

[30] Degler, *In Search of Human Nature* (cit. n. 3).

tural influences.[31] The turn to culture and society as primary determinants of human behavior was thus an international trend that played itself out in different political contexts in various ways.

During the 1960s, East Germany underwent a political, economic, and scientific renewal. This spirit of reform was reflected in the country's commitment to "the new economic system"—a concept based on advancing socialism by taking advantage of science, technology, and economic rationality.[32] Psychologists and psychiatrists were one professional group recruited to further "scientific socialism." Their aim was to reform the psychological sciences as well as psychiatric institutions. As part of the commitment to establish "scientific socialism," psychologists and psychiatrists also critically reexamined various psychodiagnostic, psychometric (intelligence and personality traits testing), and depth psychological testing methods. Such diagnostic testing methods were critiqued as bourgeois and unscientific[33] as they were thought to predetermine individual characteristics.[34] The use of these methods became politically suspect, and they were therefore marginalized in practice.[35] The goal was to reconstitute them in line with the normative constituents of "the socialist personality" and its essential malleability and transformability by given historical and social circumstances.[36]

The Therapeutic Turn in Socialist Psychology

At a 1963 symposium about psychiatric rehabilitation in the town of Rodewisch, participants drew up what became known as the Rodewischer Theses.[37] These delineated how psychiatric institutions were to be transformed. The impetus for change in East German psychiatric institutions paralleled similar reforms in West Germany at the time. Indeed, the Rodewischer Theses borrowed "ideas of West German advocates of social psychiatry."[38] The attempt to be more responsive to people's individual rehabilitative needs by attending to the social causes of mental illness was thus part of an international trend in social psychiatry, although it took a particular form in the institutional context of the GDR.

Indeed, East Germany's ministry of health was very concerned about remaining competitive with West German clinics and treatments.[39] The reformist aims included:

[31] James A. Holstein and Jaber F. Gubrium, *The Self We Live by* (Oxford, 2000); Charles Lindholm, *Culture and Identity* (Boston, 2001).

[32] Dietrich Staritz, *Geschichte der DDR* (Frankfurt, 1996).

[33] Report by D. Müller-Hegemann, 22 June 1953, Bericht über die Münchner Neurologen-Psychiater-Tagung vom 26–29 Aug. 1953, Bundesarchiv DQ-1/5699, Abteilung Deutsches Reich und DDR, BAB-L.

[34] Busse, "Gab es eine DDR-Psychologie?" (cit. n. 28), 57; see also Dieter Feldes, Hannelore Weise, and Klaus Weise, "Psychologische und soziale Aspekte der psychiatrischen Diagnostik und Klassifikation," in *Sozialpsychiatrische Forschung und Praxis,* ed. Otto Bach et al. (Leipzig, 1976), 66–90.

[35] As Lothar Sprung and Helga Sprung point out, psychodiagnostic research was long underfunded. It was not until the early 1980s that a psychodiagnostic center was established in Berlin. "Geschichte der Psychodiagnostik in der Deutschen Demokratischen Republik—Ausbildung, Weiterbildung, Forschung, Praxis," *Psychologie und Geschichte* 7 (1995): 115–40.

[36] Ibid.; Lothar Sprung and Helga Sprung, "'Ein Zeitalter wird besichtigt'—Psychologie in Deutschland im 20. Jahrhundert," *Psychologie und Geschichte* 8 (2000): 360–96.

[37] Siegfried Schirmer, Karl Müller, and Helmut F. Späte, "Brandenburger Thesen zur Therapeutischen Gemeinschaft," *Psychiatrie, Neurologie und medizinische Psychologie* 28 (1976): 21–5.

[38] Eghigian "Was There a Communist Psychiatry?" (cit. n. 18), 365.

[39] From Krankenhaus für Neurologie und Psychiatrie, 28 Dec. 1964, Dr. Ulrich, Ärztlicher Direktor to Rat des Bezirks Potsdam, Abt. Gesundheits-und Sozialwesen, Herrn Bezirksarzt Dr. Richard,

transforming psychiatric clinics by providing rehabilitation programs, rather than offering care solely for patients with chronic and debilitative diseases; shortening patients' stays at psychiatric institutions; modernizing clinical facilities; increasing outpatient psychiatric care; and offering various therapeutic treatment methods (ranging
from occupational to sports and group therapies). These changes were to cater more
effectively to the needs of the mentally ill while also being part of a campaign to surpass the West German health care system.[40] The ascent of psychotherapeutic methods
within the East German health care system not only mirrored its adaptation of, and
ideological competition with, West German social psychiatry but also went hand in
glove with the socialist vision for transforming society and its people.[41]

In conjunction with the therapeutic turn in East German psychiatry, psychiatrists
and psychologists started to rethink psychodiagnostic concepts and methods. Here I
will focus solely on how concepts of "neuroses" were reconceptualized so as to take
account of "scientific socialism."

In 1968, the East German psychiatrist Christa Kohler first attempted to define concepts of "neuroses" from a Marxist perspective.[42] She pointed out that the way neurotic pathologies are conceptualized is inevitably intertwined with the philosophy, the
ideology, and the "image of man" of a given society.[43] Kohler attacked previous conceptions of neuroses put forth by such diverse proponents as Sigmund Freud, Alfred
Adler, Carl G. Jung, and Harald Schultz-Hencke as being infused with bourgeois conceptions of human nature. She maintained that their conceptions tended to be either
ahistorical, spiritualistic, agnostic, individualistic, or biologically determinist. All of
them, she argued, negated social determinants of behavior and failed to put forth a
coherent concept of human nature.[44]

What then, according to Kohler (1968), should be the concept of human nature to
inform psychological research? Kohler proposed, in accord with Marxist-Leninist
theory, that human nature was the product of the "dialectics" between personality, its
genetic make-up, society, and the environment. In other words, human beings were
socially determined, but social factors were ameliorated by people's genetic make-up.
Kohler argued that this "image of men"[45] called for a new interdisciplinary research
program built upon Marxism-Leninism. This is especially pertinent as individual
characteristics that might thrive under capitalism, such as communication difficulties
and egocentrism,[46] underlay neurotic illnesses. These were problems that could be
eliminated in a socialist society that fostered, nurtured, and enhanced personalities

---

Bundesarchiv DQ1-6076, Abteilung Deutsches Reich und DDR, BAB-L; Memorandum 15 July 1964
von Dr. Ulrich, Ärztlicher Direktor Neuruppin, Bundesarchiv DQ1-6676, Abteilung Deutsches Reich
und DDR, BAB-L.

[40] Psychiatrie, Bundesarchiv DQ1-3343, Abteilung Deutsches Reich und DDR, BAB-L.

[41] Schirmer, Müller, and Späte, "Brandenburger Thesen zur Therapeutischen Gemeinschaft" (cit.
n. 37); Greg Eghigian, "The Psychologization of the Socialist Self: East German Forensic Psychological Science and Its Deviants, 1945–1975," *German History* 22 (2004): 181–205.

[42] Werner König, "Zur Notwendigkeit weiterer Auseinandersetzungen mit der Psychoanalyse und
anderen psychotherapeutischen Schulen," in *Neurosen: Ätiopathogenese Diagnostik und Therapie,*
ed. Kurt Höck, Hans Szewczyk, and Harro Wendt (Berlin, 1973), 62.

[43] Christa Kohler, "Der Einfluss des Menschenbildes auf die Neurosentheorie," in *Beiträge zu einer
allgemeinen Theorie der Psychiatrie,* ed. Lothar Pickenhain and Achim Thom (Jena, Germany, 1968),
153–4.

[44] Ibid., 164; see also Höck, Szewczyk, and Wendt, *Neurosen: Ätiopathogenese Diagnostik und
Therapie* (cit. n. 42).

[45] Kohler, "Der Einfluss des Menschenbildes auf die Neurosentheorie" (cit. n. 43), 154.

[46] Ibid., 173–9.

with socialist sensibilities. Therefore a new research program should focus on how and to what extent social and individual factors could cause neurotic pathologies in a socialist society.

The concepts outlined by Kohler in 1968 were taken up at a symposium of the Society for Medical Psychotherapy of the GDR (Gesellschaft für Ärztliche Psychotherapie der DDR) in Bad Elster in 1969. In an edited volume titled *Neuroses* (*Neurosen*), the institutionally well-established and respected psychiatrist Kurt Höck put forth a new comprehensive definition of neurotic disorders.[47] For him: "a theory of neurosis" must account for "the specific features of personality under the conditions of socialism and its relation to the environment."[48]

Höck argued that dialectical materialism highlighted in a new way the relationship between humans and their environment. In his theory of neurosis, and its proposed diagnosis and treatment, he took into account the socialist conception of human nature as a biosocial unit.[49] He proposed that dialectical and historical materialism in conjunction with Pavlovian research into reflex conditioning could map out the social and biological aspects of personality.

Höck defined neurosis as a disorder that could produce both psychological and somatic symptoms. The symptoms were thought to be triggered by an experience or event that affect the relation between humans and their environment.[50] Höck drew attention to the a priori assumptions that informed this definition of neurosis. These included the beliefs that there exists a unity between "the human-environment-system," and that "external" social conditions largely determined "internal" biological human characteristics.[51] By taking account of such Marxist assumptions, he located his theoretical contribution explicitly in the burgeoning literature on the new and innovative socialist psychology of the time.

Höck proceeded to categorize neuroses into milder "neurotic reactions" and more severe "neurotic developments."[52] "Neurotic reactions" were subdivided into "lasting affective reactions" and "functionally fixated neurotic reactions."[53] Both neurotic reactions were thought to have external/environmental/social causes. However, the "functionally fixated neurotic reactions" became (in line with Pavlov's findings on conditioned reflexes) conditioned/fixated into behavioral patterns.[54] The more severe

[47] Höck, Szewczyk, and Wendt, *Neurosen* (cit. n. 42). Kurt Höck was an institutionally and politically well-established psychiatrist who, at the time, directed the psychotherapeutic department of a prominent Berlin clinic: das Haus der Gesundheit. Höck was asked to take over the Berlin clinic in the 1950s to eradicate the practice of dynamically oriented individual therapies. These were seen as too costly, time consuming, and inappropriate as they did not keep up with what was then perceived as the latest developments in psychology: the Pavlovian approach and biologically based treatment methods. Upon his arrival at the clinic, Höck introduced biologically based treatments such as *autogenous training* (a form of hypnosis) and initiated various group therapies that went hand in hand with an ideological commitment to enhance the socialist collective. See Leuenberger, "Socialist Psychotherapy and Its Dissidents" (cit. n. 2); Petra Sommer, "Kurt Höck und die psychotherapeutische Abteilung am 'Haus der Gesundheit' in Berlin—institutionelle und zeitgeschichtliche Aspekte der Entwicklung der Gruppenpsychotherapie in der DDR," *Gruppenpsychotherapeutische Gruppendynamik* 33 (1997): 130–47.

[48] Kurt Höck and Werner König, *Neurosenlehre und Psychotherapie* (Jena, Germany, 1979), 31.

[49] Ibid., 34

[50] Ibid., 36.

[51] Höck, Szewczyk, and Wendt, *Neurosen* (cit. n. 42).

[52] Kurt Höck, "Zur Definition und Klassifikation der Neurosen," *Psychiatrie, Neurologie und medizinsche Psychologie* 8 (1976): 484.

[53] Ibid., 484.

[54] Höck and König, *Neurosenlehre und Psychotherapie* (cit. n. 48), 44.

classification of "neurotic developments" were, however, caused by "intrapsychical processes" that stemmed from maladjustments during childhood.[55] These disorders were further categorized into "primary developmental disorder" and "secondary developmental disorder."[56] The primary disorder was acquired and "fixated"[57] during early childhood and consisted of "inadequate norms and value systems in terms of wrong expectations, misconceptions and maladjustments."[58] Such an inadequate socialization supposedly caused deficient personality structures that affected people's ability to handle psychological stress. The "secondary developmental disorder" was caused by biological or social events that disturbed the "person-environment" relationship during adult life.[59] The aim of treating all these four neurotic disorders was to "liberate" patients from their neurosis and to reestablish a "healthy dialectics" between them and their environment.[60]

From Tidy Theories to Messy Practices

Kurt Höck's neurotic nosology was to become what one interviewee called, the "typical GDR diagnosis."[61] These four neurotic categories were influential on how practitioners "diagnosed . . . in practice."[62] They were written into patient's medical files and referral letters, and they were used in publications. Indeed, articles published at the time attest to the widespread use of these classificatory practices.[63] One interviewee puts it as follows: "[If] one was to write an official article then it was always written within the official framework of the GDR's own neurotic classifications. They were developmental disorders."[64]

Although they relied on theories and diagnostic categories based on socialist psychology, East German practitioners kept up international contacts with colleagues and international organizations. The Ministry of Health, in conjunction with professional societies (such as the Society for Medical Psychotherapy of the GDR), frequently sent delegates to international conferences to keep abreast of international and West German professional advances. The aim was to compete with professional developments elsewhere and to persuade national and international audiences of East Germany's successes in conceptualizing and institutionalizing psychiatric, psychological, and psychotherapeutic care.[65] As a result, some practitioners were well aware of international, particularly West German, classificatory systems of mental disorders based upon other theories of neurosis.

---

[55] Höck, "Zur Definition und Klassifikation der Neurosen" (cit. n. 52), 485.

[56] Ibid., 485–6.

[57] Höck and König, *Neurosenlehre und Psychotherapie* (cit. n. 48), 30.

[58] Ibid., 47; Höck, Szewczyk, and Wendt, *Neurosen* (cit n. 42).

[59] Höck and König, *Neurosenlehre und Psychotherapie* (cit. n. 48).

[60] Ibid., 32.

[61] Interview (H) 2003.

[62] Interview (St) 2003.

[63] See Alfred Katzenstein and Achim Thom, "Die historische Leistung und die Grenzen des Werkes von Sigmund Freud (1856–1939)," *Psychiatrie, Neurologie und medizinische Psychologie* 34 (1982): 68–78; Christina Schröder and Werner König, "Die Integration der Psychotherapie in die klinische Medizin in Deutschland—der Beitrag von J. H. Schultz," *Psychiatrie, Neurologie und medizinische Psychologie* 36 (1984): 586–90.

[64] Interview (St) 2003.

[65] Ministerium für Gesundheitswesen, Ausgewählte Ergebnisse der Entwicklung des sozialistischen Gesundheitswesen, Sept. 1963, Referentenmaterial, Bundesarchiv Q-1/DQ1/1836, Abteilung Deutsches Reich und DDR, BAB-L.

In West Germany, definitions of neurosis were set within a very different institutional and intellectual context. Psychoanalytic training had became formalized and professionalized, and psychoanalysis was institutionalized as part of the health services.[66] By 1967, the state health insurance system was reimbursing psychoanalytical and depth-psychological treatments. Psychoanalysis was also popularized by the renowned German psychoanalyst Alexander Mitscherlich. Thomas Müller and Desiree Ricken point out that, at that time, psychological themes were also increasingly taken up in the popular media. This correlated with a rise in neurotic diagnosis between 1950 and 1970 in clinical practice. Unlike in East Germany, where he was decried as bourgeois and unscientific, Sigmund Freud and his legacy infused developments in West German psychotherapy. Additionally, professional and personal contacts with Anglo-Saxon practitioners also informed West German psychotherapeutic practices.[67]

Definitions of neurosis in West Germany, like those in the East, also accounted for individual character, biological make-up, and the environment. Given the wide range of theoretical approaches in West German psychotherapeutic theory, however, there was little agreement on the exact etiology of neuroses.[68] However, Roderich Hohage contends that practitioners did agree that "the causes of neurotic disorders lay within childhood."[69] The predominant assumption was that, during early childhood, trauma and problematic parent-child relationships gave rise to "unconscious conflicts." Besides such conceptual differences over the etiology of neurosis between East and West German practitioners, West German psychotherapeutic theory also focused on a different set of concerns than those elaborated in Kurt Höck's approach. The emphasis was on the notion of the unconscious, sexual and libidinal desires, psychological resistance, and repression.[70]

Concurrent with the ascent of psychoanalytic ideas in West German psychotherapy, psychoanalytic and psychodynamic ideas experienced a renaissance internationally as well. These approaches informed and were legitimized by *The Diagnostic and Statistical Manual of Mental Disorders: DSM-II* (published by the American Psychiatric Association [APA] in 1968). It was not until 1980 that within its new expanded edition—the *DSM-III*—the psychodynamic tradition was largely abandoned in favor of a biopsychiatric approach.[71] West German practitioners increasingly used the language and terminology of the *DSM* to satisfy various bureaucratic requirements of the statutory health insurance. These developments were also increasingly to affect East German psychotherapeutic practices.

---

[66] Annemarie Dührsen, *Analytische Psychotherapie in Theorie, Praxis und Ergebnissen* (Göttingen, Germany, 1972). Psychoanalytic practitioners remained ambiguous about the advantages of turning psychoanalysis into an institutionalized profession. Many argued that it hereby had lost its potential for social criticism and reform. See Margarete Mitscherlich-Nielsen and Detlef Michaelis, "Psychoanalyse in der Bundesrepublik," *Psyche* 38 (1984): 577–84; Johannes Cremerium, "Die Präsenz des Dritten in der Psychoanalyse: Zur Problematik der Fremdfinanzierung," *Psyche* 35 (1981): 1–41.

[67] For an analysis of how psychoanalysis was adopted and shaped by various intellectual traditions in West Germany, see Thomas Müller and Desiree Ricken, "Alexander Mitscherlichs 'politische' Psychoanalyse, seine Beziehung zur Humanmedizin und die Wahrnehmung der bundesdeutschen Öffentlichkeit," in *Geschichte der Psychoanalyse: Tel Aviv Jahrbuch für Deutsche Geschichte,* vol. 32, ed. Moshe Zuckermann (Göttingen, Germany, 2004), 219-257.

[68] Roderich Hohage, *Analytisch orientierte Psychotherapie in der Praxis* (Stuttgart, 1997).

[69] Ibid., 36.

[70] See Dührsen, *Analytische Psychotherapie in Theorie, Praxis und Ergebnissen* (cit. n. 66); Franz R. Faber and Rudolf Haarstrick, *Kommentar: Psychotherapie-Richtlinien* (Neckarsulm, Germany, 1996).

[71] Kirk and Kutchins, *The Selling of DSM* (cit. n. 6).

After the 1970s, growing economic and political cooperation between the two German states (the result of *Ostpolitik* as propagated by West Germany's Social Democratic chancellor Willy Brandt) brought about a renewed openness to West German and international influences. Western literature was increasingly cited alongside conventional Soviet sources, and some formerly repudiated Western psychodiagnostic tools and tests (such as Minnesota Multiphasic Personality Inventory) were adopted in clinical practice.[72] Concurrent with this détente, psychotherapeutic practitioners began to use international classificatory systems for mental disorders, such as *DSM* and the *International Classification of Disease* (*ICD*) (published by the World Health Organization, WHO), in their clinical practice. As one interviewee reported:

> [In] the 1980s there was a change in the international classification. Early disorders were defined as narcistic personality disorder or as borderline disorder and this was also taken on . . . but . . . that was unofficial—between colleagues one could maybe use the specialist terminology of borderline or narcistic personality disorder. But it wasn't recorded like that. Instead they were [recorded as] primary neurotic developmental disorders.[73]

Accordingly, despite the fact that clinicians used *DSM* classifications in practice, for official documents they continued to use Kurt Höck's terminology. Interviewees also maintained that other classificatory systems, such as Freudian psychoanalysis, were invoked in practice: "one could also use oral, anal, . . . genital personality according to Freud. But that was not official."[74]

Arguably, the kind of classificatory system used and how it was implemented depended on local clinical practices. The classificatory preferences of colleagues, heads, and supervisors, as well as of clinics' international networks, were decisive in this respect: "Clinics could always decide themselves how they codified and this depended on the head and what he preferred . . . the diagnoses were decided during visitations by the head."[75]

Besides the role of colleagues and superiors, the geographic location of a particular clinic and its networks to other national and international clinics affected diagnostic preferences. For instance, in a rural northern clinic one interviewee observed: "We didn't have the term borderline personality disorder in the GDR, or maybe we just didn't have it in [name of town]. That was . . . codified under primary developmental disorder."[76] By contrast, interviewees (from a Berlin clinic) reported that they were eager to keep up with West German developments and used the *DSM* classificatory schema in practice.[77]

These local variations in the use of classificatory systems were often linked to a consensus built up among like-minded colleagues at a clinic. This, in turn, was to a large extent determined by what "the head" was doing. The department heads may or may not have been party members, but they could substantially influence the intellectual direction of the whole unit. As one interviewee reported: "There was the type of SED [member of the German Socialist Unity Party] head who was a protective

---

[72] Eghigian, "Was There a Communist Psychiatry?" (cit. n. 18), 366.
[73] Interview (St) 2003.
[74] Ibid.
[75] Interview (H) 2003.
[76] Ibid.
[77] Interview (Hh) 2003.

shield for others, he warded off what came—and there were others who conformed and pressurized others and checked that they all were politically flawless."[78]

Heads of medical units could thus shield others from political intervention or enforce the party line. How staff classified mental disorders in various clinics was therefore crucially determined by "the categories of the head doctor."[79] Local hierarchical arrangements could thereby dictate to what extent staff conformed or diverged from official practices of categorization. As Mitchell Ash points out: "the traditional top-down, authoritarian structure of German academic research institutions"[80] was thus also carried over into East Germany's psychiatric institutions.

Additionally, depending on practitioners' status and seniority, they also enjoyed remarkable independence in clinical practice. As state employees, certified practitioners were under no economic pressure and were not formally accountable for their clinical activities. Treatment regimes were at the discretion of the treating physicians and did not need approval from outside agencies (e.g., the health insurance provider). Psychological practice, therefore, could thus form a "niche."[81] It not only provided a space to experiment with various therapeutic approaches but also facilitated East Germany's thriving black economy. As one respondent pointed out: "There were practitioners who, when someone rang downstairs, threw sick leave certificates out the window if it was the plumber whom he needed urgently."[82] In other words, some practitioners would freely give out sick leave certificates so the plumber could get off work to come around to their houses to do some long needed repairs for some extra income.

Depending on local, social, and institutional contexts, people could thus resist, recreate, and improvise a set of informal practices lying outside the formal command structure. Clinical practice could constitute a niche in which practitioners could walk the fine line between dogma and dissidence. Professional colleagues and heads could crucially shape how formal theoretical precepts were incorporated in practice. In addition, various networks among colleagues that spanned geographic, disciplinary, and ideological boundaries assured the influx of West German and international classificatory practices (e.g., Berlin's Charité clinic had extensive international networks and used the *DSM* classification system of disease).[83]

Arguably, East German developments in the psychological sciences must be understood in relation to the international context as well as to local factors.[84] Scholars of East German science[85] disagree over whether East German psychology was in line

---

[78] Interview (G) 2001.

[79] Interview (H) 2003.

[80] Mitchell Ash, "Kurt Gottschaldt and Psychological Research in Nazi and Socialist Germany," in Macrakis and Hoffmann, *Science under Socialism* (cit. n. 16), 298.

[81] Günther Gaus, *Wo Deutschland liegt: Eine Ortsbestimmung* (Munich, 1986); Macrakis and Hoffmann, *Science under Socialism* (cit. n. 16); Fulbrook, *The Two Germanies, 1945–1990* (cit. n. 12); Staritz, *Geschichte der DDR* (cit. n. 32); see also James C. Scott, *Seeing Like a State: How Certain Schemes to Improve the Human Condition Have Failed* (New Haven, Conn., 1988).

[82] Interview (G) 2001.

[83] The Charité (Berlin) was nationally and internationally recognized as one of the best research clinics in East Germany. It is also there that Professor Karl Leonhard (*Individualtherapie der Neurosen* [Berlin, 1981]) developed a nosology for severe mental disorders. How Leonhard established his diagnostic schema and its relationship to political ideology (e.g., he was branded a biological determinist and eventually integrated behavioral analysis and Pavlovian theories into his therapeutic approach) is a matter for further investigation.

[84] See Busse, "Gab es eine DDR-Psychology?" (cit. n. 28).

[85] See Sprung and Sprung, "Geschichte der Psychodiagnostik in der Deutschen Demokratischen Republic" (cit. n. 35); Sprung and Sprung "Ein Zeitalter wird besichtigt" (cit. n. 36); Busse, "Gab es eine

with international developments or had become a uniquely socialist science. Certainly, as Slava Gerovitch has documented, the Eastern European scientific community was aware of, and was competing with, Western scientific approaches. The official aim was to either "criticize and destroy" or "overtake and surpass" Western science.[86]

International scientific developments did not, therefore, bypass East German science. For instance, the turn to culture, environment, and society as a way of understanding human similarities and differences in the social and behavioral sciences was an international trend that had been underway since the 1920s. The power of the cultural paradigm was also dramatically enhanced by the disastrous consequences of the rise of German eugenics as a justification for exterminating people based on their race and biology during WWII.[87] In the postwar period, biologically and racially based theories became temporarily eclipsed by the cultural turn (only to rise again in the 1950s and 1960s in terms of probabilistic, but not deterministic, genetic theories). Particularly in Germany, the postwar period was a period of renewal. Socialism and democracies were driven by visions of a "good" society,[88] and scientists were predisposed to favor change, progress, social improvement, and reform.[89] Culture, environment, and society as primary determinants of human behavior could potentially bring about desired changes in democratic as well as socialist societies. East German psychologists' commitment to environment and culture in bringing people together or setting them apart was thus very much in keeping with intellectual sentiments at the time.

But science is always also produced locally. Science studies has shown how science is contingent upon often very local cultural, material, and social resources. To advance their science and resolve scientific controversies, scientists sometimes have to establish alliances with funding agencies and political actors. They need to "mobilize material and financial as well as conceptual, rhetorical, and ideological support."[90]

The example of the remaking of neurotic categories by Kurt Höck reveals how these local resources play out. Höck had the institutional status and resources to produce a shift in the way neurotic disorders were understood. As a socialist and director of a leading clinic for psychotherapy, he had the clout and resources necessary to bring about changes in the psychological sciences.[91] He also cleverly used Marxism-Leninism to make the rhetorical case for a new approach to neurotic disorders. His definition of neuroses addressed the philosophical concerns of dialectical materialism by attending to the "dialectics" between humans and their environment. In his treatises, Höck acknowledged the importance of the work of Ivan Pavlov by explicitly drawing on Pavlovian theory, methods, and terminology.[92] His approach also paid tribute to the

---

DDR-Psychology?" (cit. n. 28); Busse, "Von der Sowjetwissenschaft lernen" (cit. n. 28); Joachim Kocka, *Vereinigungskrise: Zur Geschichte der Gegenwart* (Göttingen, Germany, 1995).

[86] Slava Gerovitch, *From Newspeak to Cyberspeak: A History of Soviet Cybernetics* (Cambridge, Mass., 2002), 17.

[87] Degler, *In Search of Human Nature* (cit. n. 3).

[88] Talcott Parsons, "The Symbolic Environment of Modern Economies," in *Economic Sociology,* ed. Richard Swedberg (Cheltenham, UK, 1996), 218–35.

[89] Degler, *In Search of Human Nature* (cit. n. 3).

[90] Ash, "Kurt Gottschaldt" (cit. n. 80), 289–90.

[91] Leuenberger, "Socialist Psychotherapy and Its Dissidents" (cit. n. 2).

[92] Höck aligns his work with H. Schultz-Hencke who reconceptualized the relationship between physical and psychological processes in terms of Pavlovian theory (H. Schultz-Hencke, "Arzt und Psychotherapie," *Das Deutsche Gesundheitswesen* 1 (1946): 247–52); see also Kurt Höck, ed., *Zielstel-*

importance ascribed to social determinants of human behavior in Marxist-Leninist theory.

By linking neuroses to the philosophical concerns of material dialectics, Höck bowed "toward the political discourse."[93] He used various locally contingent rhetorical, institutional, and cultural resources to advance his science. His classification schema was part of a political system and a socialist vision. However, his grand vision often faltered as it encountered clinical practice, medical hierarchies, and national and transnational professional networks.

FROM "SOCIALIST PERSONALITIES" TO A DEMOCRATIC CITIZENRY

During the 1980s, communist countries from the Soviet Union to Hungary increasingly adopted liberalizing political and economic policies. The East German government was under pressure to follow suit. By the fall of 1989, there were regular political protests, and East Germans left in droves, packing cars and trains to the Czech Republic and Hungary, which had both opened their borders to the West. On November 9, the Central Committee of the Communist Party met to discuss "provisional regulations"[94] that would permit East Germans to travel to the West. They were to be made public the following day. However, East Germany's political leader, Egon Krenz, asked Gunther Schabowski (a Central Committee member who had not been present at the meeting) to mention these new travel regulations at a press conference that evening. When quizzed by a journalist as to when the new rules would come into effect, Schabowski hesitated and then said "immediately!"[95] There was a run on the border, thousands of people crowded the checkpoints, and after stamping a few passports, the overwhelmed guards opened the barriers. By midnight, all border crossings within Berlin, and an hour later all checkpoints to West Germany, were open.

After the Berlin wall had crumbled, neoliberal reforms were instituted to liberalize and privatize the East German economy and to restructure its political system. Pre-1989, East Germany might have justifiably been a "socialist laboratory,"[96] but now it was to become yet another laboratory for neoliberal economic and social reforms.

Scholars have pointed to various economic developments that in various degrees facilitated the swift adoption of the neoliberal model. What promoted the neoliberal restructuring process in East Germany included: transnational networks composed of Eastern European and American economists before 1989,[97] as well as the support of international agencies, Western governments, and Western advisers to the transitioning economies.[98] In addition, by rapidly adopting already established and legitimate West German institutions and policies, and transforming its economic and political

---

lungen und Entwicklungen der Psychotherapie in der DDR—ausgewählte Schriften von Kurt Höck anlässlich seines 65. Geburtstages: Psychotherapie-Berichte 31 (Berlin, 1985); Höck, *Psychotherapie in der DDR* (cit. n. 25).

[93] Ash, "Kurt Gottschaldt" (cit. n. 80), 295.

[94] Staritz, *Geschichte der DDR* (cit. n. 32), 381.

[95] Ibid., 381.

[96] Johanna Bockman and Gil Eyal, "Eastern Europe as a Laboratory for Economic Knowledge: The Transnational Roots of Neoliberalism," *American Journal of Sociology* 108 (2002): 310–52.

[97] Ibid.

[98] Lawrence P. King, *The Basic Features of Post-Communist Capitalism in Eastern Europe* (London, 2001); Eric Hanley, Lawrence King, and Istvan Toth Janos, "The State, International Agencies, and Property Transformation in Postcommunist Hungary," *Amer. J. Sociol.* 108 (2002): 129–67.

institutions accordingly, East Germany gained legitimacy for international financial organizations and investors.[99]

As part of this whole-scale institutional transfer of West German political and economic institutions,[100] the East German health care system was overhauled. The institutional structure and professional standards of the West German health care system were successively introduced.[101] The institutional transfer was, at the time, crucially facilitated by the support of powerful social interest groups and professional lobbies. These had overridden debates about the shortcomings of the West German health care system and the possibility of upholding some East German institutional structures and services.[102] East German psychologists, psychiatrists, and psychotherapists were now thrown into a laboratory of a different kind. They were novices in dealing with what formerly were West German institutional landscapes. Their experiences reveal some of the taken-for-granted practices upon which the West German psychological profession had been built.

Most important, East Germany's psychological services were increasingly transferred from the state to the private sector. This meant practitioners worked under a different set of economic constraints and possibilities. For instance, they were to largely rely on the state health insurance system to cover their services. Practitioners were hereby subject to specific bureaucratic guidelines as to what constituted a psychological problem and its appropriate therapeutic treatments. These guidelines, in turn, became the official arbitrator of which psychological theories and practices were seen as legitimate. Reimbursable therapeutic methods included: *behavioral therapy*—a short-term, conflict-oriented therapy that attempts to alter maladaptive behavior according to learning theoretical precepts; *analytical therapy*—a long-term therapy that is to reveal repressed and unconscious neurotic conflicts through the processes of "regression," "transference," and "resistance"; and *depth-psychological psychotherapy*—a short-term psycho-dynamically oriented conflict-oriented therapy that addresses a specific unconscious conflict as the cause of the psychological illness.[103] As a result of these new stipulations, East German practitioners had to fulfill certain professional qualifications to continue practicing. Although they could get their treatments reimbursed under the categories of behavioral therapy and depth-psychological therapy, their qualifications were not thought to be sufficient for the practice of analytical therapy. Such guidelines and stipulations often encouraged practitioners to forgo previously used therapeutic methods in favor of approaches that were institutionally legitimized at the time, such as psychoanalytically informed approaches (that had, as mentioned earlier, been established within the West German mental health care system since 1967).

---

[99] Walter W. Powell and Paul J. DiMaggio, eds., *The New Institutionalism in Organizational Analysis* (Chicago, 1991); Sarah L. Babb, *Managing Mexico: Economists from Nationalism to Neo-Liberalism* (Princeton, N.J., 2001).

[100] Hans Joas and Martin Kohli, eds., *Der Zusammenbruch der DDR* (Frankfurt, 1993).

[101] See Philip Manow, *Gesundheitspolitik im Einigungsprozess* (Frankfurt, 1994). Paul DiMaggio and Walter Powell (*New Institutionalism in Organizational Analysis* [cit. n. 99]) also have pointed out that during uncertain times, organizations may often enhance their legitimacy by replicating already established procedures. In this case, there was significant pressure on East German governmental and nongovernmental organizations to replicate the already established structure of comparatively powerful and successful West German institutions.

[102] Manow, *Gesundheitspolitik im Einigungsprozess* (cit. n. 101).

[103] Faber and Haarstrick, *Kommentar: Psychotherapie-Richtlinien* (cit. n. 70); see also Christine Leuenberger, "The End of Socialism and the Reinvention of the Self: A Study of the East German Psychotherapeutic Community in Transition," *Theory and Society* 31 (2002): 257–82.

Indeed, as part of the bureaucratic overhaul, East German practitioners, like their West German counterparts, were now required formally to submit treatment plans for approval by therapeutic referees (qualified professionals who monitored whether the proposed course of treatment satisfied the stipulations set forth by the health insurance system) before treatment could commence. These requests were to be framed as lengthy discursive reports about patients' symptoms, states of health, and biographies. The treatment plans also proposed a diagnosis, a course of treatment, and a prognosis. They were to be couched within psychoanalytically informed terminology and methods. As one interviewee pointed out:

> Because we have to justify this analytical model to insurance or the referees in each single case . . . we are all very much forced to take into account this way of thinking and somehow identify with it and I think . . . you are forced to get a bit of training in this in order to get the reports through.[104]

Arguably, it was such bureaucratic requirements that inadvertently led to an ascent of psychoanalytical thinking, thereby propelling "individualizing" approaches to human nature.[105] During the immediate postwall period, numerous treatises were published decrying the supposedly collectivist-oriented East Germans[106] and proposing the necessity for them to individualize. Psychologists, psychotherapists, and self-help organizers attested to the need to reform East Germany's culture of dependency into a culture that was to encourage individuation, individual autonomy, and independence.

The neoliberal economic model that had remade East Germany's economic and political landscape seemed to have found its counterpart in psychological practices. There, individualizing forms of knowledge could help remake a people into neoliberal subjects who could independently and self-reliantly navigate the newly emerging market economy. As Nikolas Rose points out, individuals who assume themselves to be self-contained, independent, and autonomous decision makers may enhance their ability to advance economically in a competitive market economy and integrate into a neoliberal democratic citizenry.[107] The fact that therapeutic practitioners could now find economic reward and therapeutic practice to be fitting partners in the new capitalist dance had, however, less to do with political stipulations (as in former socialist East Germany) than with West Germany's institutional inertia and bureaucratic accounting requirements.

Concurrent with the ascent of "individualizing forms of knowledge" within the East German health care system, classificatory schemes such as the APA's *Diagnostic and Statistical Manual of Mental Disorders,*[108] as well as WHO's *International*

---

[104] Interview (Hh) 1997.

[105] Rose, "Individualizing Psychology" (cit. n. 1); Susanne Kirschner, "The Assenting Echo: Anglo-American Values in Contemporary Psychoanalytic Developmental Psychology," *Social Research* 57 (1990): 821–57.

[106] See Hans-Joachim Maaz, *Gefühlsstau: Ein Psychogramm der DDR* (Berlin, 1990); Maaz, *Das Gestürzte Volk: Die Verunglückte Einheit* (Berlin, 1991); Reimer Hinrichs, "Patient DDR," in Stefan Welzk, ed., *Abriss der DDR* (Berlin, 1990), 57–66; Brigitte Rauschenbach, ed., *Erinnern, Wiederholen, Durcharbeiten: Zur Psycho-Analyse deutscher Wenden* (Berlin, 1992); Wolfgang Möller, "Entfremdung. Eine Heilungsgeschichte," in Welzk, *Abriss der DDR* (Berlin, 1990), 67–73.

[107] Rose, "Individualizing Psychology" (cit. n. 1).

[108] In the United States, the earliest classification system for mental disorder was developed for the census by the federal government. In the 1840 census, there was just one mental disorder: idiocy, which included insanity. The categories, however, multiplied with every census. By 1918, the American Medico-Psychological Association (forerunner of the American Psychiatric Association) in

Classification of Disease[109] had to now be used to categorize what were taken to be mental disorders into a reimbursable nosology. Practitioners used these categories to specify a diagnosis on insurance forms. This ensured that the statutory health insurance would cover the treatment.[110] East German practitioners therefore adopted various, often contradictory, classificatory practices that stemmed from either psychoanalytic approaches or the latest versions of the *DSM* and *ICD*. While many had long been familiar with these classificatory schemas (such as *DSM* and *ICD*) because they had worked in clinics where "they took pride in not being behind the times and having the same standard as West Germany,"[111] others encountered them only after 1989. "Only in the transition [in 1989] did I get to know about this American classification [*DSM*] and also the *ICD-10* [the then latest version of the *ICD* manual] of the WHO. That I only got to know about now."[112]

However, in practice, various classificatory systems continued to be used, including Kurt Höck's nosology. As one interviewee observed: "years after the transition some physicians wrote down primary developmental disorder because they once learned it this way and perhaps found that also quite useable and practicable."[113]After 1989 just as before, tidy theories and classificatory stipulations became mangled in

---

cooperation with the National Committee for Mental Hygiene published the first standardized psychiatric nosology, the *Statistical Manual for the Use of Institutions for the Insane,* which contained twenty-two categories based on biological and somatic disorders. This manual went through ten editions between 1918 and 1942 and retained its somatic orientation throughout. The first major change of nosology was embodied in APA's publication in 1952 of *DSM-I.* It contained just over 100 diagnostic categories. When *DSM-II* was published in 1968, the number of categories had expanded to nearly 200 and stood in the psychodynamic and psychoanalytic tradition. By 1980, these had expanded to approximately 265 categories in *DSM-III.* This manual was a radical departure from the psychodynamic tradition in favor of a psychiatric approach reimbursable by health insurance. *DSM-III-R* (published in 1987) (with approximately 300 mental disorders) and *DSM-IV* (with up to 357 categories) proceeded within the same psychiatric tradition. (Note: As there is disagreement over the exact number of categories for mental disorder in the various updates [compare Kirk and Kutchins, *The Selling of DSM* (cit. n. 6); and Eaton, *Sociology of Mental Disorders* (cit. n. 10)] I only give approximate figures.) As medical sociologists point out, the dramatic increase in these categories is yet another example of the increasing medicalization of wider and wider ranges of behavior. See Peter Conrad, "Medicalization and Social Control," *Annual Review of Sociology* 18 (1992): 209–32.

[109] *ICD* originated in the late nineteenth century. It was first published in 1893 as the *Bertillon Classification* or *International List of Causes of Death.* It has been revised roughly every ten years since then. *ICD* is vital for assembling, standardizing, and aggregating medical information (including mortality as well as the mental and physical health of populations) for an international public health bureaucracy. This facilitates the evaluation of epidemiological trends and national and international health care. *ICD*'s latest revision, *ICD-10,* is very descriptive of symptoms. It has disease categories such as "nail biting" and "hair pulling." By being purely descriptive, its proponents claim, it is more objective and less theory-laden. See Bowker and Star, *Sorting Things Out* (cit. n. 5), for an in-depth description and history of *ICD; International Classification of Disease (ICD),* http://www.who.int/whosis/icd10; Volker Faust, *Psychosoziale Gesundheit: Die Neurosen—einst und heute* (2003), http://www.psychosoziale-gesundheit.net/psychiatrie/neurosen/html.

[110] The necessity to categorize patients' behavior and symptoms into different categorical schemes for submitting treatment plans and requesting reimbursement for treatments stems from two longstanding traditions in the German psychological sciences. On the one hand, there is the legacy of Emil Kräpelin, who in the late nineteenth century identified and categorized clinical signs and disease to constitute a classificatory scheme that would satisfy scientific pretensions. On the other hand, Sigmund Freud instigated the practice of describing psychological states as discrete entities (Kirk and Kutchins, *The Selling of DSM* [cit. n. 6]). German practitioners continue to use both approaches for understanding disease as they need to request reimbursement for services rendered by using *ICD* categories, but submit treatment plans in which they describe patients' psychological state phenomenologically.

[111] Interview (Hh) 2003.

[112] Interview (S) 2003.

[113] Interview (Hh) 2003.

therapeutic practice.[114] Pre-1989, local cultural circumstances largely determined which classificatory system was used in practice. In the post-transition period, the use of different classificatory schemes was yet again largely dependent on one's place of work and on the theoretical and methodological preferences of heads and staff.

By 2000, however, new rules stipulated that to get treatments reimbursed by the state health insurance system, East (and West) German practitioners were obligated to use *ICD-10* categories for classifying physical and mental conditions. At that time, practicing clinicians could attend courses offered as part of an effort to familiarize them with *ICD-10* categories. Students in training, however, were now only taught *ICD-10* diagnostic categories. Consequently, as one respondent put it, these students have little or no knowledge of any other (what she referred to as) "strange classificatory schemes."[115] The reason: "*ICD-10* is now the strict foundation for applying, for instance, for therapeutic services to be covered by the health insurance. There is a number for it without the background [of the disorder] being spelled out exactly."[116]

Practitioners have, according to one interviewee, "to settle for an *ICD-10* diagnosis" as the reimbursement claim to the health insurance; otherwise the claim will "come back."[117] Consequently, the category of neurosis has also largely been displaced by other categories of mental disorders such as "borderline" or "personality disorder." As one respondent put it in 2003: "The term 'neurosis' doesn't exist any longer; these are now all 'personality disorders.'"[118]

The concept of "neurosis" has been around for more than 200 years (the term was first coined by a Scottish doctor, William Cullen, in 1769) and has throughout its history been conceptualized in different ways.[119] It was appropriated by Kurt Höck and integrated into a socialist psychology. In *DSM-II* (1968), neurosis was defined as a category applicable to depressed individuals who suffered from conscious or unconscious anxieties and distress.[120] Despite its many meanings and long history, the concept of "neurosis" fell out of favor in the United States with the publication of the updated *DSM-III* in 1980. Consequently, what formerly was described as a neurosis was increasingly categorized as an anxiety or depressive disorder.[121]

According to interviewees, it is the concept of "personality disorder" that has now eclipsed the use of the term "neurosis" and has turned into somewhat of "a diagnostic preference"[122] among German practitioners.[123] Practitioners find "an inflation of borderline and personality disorders. There were suddenly so many patients with this diagnosis after the transition. I think we would have seen that differently in the past."[124]

---

[114] Andrew Pickering, *The Mangle of Practice* (Chicago, 1995).

[115] Interview (St) 2003.

[116] Ibid.

[117] Interview (H) 2003.

[118] Interview (St) 2003.

[119] Faust, *Psychosoziale Gesundheit* (cit. n. 109).

[120] Eaton, *Sociology of Mental Disorders* (cit. n. 10).

[121] See David Healy, "Shaping the Intimate: Influences on the Experiences of Everyday Nerves," *Social Studies of Science* 34 (2004): 219–45.

[122] Jan Goldstein, *Console and Classify: The French Psychiatric Profession in the Nineteenth Century* (Chicago, 1988), 330.

[123] There are many different types of personality disorders that describe a range of maladaptive patterns of behavior that can include paranoia, excessive mood swings, and/or shyness. They can also characterize explosive, hysterical, or antisocial personalities that can cause problems for others. Eaton, *Sociology of Mental Disorders* (cit. n. 10), 17.

[124] Interview (H) 2003.

Such shifts suggest that changes in diagnostic practices are not due to a more accurate depiction of the patients' mental state, nor is it the result of new research. Rather, to "name" a disorder is to slot a diagnosis within a prevailing medical discourse or categorization scheme. It is then the "relationships of distinction, opposition, super- and subordination, etc., to other categories"[125] gathered from a particular discourse that provide a way to make sense of psychological phenomena.

Furthermore, once a dominant medical discourse or classificatory system becomes established, practitioners use it as a pamphlet to the mind. Indeed, as Jan Goldstein points out, once a sufficiently large number of physicians eager to keep up with "the latest advances in medical knowledge" share a disease category, they can produce an "epidemiological trend."[126] In other words, their clinical perceptions are shaped to expect to find the disease they are looking for (whether it be hysteria, personality disorder, attention deficit disorder, or neurosis).

How important, however, are diagnostic categories now, and to whom do they matter? As we've seen for East German practice before the transition in 1989, diagnostic categories were used to make sense of a patient's condition and to propose a course of treatment, and the category was recorded in the patient's medical files and in referral letters. However, in the post-transition period, diagnostic categories have acquired a new significance as they are required for getting reimbursement from the state health insurance system and for approval of proposed treatments. As pointed out earlier, this requires a dual system whereby certain diagnostic categories (*ICD-10*) are appropriate for requesting reimbursement but are illegitimate for the purposes of gaining approval from therapeutic referees for proposed treatment plans. As interviewees point out:

> This is a little double-tracked because those are two completely different systems in which they enter. The referee wants to know whether I have understood the case and whether I have a feasible theory why this patient got ill and what I intend to do about it. And the state health insurance system only wants to know what sort of cases are being treated.[127]

As the referee expects a lengthy psychoanalytically informed description of the patients' symptoms, diagnoses, and prognoses, writing *ICD-10* diagnostic categories into these reports would be inappropriate as the referee would "feel as though he was getting covered with shit."[128] However, *ICD-10* diagnostic categories are used to get reimbursement for treatments from insurance. Just as in the United States or other West European health care systems, these categories have become "business acts."[129] Their biomedical language has become the "lingua franca of the medical insurance companies."[130] As one interviewee put it:

> This [*ICD-10*] has been imposed on us by the health insurance. They want to have comparability. They want to have an overview . . . over the cases treated. . . . So that this can somehow be processed further.[131]

[125] Danziger, *Naming the Mind* (cit. n. 3), 7.
[126] Goldstein, *Console and Classify* (cit. n. 122), 330.
[127] Interview (Hh) 2003.
[128] Interview (H) 2003.
[129] Kirk and Kutchins, *The Selling of DSM* (cit. n. 6), 233.
[130] Bowker and Star, *Sorting Things Out* (cit. n. 5), 47.
[131] Interview (Hh) 2003.

Thus in both parts of Germany two separate diagnostic classification schemes are used to get therapeutic services approved, extended, and reimbursed. The practice of classifying and diagnosing patients' conditions has become a complex administrative task that involves negotiating with people from various institutional contexts that fall outside the parameters of therapeutic practice. East German psychotherapists maintain that such involvement of outside agencies in their therapeutic practice is a new experience for them. Therefore they pointedly reflect on its consequences for clinical practice:

> Now . . . the financial sponsor becomes visible and becomes also a contact . . . For instance it's in the interest of the health insurance system that patients aren't in hospital for so long anymore, and you have to prove in increasingly short intervals why they still need to be there . . . That is also one reason why you would write down a severe diagnosis. . . . The contrast to out-patient treatment is that there you write down a mild diagnosis, and you try to avoid severe diagnosis . . . So in both cases it is a financial matter.[132]

This entails that in a clinical context practitioners may "overdiagnose" to get a proposed course of treatment reimbursed. By contrast, they may also "underdiagnose" to avoid stigmatizing the client. Medical sociologists[133] point out that the practice of overdiagnosing and underdiagnosing is a widespread phenomenon, often carried out for various political and instrumental purposes other than clinical diagnosis. Far from constituting a failure of technical rationality, such practices actually are the mark of effective bureaucrats who can creatively manage and selectively present information. The increasingly routine use of underdiagnosing among East German practitioners is evident from the following quote:

> My colleague . . . writes only one diagnosis. There is something called "unspecific neurosis" that is what she always uses . . . She always does this because she thinks that it is not the health insurances' business what [patients] suffer from.[134]

For these practitioners, the category of "unspecific neurosis" is neither stigmatizing nor informative about a patient's condition. It is a wastepaper basket for symptoms that include tiredness after mental exertion, loss of concentration, burnout syndrome, emotional desensitivity, and a sense of unreality.[135]

Stuart Kirk and Herb Kutchins show how in clinical practices in the United States diagnosing symptoms and problems can be done "ritualistically."[136] Faced with the bureaucratic necessity of labeling a patient's condition in line with *ICD*'s range of available categories, East German practitioners also find that diagnosing turns into somewhat of a ritual. "I never put that much meaning into it [the diagnostic categories] . . . I know a couple of numbers by heart—most gets decided pragmatically anyhow . . . when I do my accounts."[137]

At present, disease categories are not only useful for deciding a course of treatment

---

[132] Interview (H) 2003.
[133] Kirk and Kutchins, *The Selling of DSM* (cit. n. 6).
[134] Interview (H) 2003.
[135] See "Neurosen: Einteilung der Neurosen nach ICD 10 (Auszug)," http://www.wartezimmer.de/deutschland/index.shtml.
[136] Kirk and Kutchins, *The Selling of DSM* (cit. n. 6), 235.
[137] Interview (H) 2003.

but also vital when dealing with outside agencies. As is true for any modern health-care bureaucracy, classificatory schemes can at different times be linked to external administrative, governmental, and legal requirements and needs (e.g., health insurance, organizational, and caseload constraints).[138] As Goeffrey Bowker and Susan Leigh Star point out, such external institutional requirements are often more decisive for the success of particular disease categories than the desire by clinicians to provide a nosology.[139]

The reduction of mental disorders to *ICD-10* diagnostic categories also has the usual sorts of administrative advantages and conveniences. It makes clinical work visible, legitimate, and comparable across different sites; it enables cost of service and resource needs to be determined; and it classifies and categorizes people in line with organizational demands and available programs. Thus the use of classificatory schemes in East Germany (as elsewhere) is most adapted to the functioning of the modern bureaucratic state.[140]

In the therapeutic encounter, however, practitioners can have a different and often contradictory set of concerns. They are "conditioned by training and reinforced by the exigencies of practice."[141] For practitioners trained in psychodynamic and psycho-analytic approaches, diagnosing plays "not such a big role."[142] Rather than letting classificatory manuals determine how they categorize behavior, they self-consciously distance themselves from what they take to be reductionist definitions of human behavior. They attend to how "transference" and "countertransference" reveal the "specific psychodynamics"[143] of the client. In other words, they are concerned with "what happens with me, what does he do to me."[144] They see their task as listening, deciphering, and interpreting psychological phenomena within the psychoanalytic tradition.

While contingencies of clinical practice temper the interpretative force of new classificatory schemata, the increasing shift of therapeutic services from the state to the private sector has also affected the therapeutic relationship as well as clinicians' diagnostic preferences. The entry of the "psy" sciences into the East German market-

[138] See Kirk and Kutchins, *The Selling of DSM* (cit. n. 6); Elisabeth C. Cookey and Phil Brown, "Spinning on Its Axes: DSM and the Social Construction of Psychiatric Diagnosis," *International Journal of Health Services* 28 (1998): 525–54.

[139] Geoffrey C. Bowker and Susan L. Star point out that advocacy groups, social movements, health-related organizations, pharmaceutical companies, academic researchers, and clinicians can all be central in creating specific diagnoses (Bowker and Star, *Sorting Things Out* [cit. n. 5], 560). For instance, in the case of attention deficit hyperactivity disorder (ADHD), the popular media, professional and academic publications, pharmaceutical companies, and various health insurance policies, all facilitated its legitimation as a disease category. Conrad and Potter argue that therefore there is a "feedback loop among professionals, claims-makers, media, and the public in terms of the creation, expansion, and application of illness categories." (Peter Conrad and Deborah Potter, "From Hyperactive Children to ADHD Adults: Observations on the Expansion of Medical Categories," *Social Problems* 47 (2002): 575.) See also Hacking, "The Invention of Split Personality," "Making Up People," and *Mad Travellers* (all cit. n. 8); Hacking "World Making by Kind Making," and *Rewriting the Soul* (both cit. n. 6); Kirk and Kutchins, *The Selling of DSM* (cit. n. 6); David Healy, *The Creation of Psychopharmacology* (Cambridge, Mass., 2002); Cookey and Brown, "Spinning on Its Axes" (cit. n. 138).

[140] Kirk and Kutchins, *The Selling of DSM* (cit. n. 6); Bowker and Star, *Sorting Things Out* (cit. n. 5); Cookey and Brown, "Spinning on Its Axes" (cit. n. 138); Theodore M. Porter, *Trust in Numbers: The Pursuit of Objectivity in Science and Public Life* (Princeton, N.J., 1995).

[141] Lunbeck, *Psychiatric Persuasion* (cit. n. 11), 116.

[142] Interview (Hh) 2003.

[143] Ibid.

[144] Interview (Hh) 1997.

place throughout the 1990s introduced various choices for consumers ranging from self-help groups to new age movements to different forms of therapy for dealing with the daily exigencies of life.[145] As a result of market competition, practitioners are now poignantly aware that "diagnosis always arises in a relationship"[146] with their clients. While in the former GDR, psychiatric clinics were institutions of last resort for people suffering from chronic and debilitating diseases and psychiatrists had the institutionally legitimized power to define and categorize patients' conditions, private clinical practices operate in a competitive environment in which practitioners need to sustain and attract a clientele.[147] The logic of the market inadvertently enforces the medicalization of a wider and wider range of "normal" behaviors and conditions, making them subject to psychological interventions.[148] Consequently, severe diagnoses have been eclipsed by milder psychological conditions ranging from conflicts in early childhood to various borderline and personality disorders.[149] At the same time as the quality of social interaction and other ad hoc features of the therapeutic interchange[150] codetermine the diagnostic categories decided upon, they also ultimately turn into administrative tools to fulfill various institutional requirements that lie beyond the immediacy of the therapeutic encounter.

CONCLUSION

How have concepts of "normal" and "pathological" selfhood become a political and scientific project caught up in the ideological battles of the twentieth century? What

---

[145] Some argue that the entry of therapeutic services into the East German marketplace has brought about a "psychoboom." See, e.g., Andreas Peglau, "'Vorsicht, Psychoboom!' Ratschläge für Psychotherapie-Interessierte," in *Weltall, Erde, . . . Ich*, ed. Andreas Peglau and Ich e.V. Berlin (Berlin, 2000), 201–2; Bierhoff Burkhard, "Organisation und Charakter: Sozialpsychologische Aspekte zur Analyse von Sekten und Psychogruppen," in *Die neuen "Seelenfänger": Religion zwischen Abhängigkeit und Selbstbestimmung*, ed. International Erich-Fromm-Gesellschaft (Tübingen, Germany, 1998), 33–60. "Wucherndes Dickicht," *Spiegel Online* 25 (15 June 1998), http://www.service. spiegel.de/digas/servlet/find. Thiel Wolfgang (2000), "Selbsthilfegruppen in den neuen Bundesländern zehn Jahre nach der Wiedervereinigung," http://fesportal.fes.de/pls/portal30/docs/Folder/BUERGER GESELLSCHAFT/Thiel8.htm; see also Peglau, "'Vorsicht, Psychoboom!'"; Miriam Gebhardt, *Sünde, Seele, Sex* (Munich, 2002).

[146] Interview (H) 2003.

[147] While at present ever wider ranges of behavior are medicalized, they fall within milder diagnostic categories than was common in the GDR's centralized health care system that was geared toward the treatment of severe mental disorders. Despite efforts at psychiatric reform in the 1960s and the 1970s, a bankrupt East German state lacked the resources for improvements. Psychiatric services therefore continued to cater to patients with chronic and deteriorating disorders in large, dilapidated, and "asylum-like" state-run hospitals. D. Müller-Hegemann to Ministry of Health, letter, 21 Feb. 1966, Bundesarchiv DQ-1/6676, Abteilung Deutsches Reich und DDR, BAB-L.

[148] Medical sociologists have long recognized that market mechanisms enforce medicalization. See, e.g., Conrad, " Medicalization and Social Control," (cit. n. 108); Renee Fox, "The Medicalization and Demedicalization of American Society," in *Sociology of Health and Illness,* ed. Peter Conrad and Rochelle Kern (New York, 1981), 527–34; Robert A. Aronowitz, *Making Sense of Illness* (Cambridge, 1998); Healy, "Shaping the Intimate" (cit. n. 121).

[149] Interview (G) 2001; interview (F) 1997; interview (M) 1997.

[150] In his sociological study of psychiatric intake interviews, Jörg Bergmann points to the link between the quality of social interaction and the severity of the diagnoses. Others also point out that the relative severity of a diagnosis depends on factors such as the institutional context in which the treatment is taking place as well as feelings of sympathy between clients and their therapists. Jörg Bergmann, *Interaktion und Exploration: Eine konversationsanalytische Studie zur sozialen Organisation der Eröffnungsphase von Psychiatrischen Aufnahmegesprächen* (PhD diss., Univ. Konstanz, Konstanz, Germany, 1980). See also Hacking, *Mad Travellers* (cit. n. 8); Lunbeck, *Psychiatric Persuasion* (cit. n. 11); Wirth-Cauchon, *Women and Borderline Personality Disorder* (cit. n. 9).

role have cultural institutions, politics, and science played in making human beings visible, understandable, and treatable in pre-and post-1989 East Germany? Certainly shifts in the category of "neurosis" as a way of understanding selfhood reveal how aspects of the self became entangled with wider social, institutional, political, and cultural forces.

The macro shifts in the political and administrative culture of East Germany reveal how as part of such a large-scale social and political change, diagnostic preferences can change as well. These preferences are embedded in particular cultures of practices that make them seem natural and inevitable as a tool for understanding and interpreting psychological issues. Rather than representing the actual inner workings of the psyche, diagnostic categories involve what Kurt Danziger calls the "naming" of specific phenomena that make sense only within a given discourse or classificatory schema. These are sustained by a scientific culture that determines "framework of conceptions" that enable particular ways of seeing.[151]

Whether we look at the categories of neurosis before the transition from state socialism in 1989 or after, at both times they were historical, political, and institutional artifacts that spoke to the organization of a society at a particular time and place. Before 1989, Marxist psychology and psychodiagnostics were reconceptualized and often molded by nonpsychologists.[152] Practitioners could draw on various political, philosophical, and scientific trends to reconceptualize mental health and pathologies. Psychological concepts could thereby become embedded in the ideological visions of a "good" society.[153] The concept of neurosis, in particular, was redefined and established in the academy as a tool to enhance "socialist personalities." However, this supposedly homogenous conception of neurosis was undercut by cultures of institutional practices that varied across localities and regions. The professional ties of particular clinical institutions to national and international networks also affected the use and dissemination of various international classificatory schemes of mental disorder (such as *DSM* and *ICD*) across the former GDR.

After 1989, the institutional transfer of the West German health care system into East Germany legitimized a set of diagnostic categories that also became an administrative necessity in dealing with the health care bureaucracy. To navigate and sustain their therapeutic careers, practitioners frequently adopted "individualizing" therapeutic approaches that had become the manual to the inner workings of the individual psyches while, at the same time, becoming the "lingua franca"[154] of administering "psy" services.[155] The rise of certain diagnostic categories and preferences also went hand in glove with the transfer of a market-oriented economic and political system that was tied into global networks and institutions such as the World Health Organization. Just like the remaking of psychological categories under socialism,[156] the dissemination of international categories of disease such as *ICD* throughout East German clinical practices was not driven by clinicians. But unlike socialist diagnostics, the use of such international classificatory schemata was less tied to politics and sci-

---

[151] Danziger, *Naming the Mind* (cit. n. 3), 3.
[152] Busse, "Von der Sowjetwissenschaft lernen" (cit. n. 28).
[153] Parsons, "Symbolic Environment of Modern Economies" (cit. n. 88).
[154] Bowker and Star, *Sorting Things Out* (cit. n. 5), 47.
[155] Nikolas Rose, "Assembling the Modern Self," in *Rewriting the Self,* ed. Roy Porter (London, 1997), 224-48.
[156] Busse, "Von der Sowjetwissenschaft lernen" (cit. n. 28).

ence than to various exigencies determined by a public health bureaucracy linked to modern states and to the needs of medical insurance companies.[157]

Nikolas Rose points out that socialism and its commitment to the welfare of its citizens from the "cradle to the grave" was grounded in a particular way of thinking about politics, power, and individual subjects forged in the nineteenth century.[158] The nation-state was thought to enact binding laws across its territories and took on the role of a moral gatekeeper that also devised interventionist social engineering policies. Individual subjects were understood in relation to "state and politics."[159] While the state structured "from above," the people could often enact "an anti-logic" from below[160] by conforming to or resisting political powers. At the same time, there were spaces immune from political intervention. Political control over activities was insulated by claims to professional autonomy and trust in expertise.[161] This enabled East German practitioners to transform their professional practices into "niches" that could accommodate or resist state-enforced regulations without oversight or political intervention.

However, the transition from socialism to a market economy brought about a new "relation between government, expertise and subjectivity."[162] Various institutional actors who enforce the logic of different programs, associations, and institutions have eclipsed big politics, big government, and interventionist social engineering. Individual subjects, rather than being hostage only to the state and politics, are governed by an array of obligations, programs, and techniques embedded in various institutional frameworks. Voluntary organizations, associations, and professional organizations of all kinds enforce certain commitments to transparency, efficiency, and professional accountability that are inscribed in institutional practices.

Michael Powers argues that in advanced liberal governments accounting practices and audits in service for particular programs have become a central mechanism for governing.[163] Auditing sets "objectives, proliferating standardized forms, generating new systems of record keeping and accounting," and "governing paper trails." Auditing renders decisions visible and amenable to evaluation and creates "accountability to one set of norms."[164] Auditing and evaluation practices hereby become forms of social control that enforce particular practices and ways of knowing. As one interviewee pointed out: "because we have to justify this analytical model to the insurance . . . we are all forced to take into account this way of thinking and somehow identify with it."[165]

Subjects thereby become "enwrapped in webs of knowledge . . . through which their actions can be shaped and steered and by means of which they can steer themselves."[166] The manner of governing individuals and their knowledge and practices have thus changed. Instead of accounting for an overt political agenda in theory and practice, therapeutic practitioners are continuously held accountable to numerous

---

[157] Bowker and Star, *Sorting Things Out* (cit. n. 5).
[158] Nikolas Rose, *Powers of Freedom* (Cambridge, 1999), 139.
[159] Ibid., 3.
[160] Ibid., 277.
[161] Ibid., 165.
[162] Ibid., 141.
[163] Rose, *Powers of Freedom* (cit. n. 158), 154. Michael Powers, *The Audit Society: Rituals of Verification* (Oxford, 1997).
[164] Rose, *Powers of Freedom* (cit. n. 158), 154.
[165] Interview (Hh) 1997.
[166] Rose, *Powers of Freedom* (cit. n. 158), 147.

agencies for their actions and activities, evaluating them in light of certain normative and theoretical standards. Such external exigencies, however, may set the formal stipulation for the knowledge and record-keeping practices of therapeutic practitioners, but they do not determine what happens in the therapeutic encounter, which remains, as it always has, ad hoc and messy.

In Germany, the transition from state socialism in 1989 is commonly thought to have fundamentally changed East German society, politics, and economy. Indeed, the rapid shift to a market-driven economy has entailed an array of institutional and cultural transformations. Expert knowledge systems have been replaced. Neoliberal policies have been enforced and totalitarian politics have become eclipsed by an "audit society."[167] But as interviewees maintain: "in some way what was there in the GDR, has suddenly now returned again, under different circumstances."[168]

Indeed, a set of dilemmas faced by people from day to day has hardly changed at all. People grapple with similar tensions arising between theory and practice, formal stipulation and informal improvisation, and divergences among local, regional, national, and international institutional cultures. The history of diagnostic categories thus resides within a complex web of meanings and localities. Both before and after 1989, then, diagnostic categories always were and are part of various political, scientific, and administrative cultures while standing in a tenuous and fragile relationship to local, institutional, and therapeutic arrangements and practices.

[167] Powers, *The Audit Society* (cit. n. 163).
[168] Interview (P) 2003.

The Chimera of Liberal Individualism:

How Cells Became Selves
in Human Clinical Genetics

*By Aryn Martin**

ABSTRACT

Using Ian Hacking's notion of "making up people," this paper argues that human chimeras—people who contain more than one genetically distinct cell population—have been made up. As with multiple personality, the discourse surrounding the phenomenon of chimerism offers a novel vantage point for examining the sociopolitical processes of subject formation. Evidence from archives, interviews with cell scientists, and popular sources will show that, in a strange leap that has come to seem self-evident, journalists, laypeople, and even scientists have come to equate genomes with selves and hence conclude that chimeras are more than one person. Thus far, the challenge that chimeras pose to the simple alignment of genome-body-person has been limited both by relegating chimeras to freak show status and by liberal institutions' demands that individuals be singular.

INTRODUCTION

Chimera *n*
1. a female fire-breathing monster in Greek mythology, typically represented as a combination of a lion's head, goat's body, and serpent's tail

chimera or chimaera *n*
1. a figment of the imagination, for example, a wildly unrealistic idea or hope or a completely impractical plan
2. an organism, or part of one, with at least two genetically different tissues resulting from the mutation, the grafting of plants, or the insertion of foreign cells into an embryo[1]

Nikolas Rose argues that "[t]he conceptual dispersion of 'the self' appears to go hand in hand with its 'governmental' intensification."[2] In other words, the more social

*Department of Sociology, York University, 2150 Vari Hall, 4700 Keele St., Toronto, Ontario, Canada M3J 1P3; aryn@yorku.ca.

Thanks to all the participants of the conference "The Self as Scientific and Political Project in the Twentieth Century: The Human Sciences between Utopia and Reform," Pennsylvania State University, October 2003, for their feedback and encouragement, especially to Greg Eghigian, who was an excellent host, and to Christine Leuenberger for inviting me to go in the first place. Thanks also to Michael Lynch, Stephen Hilgartner, and Rachel Prentice for their roles in shaping this work. The research for this piece was supported by NSF award # 0432120. All interviews were conducted in confidentiality, and the names of interviewees are withheld by mutual agreement.

[1] *New Oxford American Dictionary,* 2nd ed. (Oxford, 2005).
[2] Nikolas Rose, *Inventing Our Selves: Psychology, Power, and Personhood* (Cambridge, 1998), 170.

theorists decree the fragmentation and multiplicity of the self, the greater is governmental investment in its unitary and coherent existence. While the self has, in recent
decades, been an object of study for social and human scientists, this article examines
a case in which *biomedical scientists* characterize a fragmented self. An immunologist described to me a dispersion of the self that is more material than conceptual:

> So my worldview is that I think we're probably all chimeras and that we're probably all
> not just circulating chimeras, we're tissue chimeras, too. It's just a different kind of view
> of the self. I mean, the self's not a clone of one. It actually, intrinsically, the thing we call
> a self, has these minor populations.[3]

Current biology posits that, very rarely, human chimeras develop spontaneously
when fraternal twin embryos fuse or when one twin absorbs the other. Smaller degrees of chimerism, called *microchimerism,* arise from pregnancy, from transplanta-
tion, and from blood exchanged in utero during twin development. True to Rose's observation, the "discovery" and surgical design of selves that are "not a clone of one"
seems to coincide with the paradoxical intensification of modes of governance, in
late twentieth- and early twenty-first-century America (and elsewhere), that rely on
genetic oneness. That the immunologist quoted above can speak of a "different kind
of view of the self" discloses her assumption that there is a dominant view of the self
as genetically uniform in each and every cell. Is this true outside of her community
of cell scientists, and if so, since when? In this paper, I use scientific and popular accounts of chimerism to posit that the genetically uniform self came into being relatively recently, in part because of governmental intensification of "identity projects"
that make a person's genome a salient feature of his or her citizenship. Moreover,
when cells become proxies for persons, as they do in DNA profiling for administrative purposes, the proliferation of cell populations in a person necessarily prob-
lematizes the alliance of biological and political individuality required by those institutions.

In the first part of the article, I show how chimerism—as a way of being a person—
was "made up" in the latter half of the twentieth century. As with Ian Hacking's example of multiple personality, chimerism has a traceable genealogy of individual
cases and allied professionals.[4] In 1953, the first human chimera, Mrs. McK, was found
to contain blood that "belonged to" her twin. However, the category of chimerism
wasn't known to the public until the turn of the twenty-first century; real and fictitious
cases of chimerism are now making the news. I suggest that cellular multiplicity—
while a fascination to blood scientists in the 1950s—lacked public or political resonance because at that time the self was not coupled to a singular genome. For
chimerism to provoke angst about proper personhood, a genome had to first become
a thing one has.

Later in the paper, I review a number of genetic identity projects employed by gov-

---

[3] Interview with immunologist, 18 Sept. 2004.
[4] Ian Hacking, *Rewriting the Soul: Multiple Personality and the Sciences of Memory* (Princeton, N.J,
1995). Hacking uses the term "multiples" and eschews the term "multiple personality disorder," although this is the diagnostic category by which multiples are more commonly known. Hacking first
describes the concept in Hacking, "Making Up People," in *Reconstructing Individualism: Autonomy,
Individuality, and the Self in Western Thought,* ed. Thomas C. Heller, Morton Sosna, and David E.
Wellbery (Palo Alto, Calif., 1986), 222–36.

ernments and courts since the 1980s that have contributed to the assumption that a person does and should have an identifiable, traceable, storable individual genome. Finally, I explore the recent proliferation of real and fictitious chimera cases. Commentators—including journalists, geneticists, and the public—project a genetic notion of the self on to chimeras and characterize them as more than one person. We will see that although citizens and scientists speak of chimeras as though they are multiple, the challenge to the unified subject is ultimately ephemeral. The machinery of individualism quickly snaps back into place to manage the potential for selves to exceed bodily boundaries.

MULTIPLES

Ian Hacking has described multiple personality as a historically transient mental disorder.[5] It came into being, with a patient known as Felida X, in late nineteenth-century France, and it has achieved immense popularity in the United States since the 1970s. Multiple personality disorder is an example of "making up people," an idea that Hacking describes thus: "Inventing or molding a new kind, a new classification, of people or of behavior may create new ways to be a person, new choices to make, for good or evil. There are new descriptions, and hence new actions under a description."[6] People are made up through a number of avenues, in Hacking's account. Sometimes official statistics that record births, deaths, and diseases create new kinds of people, stored in ledgers and files. Sometimes making up people involves "semantic contagion," in which ways of talking about things, or imposing narratives retrospectively, are disseminated through official and unofficial channels and are adopted by people as possibilities for being.[7] Furthermore, multiples were (and are) made up through situated interviews, interactions, and therapies with experts who already conceive of the possibility of being multiple or split.

Hacking gives the following example of a person being made up in such an interaction. Janet is a late nineteenth-century French psychiatrist, and Lucie is his patient:

Janet: Do you understand me?
Lucie (writes): No.
J: But to reply you must understand me!
L: Oh yes, absolutely.
J: Then what are you doing?
L: Don't know.
J: It is certain that someone is understanding me.
L: Yes.
J. Who is that?
L. Somebody besides Lucie.
J. Aha! Another person. Would you like to give her a name?
L. No.
J. Yes, It would be far easier that way.

---

[5] Hacking, *Rewriting the Soul* (cit. n. 4).
[6] Ibid., 239.
[7] Ibid., 238, 256–7.

L. Oh well. If you want: Adrienne.
J. Then, Adrienne, do you understand me?
L. Yes.[8]

In this example, two kinds of people are made up. One new "person" is Adrienne, who is conjured in the interaction between doctor and patient. The simple practice of naming gives Adrienne an anchor for future characteristics and behaviors. She can now appear and reappear in Janet's notes and in Lucie's mind and actions. The other kind of "person" made up in this exchange, and in ongoing treatment, medical records, and Janet's publications, is Lucie, a patient with a disorder called *dédoublement* (the nomenclature for what later became multiple personality disorder). This second meaning, the coming and going of categories, and consequently people who "are" that kind of person, applies also to Hacking's examples of consumptives, homosexuals, and perverts. Examples explored by other historians of medicine could include anorexics[9] and hysterics.[10]

In the case of multiple personality, Lucie exists *as a patient* in order to contain the disruption that ensues when both Lucie and Adrienne inhabit Lucie's body. In other cultures, Hacking points out by drawing upon anthropologist Mary Douglas's work, "people are happy to have four selves."[11] The liberal democratic institutions that require autonomous, properly separated individuals could not long tolerate a disruption in the chain that links a single psyche to a single subject. In its heyday—the 1980s—multiple personality was offered as a defense in numerous court cases in the United States.[12] Although expert witnesses testified, and details of the cases corroborated the separation of personalities into distinct intentional actors (called "alters"), the defense did not usually succeed in the courts. One judicial approach—the unified approach—demands the unification of the body and the culpable person:

> The unified approach reasons that, regardless of the number of personalities, one body contains only one person. Thus, if one person committed a crime, then only one person is liable for that crime. Courts that use this approach do not consider separately the mental state of the alters and the host.[13]

In one case, for example, an alter named John Gustard was prohibited from testifying, although his testimony would have allegedly exculpated the defendant (his host personality), James Woodard.[14] I will explore how chimeras fare in court later in the

[8] Pierre Janet, "Les actes inconscients et le dédoublement de la personnalité pendant le somnambulisme provoqué," *Revue Philosophique* 22 (1886): 581. Quoted in Hacking, *Rewriting the Soul* (cit. n. 4), 224–5.

[9] For a detailed history of anorexia, see Joan Jacobs Brumberg, *Fasting Girls: The Emergence of Anorexia Nervosa as a Modern Disease* (Cambridge, Mass., 1988).

[10] See, for a brief discussion of hysteria, Michel Foucault, *The History of Sexuality: An Introduction,* vol. 1 (New York, 1978).

[11] Hacking, *Rewriting the Soul* (cit. n. 4), 146. Hacking is referring to Mary Douglas, "The Person in an Enterprise Culture," in *Understanding the Enterprise Culture: Themes in the Work of Mary Douglas,* ed. S. H. Heap and A. Ross (Edinburgh, 1992), 41–62.

[12] These cases are summarized in Juliette K. Orr, "Multiple Personality Disorder and the Criminal Court: A New Approach," *Southwestern University Law Review* 28 (1999): 651–76.

[13] Ibid., 657.

[14] Orr's example, which I am borrowing, is from *State v. Woodard,* 404 S.E.2d 6 (N.C. Ct. App. 1991).

article; first, however, I will make the case that, like multiples, chimeras have been made up.

MRS. McK: THE FIRST HUMAN CHIMERA

Prior to 1953, it was impossible to be a human chimera. Plants and animals could be chimeric, but people could not. That year, a Mrs. McK donated blood at a northern English clinic.[15] Her blood appeared to be a mixture of two types of blood, O and A. Stumped as to the explanation, the local clinic doctor, Ivor Dunsford, sent a specimen of the blood to Robert Race and Ruth Sanger, specialists at the Medical Research Council (MRC) Blood Group Unit in London. Race could also separate the blood into two types. While puzzling over the explanation, he remembered that such a situation had been found in fraternal twin cows. Race asked Dunsford whether Mrs. McK had been a twin, proposing that the twin's blood had crossed over during gestation and continued to circulate in Mrs. McK even now, decades later. Dunsford reported that the twin hypothesis was correct; Mrs. McK's twin brother had died at the age of three. At this point in the investigation and in the laboratory notes, Mrs. McK's blood went from being marked as her own (though classed into "Mrs. McK I" and "Mrs. McK II") to being hers and her brother's, both in pedigrees and in the ways in which doctors talked about her/them. The next series of techniques were aimed at finding out which blood was her "truly begotten blood" and which was her brother's. Race wrote, in a letter to renowned immunologist Peter Medawar, "[I]sn't it extraordinary to be able to group fully a person who has been dead for 30 years!"[16] Medawar, in turn, wrote the following passage about this case in his essay "The Uniqueness of the Individual":

> There is no telling how long Mrs McK will remain a chimera, but she has now been so for twenty-eight years; probably, in the long run, her twin brother's red blood cells will slowly disappear, and so pay back the still outstanding balance of his mortality.[17]

In the first sense of making up people, then, Mrs McK's brother was newly made up, like Adrienne in the example above. He existed again, in the blood (if not in the flesh), and was given a number of biological attributes that were not assigned to him when he was alive. He was more or less conjured from Mrs. McK's body, without any access to his three-year-old corpse. In the second sense, that in which "making up people changes the space of possibilities for personhood,"[18] a new category was invented. In correspondence traveling between the investigators, the subject line was now "Mrs. McK.—the Human Chimera." Mrs McK's case was published in the *British Medical Journal,* and she became the first human "blood group chimera."[19] Whether the description had an impact on Mrs. McK's self-understanding is impossible to

---

[15] The case notes, correspondence, and laboratory notes for the Medical Research Council Blood Group Unit (hereafter cited as MRC BGU) are at the Wellcome Library for the History and Understanding of Medicine, London. The case files for Mrs. McK are indexed as SA/BGU/F20/1, Parts 1 and 2. The details of Mrs. McK's story are gleaned from these files.

[16] Robert R. Race to Peter B. Medawar, 25 April 1953, SA/BGU/F20/1, Part 1, MRC BGU.

[17] Peter Medawar, "The Uniqueness of the Individual," in *The Uniqueness of the Individual* (London, 1957), 151–2. Mrs. McK and chimeras more generally are described in Medawar's essay as exceptions that prove the rule of the immunologically unique individual.

[18] Hacking, "Making up People" (cit. n. 4), 229.

[19] I. Dunsford et al., "A Human Blood-Group Chimera," *British Medical Journal* 2(4827) (1953): 81.

know, but from the correspondence it was evidently a delight to the doctors and scientists who "discovered" her.

However, one case does not make a category. As with multiple personality disorder, the phenomenon needed to be found repeatedly to furnish a class of people rather than an isolated anomaly. When, in 1956, another pair of fraternal twins with mixed bloods was referred to Race, he wrote: "This family should certainly be published without unnecessary delay. We are often asked whether any more of these twins have turned up: we suppose a second example is awaited to lift the phenomenon out of the 'freak' category."[20] Because Race, Sanger, and the MRC blood group had established expertise and priority, when other instances of "mixed bloods" were found by other blood donation clinics, they were referred, or at least reported, to Race and Sanger. From all over England, and then from South Africa, Canada, the United States, Sweden, and elsewhere, blood samples and patient pedigrees would be sent to the MRC Blood Group Unit for confirmation or cataloging by Race and Sanger. In this way, the unit became an "obligatory passage point"[21] in the construction of the category of human chimeras. Often, when cases were referred to the unit, they would be described without reference to the word "chimera," being called instead "mosaics" or "mixed bloods." Only after a case was worked up and confirmed at the unit would it become reclassified as a chimera and added to the growing list. The definitive catalog of chimera cases was published in successive editions of Race and Sanger's classic book *Blood Groups in Man.*[22]

Blood group chimeras were the first of several subclasses of human chimeras that, since their emergence, have been described under the banner of chimerism. *Dispermy* (two sperm, one egg), *tetragametic* (two sperm, two eggs), and *genetic* chimerism were first described in the 1960s. Genetic chimeras arise when two embryos, or would-be fraternal twins, fuse early in gestation and result in one healthy person (who may, if XX/XY, have some degree of intersex development). Genetic chimeras differ from blood chimeras in that, in the former, the two cell lines are found throughout the body, while in the latter, they are confined to the blood.

In the 1960s, '70s, and '80s, cases of genetic chimerism would occasionally crop up, usually coming to doctors' attention either accidentally or because of genital ambiguities in the patient. While Race and Sanger continued to compile cases of both twin chimeras (those who shared blood in utero) and genetic chimeras (those who were born as singletons), it seems that at some point it became redundant for blood bankers and geneticists who stumbled on new cases to publish them in journal articles. The MRC archive contains much correspondence in which Race and Sanger, and later their successor Patricia Tippett, followed up on or received notice of chimera cases. In one such letter, M. Ferguson-Smith writes, "I am truly ashamed to say that I never got round to publishing the case of F.B. It is somehow never quite so exciting to write up something that has been described before."[23] Patricia Tippett published an article in 1983 that reviewed seventy cases, about half of them twin chimeras (such as

[20] Robert R. Race to Gertrude Plant, 27 Dec. 1956, SA/BGU/F20/2/1, Part 1, MRC BGU.

[21] Michel Callon, "Some Elements of a Sociology of Translation: Domestication of the Scallops and the Fisherman of St. Brieuc Bay," in *Power, Action, and Belief: A New Sociology of Knowledge?* ed. John Law (London, 1986).

[22] Robert R. Race and Ruth Sanger, *Blood Groups in Man* (Oxford, 1950, 1954, 1958, 1962, 1968, 1975).

[23] Malcolm Ferguson-Smith to Robert R. Race, 19 Feb. 1973, SA/BGU/F20/7, MRC BGU.

Mrs. McK) and the other half, dispermic chimeras.[24] Tippett concluded that "investigation of these rare people, who are natural experiments, is still entertaining and worthwhile."[25] Although Tippett's tone is somewhat disconcerting in its objectification, it illustrates that cellular or genetic multiplicity did not, at this point, seem to provoke existential speculation about the people so described. Chimerism was a "natural experiment," but it was not yet a political one. The second cell population did not create an alter in the sense of a second agent with an independent consciousness and memory, as in multiple personality.

Mrs. McK's story highlights the historicity of the category *human chimera*. "Chimerism," as a word and as a possibility, crossed over from animal species to humans (or rather "infected" humans, if we carry on Hacking's metaphor of semantic contagion) in 1953. Robert Race wrote, in a letter to a colleague, that his "only thought in using [the term "chimera"] was to get a good title for the paper."[26] In Hacking's words, "[A] kind of person came into being at the same time as the kind itself was being invented. In some cases, that is, our classifications and our classes conspire to emerge hand in hand, each egging the other on."[27] Being "made up," though, is not meant to be a denigration or trivialization of those states of being that, although historically contingent, are entirely real to those affected. Nonetheless, prior to 1953, it was impossible to be a human chimera or to (know that you) contain anyone's cells other than your own.

I'm not convinced that anyone—except maybe cytologists—did think of themselves as containing cells in the 1950s, let alone worry about whether their genomes were their own or someone else's. So while the possibility of being a human chimera was introduced in England in 1953, an immediate spate of cases did not follow. If chimeras posed any potential threat to theoretical conceptions of the biology of the individual, it was managed by "freak of nature" discourse. The chimera's namesake (the Chimera) is, after all, a mythical monster. When biologists wrote about the phenomenon, they tended to use language such as Tippett's (quoted above). For example, Curt Stern, in 1968, wrote about "those curious creatures" that "are irregular compounds, seemingly created 'against Nature.' "[28] Chimeras were peculiar, to be sure, but they did not invoke speculation about personhood. For chimeras to be disruptive to anyone's idea of persons or selves, links had to be forged between genomes and persons in the first place. As Hacking writes, "[A] movement will 'take' only if there is a larger social setting that will receive it."[29]

DNA AND POLITICAL IDENTITY PROJECTS

In this section, I briefly digress from chimeras to speculate about why, at this time and place, they seem to inspire existential uneasiness about individuation. I describe below how the American government and others have adopted genetic modes of identification and identity making in judicial, familial, and civic administration. In these interfaces, citizens are replaced by, or reduced to, a small sample of fluid or tissue from

[24] Patricia Tippett, "Blood-Group Chimeras—a Review," *Vox Sanguinis* 44(6) (1983): 333–59.
[25] Ibid., 346.
[26] Robert R. Race to R. D. Owen, 13 July 1953, SA/BGU/F20/1, Part 1, MRC BGU.
[27] Hacking, "Making up People" (cit. n. 4), 228.
[28] Curt Stern, *Genetic Mosaics and Other Essays* (Cambridge, Mass., 1968), 28.
[29] Hacking, *Rewriting the Soul* (cit. n. 4), 40.

their bodies. For the most part, this alignment of a singular genetic signature with a single body with a single citizen has come to seem obvious and trustworthy. While political identity projects are certainly not the only factors in a complex cultural trend toward genetic thinking (variously called genomania, genes'r'us, geneticization, and so on), they are, I suggest, an intriguing and novel pressure in the formation of contemporary subjects.

Since the late 1980s, the American judiciary has adopted and intensified its use of cellular DNA as a forensic guarantor of citizens' identities.[30] The chain of identity from cell to person and back again is largely taken for granted now, except when a well-resourced defense team can demonstrate technical incompetence at the level of sample handling. It is, of course, a historical accomplishment that this feature of cells should seem self-evident. Judges, in their decisions in early cases in which DNA fingerprinting featured, often included a rudimentary biology lesson. For example, the 1988 decision in *State v. Andrews,* the first case in which DNA fingerprinting admissibility was challenged, summarizes the expert evidence of a molecular geneticist from the Massachusetts Institute of Technology (MIT):

> DNA, a molecule that carries the body's genetic information, is contained in every living organism in every cell which has a nucleus (nearly all the cells of the human body). The configuration of the DNA is different in every individual with the exception of identical twins. *It is the same in all the particular person's cells,* and its characteristics remain unchanged during the life of the individual.[31]

Gradually, judges, lawyers, juries, and the public learned the simple equation: one body, many cells, one unique DNA profile. Courts and juries now routinely make judgments of culpability based on an invisible—or rather visible only when selectively processed and represented—correspondence between semen or blood and a particular body.

State adjudicators of familial relationships have adopted DNA as a tool as well. Biological indicators of paternity, such as ABO blood grouping and HLA tissue typing were used in courts prior to the so-called molecular revolution. Now DNA is considered a more accurate way to prove parentage. Critics point out that using biological indicators to create state-legitimated families precludes alternative conceptions of family arrangements that may be rooted in social relationships rather than in blood. The same could be said of DNA's entrée into immigration administration. One method of obtaining "derivative" citizenship in the United States is by proving that "a blood and a legal relationship" exist between a U.S. citizen parent and a child born abroad.[32] Petitioners have the option of providing certified DNA evidence to the U.S. embassy to prove that a parental relationship exists. The U.S. immigration Web site offers information for parents about how to undergo testing as well as about the necessity of such a measure: "The Department appreciates that this situation may be troubling to parents, but under the circumstances, it appears that there is no other way to establish

---

[30] See, e.g., Michael Lynch and Sheila Jasanoff, "Introduction: Contested Identities: Science, Law, and Forensic Practice," *Social Studies of Science* 28 (1998): 675–86.

[31] *State v. Andrews,* 533 So. 2d 841; 1988 Fla. App. LEXIS 4645; 13 Fla. L. Weekly 2364 (emphasis added).

[32] "Information for Parents on DNA and Parentage Blood Testing," http://travel.state.gov/visa/immigrants/info/info_1337.html (accessed 11 Feb. 2005).

the child's claim to U.S. citizenship."[33] Both immigration and paternity testing are conducted at people's own arrangement and expense; the information garnered by the testing is constructed as an aid to citizens rather than as a requirement. Both reduce familial relationships to matching genetic patterns.

A final type of genetic identity project is like familial testing in that it places people in or out of a category, but the group is determined by genetic commonalities other than kinship. In other words, groups are formed because people share some portion of their DNA. From the perspective of the advanced liberal state, DNA testing may appear to be an easy and certain way to determine which individuals are entitled to particular rights or benefits, such as affirmative action or Native American burial rights.[34] It may, as in the case of marriage legislation in Texas, be the final arbiter of a person's "true sex" and hence determine whom they are able to marry.[35] It may be used in medical evaluations to determine whether individuals are disabled for the purposes of the Americans with Disabilities Act and therefore entitled to its protections. These identity projects do not simply come from the state in a top-down manner but are formed at the nexus of the government, corporations offering such testing, and the groups of people or patients who, in many cases, wish to be identified. Through direct-to-consumer advertising, citizen-consumers both are enrolled in and seek out genetic information about themselves to operate within, or to subvert, institutions that have adopted administrative uses of DNA.[36]

In the landscape of selfhood since the late twentieth century, genetics may not be a constant preoccupation, but it is historically novel. While the level of "genetic literacy" at large has certainly increased, those who have had some kind of institutional confrontation with genetics—through pregnancy, genetic testing, forensic testing, immigration, or so on—are more likely to have internalized a genetic sensibility. It is neither a coincidence nor an inevitable result of biotechnological advances that genetics should become so thoroughly imbricated in governmental strategies. While the use of DNA in political identity projects such as forensics and immigration seems to be about administrative efficiency, some features of genetics lend themselves particularly well to governance at this time and place.

Rose described the political climate in the last two decades of the twentieth century in various Western states as "advanced liberalism," whereby the states' powers are no longer transparently exercised through interventionist social policies such as welfare and social security. Instead, advanced liberal democratic governments "govern through the regulated and accountable choices of autonomous agents."[37] Individuals, aided by experts, are to assume a more active, enterprising role in the management of their own lives and selves. Political theorists date this shift to around 1980, with the elections of Margaret Thatcher in the United Kingdom and Ronald Reagan in the United States, and variously call the new governing ideologies neoliberal or neoconservative. As Wendy Larner writes:

---

[33] Ibid.

[34] See, e.g., Kim Tallbear and Deborah A. Bolnick, *The Native Voice,* 3–17 Dec. 2004, D2.

[35] *Littleton v. Prange,* 9 S.W.3d 233 (1999).

[36] In this way, consumer uses of DNA testing operate like "technologies of the self." See Michel Foucault, "Technologies of the Self," in *Technologies of the Self: A Seminar with Michel Foucault,* ed. Luther H. Martin, Huck Gutman, and Patrick H. Hutton (Amherst, Mass., 1988), 16–49.

[37] Nikolas Rose, "Governing 'Advanced' Liberal Democracies," in *Foucault and Political Reason: Liberalism, Neo-Liberalism, and Rationalities of Government,* ed. Andrew Barry, Thomas Osborne, and Nikolas Rose (Chicago, 1996), 37–64, on 61.

> Neo-liberal strategies of rule, found in diverse realms, including workplaces, educational
> institutions and health and welfare agencies, encourage people to see themselves as indi-
> vidualized and active subjects responsible for enhancing their own well-being.[38]

Concordant with the privatization of responsibilities that were formerly governmental
is the individualization of risks that were, in the welfare state, managed as social risks.

The flourishing of neoliberal politics in the past few decades feeds and is fed by
genetic characterizations of health and personhood. This correlation is probably best
understood as a loose association, an enabling circularity rather than a causal relationship. Governments that "distrust collective solutions" to problems[39] favor a model that
locates risks of disease in the individual, rather than spreading risks and costs across
the society: "Collective uncertainty as regards individual risks is replaced by recog-
nizable inequality, which is transformed from a social problem into a natural fact."[40]
Enterprising and consumer-oriented citizens (allegedly one of the major effects of neo-
liberalism) may prefer this model as well—my genome, my risks, my money:

> The responsibility for the self now implicates both "corporeal" and "genetic" responsi-
> bility: one has long been responsible for the health and illness of the body, but now one
> must also know and manage the implications of one's own genome.[41]

Whether or not disease risks can be located in our genes with any degree of certainty
is an open question; however, it is this expectation that increasingly guides biotech-
nology investment and healthcare planning. This redirection of focus from social determinants of health to individual genomes is a political process. Because genome,
body, and person do not conveniently align in human chimeras, the discourse in which
they are characterized offers a crude measure of the extent to which genetic essen-
tialism has become a proxy for personhood.

THE "SEMANTIC CONTAGION" OF CHIMERISM

I will now return to the story of genetic chimeras and fast-forward a half century to
2002. Mrs. McK's contemporary counterpart, Karen, gives blood not to donate it but
so that doctors can determine her tissue type in order to properly match her to a potential kidney donor.[42] Karen, a woman in her early fifties living near Boston, has three
sons, all of whom are tested to see whether they are a tissue match. It turns out that not
only are Karen's sons not a good tissue match but two of them could not possibly be
her sons. As Karen recalls, the nurse phones and tells her:

[38] Wendy Larner, "Neo-Liberalism: Policy, Ideology, Governmentality," *Studies in Political Economy* 63 (2000): 5–25, on 13.

[39] Roxanne Mykitiuk, "Public Bodies, Private Parts: Genetics in a Post-Keynesian Era," in *Privatization, Law, and the Challenge to Feminism,* ed. Brenda Cossman and Judy Fudge (Toronto, 2002),
311–54, on 312.

[40] Thomas Lemke, "Disposition and Determinism: Genetic Diagnostics in Risk Society," *Sociological Review* 52 (2004): 550–66, on 557.

[41] Nikolas Rose and Carlos Novas, "Biological Citizenship," in *Global Anthropology,* ed. Aihwa
Ong and Stephen Collier (Oxford, 2003), 439–63, on 441. Rose and his colleague Carlos Novas have
elaborated ideas of "biological citizenship" and "somatic individuality" to get at this convergence of
politics and biology.

[42] N. Yu et al., "Disputed Maternity Leading to Identification of Tetragametic Chimerism," *New England Journal of Medicine* 346 (2002): 1545–52; National Public Radio, "Sophisticated DNA Testing
Turning Up More Cases of Chimeras, People with Two Sets of DNA," *Morning Edition,* 11 Aug. 2003.

"You know, Karen, something very unusual has happened here. We've tested your sons because they were possible donors. Your sons' blood does not match your blood. And that's an impossibility. So they couldn't be your children," is what she said to me on the phone. "These could not be your children."[43]

Upon testing biopsies of other bits of Karen's tissue—bladder, skin, hair, cheek, thyroid—her doctors decided that she is a tetragametic chimera. Dr. Margot Kruskall explains:

> In other words, she was destined originally to be fraternal twins. Both of the original fertilized eggs were girls. And so when these fertilized eggs—which presumably, because they were close to each other in her mother's womb, or for reasons we don't understand yet, fused at a very early stage of development—she developed with two separate cell lines.[44]

The case was published in May 2002 in the *New England Journal of Medicine,* and Karen decided to "go public" with an NPR interview in August 2003.

Newspapers picked up on Karen's story.[45] Some fans of the forensic television show *Crime Scene Investigation* (CSI) noted the news coverage about Karen and suggested that it would be a good story line for an episode. Indeed, the 2004 season finale of the wildly popular CSI, "Bloodlines," featured a chimera whose genetic multiplicity initially gets him off a rape charge because his blood and semen do not match.[46] Grissom, the clever forensic scientist, has a eureka moment and the camera pans to an actual article about chimeras that appeared in *Nature* in 2002, the same month Karen's case was published. When confronted, the suspect confesses his crimes and his self-knowledge:

> Grissom: You know that bone marrow donation you gave your brother? I checked your medical records. His body rejected it, and he died. My guess is that's when you first found out about your unique condition.
>
> Villain: The doctors explained it. I'm a creature of myth.
>
> Grissom: A chimera. Head of a lion, body of a goat, tail of a dragon. You're a genetic anomaly. One person, two completely different sets of DNA.

Bloggers and fans, fascinated and "creeped out" by the possibility of genetic chimeras, enthusiastically sought and shared information, often referencing the NPR interview, media reports, and scientific articles about Karen.

In 2003, a woman from Washington State, Lydia, applied to receive government assistance for herself and her three small children.[47] Routine DNA testing of the whole

---

[43] "Karen," NPR, "Sophisticated DNA Testing" (cit. n. 42).

[44] Margot Kruskall, NPR, "Sophisticated DNA Testing" (cit. n. 42).

[45] Roger Highfield, "Sons I Gave Birth to Are 'Unrelated' to Me," *The Daily Telegraph* (London), 13 Nov. 2003, 14; "When Two Become One in the Womb," BBC News, 13 Nov. 2003; Claire Ainsworth, "The Stranger Within," *New Scientist* 180(2421) (2003): 34.

[46] CBS, "Bloodlines," Episode 423 of *CSI: Crime Scene Investigation,* written by Eli Talbert and Sarah Goldfinger, directed by Kenneth Fink, first aired 20 May 2004.

[47] The details of Lydia's case are from: ABC News, "She's Her Own Twin: Two Women Don't Match Their Kids' DNA—It's a Medical Mystery," posted 15 Aug. 2006 at http://abcnews.go.com/Primetime/story?id=2315693 (accessed 25 Aug. 2006); and "*I Am My Own Twin,*" Cicada Films, first aired on Discovery Health Channel, May 2005.

family revealed that the father was a genetic match, but Lydia could not have been the children's mother. The court threatened to remove her children from her, suspecting illegal surrogacy and welfare fraud.[48] In a documentary about chimerism, her father revealed the hold DNA evidence has gained in a short time:

> I believed my daughter, but at the same time I also believed the law had something against her. I believed that somewhere, she did something, she was in trouble or something. I have always had faith in DNA.[49]

When Lydia became pregnant again, a court officer witnessed an immediate DNA test of the child at birth, which also did not match Lydia's DNA. Fortunately, attorneys on the case came across the *New England Journal of Medicine* article about Karen and proposed that Lydia might be a chimera. Testing eventually confirmed that she was chimeric, and the judge dismissed the case.

In 2005, Tyler Hamilton, a U.S. Olympic gold medal–winning cyclist, was accused of blood doping (blood transfusion to boost oxygen levels) because testing found that he carried blood other than his own. His defense? He's a chimera.[50] This defense was suggested to Hamilton by an MIT molecular biologist, David Housman, who had read about Hamilton's plight in the sports pages and contacted him with this alternative explanation for the positive test.[51] Hamilton didn't get far with the defense because a blood test ordered during the hearing did not show evidence of chimerism. He was consequently banned from the sport for two years. However, the bizarre possibility of chimerism has cast some doubt on the validity of the test itself and garnered more publicity for the condition. We are, I propose, witnessing a case of semantic contagion.

DUAL IDENTITIES

Discourse about these recent cases of chimerism reveals that genomes have come to bear some material connection to the essence of personhood, as psyches and souls did at particular historical moments. Moreover, because each cell allegedly contains a copy of the entire genetic essence of an individual, two cell lines *are rendered as though they are two people.* An article in *Nature,* titled "Dual Identities," begins with this statement: "Eight years ago in Britain, a boy was born who, genetically, was two people."[52] An article in *New Scientist* describes Karen (pseudonym "Jane"):

> In the end they discovered that Jane is a chimera, a mixture of two individuals—non-identical twin sisters—who fused in the womb and grew into a single body. Some parts of her are derived from one twin, others from the other. Jane's body was made up of two genetically distinct types of cells. There was only one conclusion: Jane was a mixture of two different people.[53]

---

[48] An interesting subtext to this story, though uncommented upon in any news media I have encountered so far, is that Lydia is white and her children racially mixed. Whereas Karen's motherhood was never seriously in doubt and prompted a medical investigation rather than a criminal one, Lydia's motherhood was immediately suspect.

[49] "*I Am My Own Twin*" (cit. n. 47).

[50] Gina Kolata, "Cheating, Or an Early Mingling of the Blood?" *New York Times,* 10 May 2005, Health and Fitness, 1.

[51] Incidentally, Housman is the same scientist quoted previously as an expert witness in the DNA fingerprinting case *State v. Andrews* (cit. no. 31).

[52] Helen Pearson, "Human Genetics: Dual Identities," *Nature* 417(6884) (2002): 10–11, on 10.

[53] Ainsworth, "Stranger Within" (cit. n. 45).

According to one of Karen's doctors, "[I]n her blood, she was one person, but in other tissues, she had evidence of being a fusion of two individuals."[54] The reporter for this article surmised: "the twin is invisible, but for chimeras the twin lives microscopically inside the body as DNA." With headlines such as "Dual Identities" and "The Stranger Within," it is precisely this multiplication of personhood that makes these cases of otherwise unremarkable healthy people newsworthy.

These samples of discourse come from a wide variety of sources that have greater and lesser degrees of epistemic authority. In multiple personality, semantic contagion involved most critically psychologists and psychiatrists. But it was also a multilayered cultural phenomenon that included books and movies, such as *The Three Faces of Eve* and *Sybil.*[55] As I mentioned above, chimeras have entered the public forum in America by way of the media, forensic crime shows, and—the new vehicle for semantic contagion—the Internet. As with multiple personality, though, doctors and scientists, and their techniques, are instrumental to the production of chimeras as well. These researchers talk about cells as representatives of the people—even *as the people*—to whom they "naturally" (i.e., genetically) belong. In fetal microchimerism research, in which investigators trace cell exchanges between mothers and their children, the conflation of cells and people routinely follows a predictably gendered motif that figures mother's cells acting as mothers are supposed to and their children's cells similarly acting in the interests of the children. A pediatrician, for example, titled an article: "So You Think Your Mother Is Always Looking over Your Shoulder? She May Be in Your Shoulder!"[56]

One laboratory director I spoke to enthusiastically elaborated when I turned the conversation to identity:

> [T]here's a saying of something like—the mother holds her children's hands for only a short time but holds their hearts forever. And there's a certain almost like a physical manifestation of that idea . . . Its like your mom's always watching you. No, she's not watching you, she's right here. You think your mother tries to instill caring and honesty and having a heart. Maybe that's because some of her cells are in my heart. . . . it almost allows you to write it down on paper. Well what is it like to be a mother and a child. Well, it's the exchange of cells.[57]

When I asked a geneticist, from the same laboratory as the previous speaker, what she thought the biological role, if any, of these cells is, she answered:

> I'd like to believe that they're there because the fetus has a vested interest in keeping its mother healthy. So if [the cells that cross the placenta] are a generation younger, and they are more plastic, and they have better regenerative properties, you know, the fetus, the neonate at that point wants its mother to be healthy.[58]

A pediatrician explained:

> What would having more of mom's cells do? Well, for the first seven years of your life, you need to be empathetic with mom; she needs to be empathetic with you. She needs to

[54] ABC News, "She's Her Own Twin" (cit. n. 47).

[55] Hacking, *Rewriting the Soul* (cit. n. 4).

[56] Judith Hall, "So You Think Your Mother Is Always Looking over Your Shoulder? She May Be in Your Shoulder!" *Journal of Pediatrics* 142(3) (2003): 233.

[57] Interview with laboratory director, 1 Feb. 2005.

[58] Interview with geneticist, 28 Oct. 2004.

understand what's your behavior because you're spending time with her and she's directing you. So that's on a behavioral, basic behavioral, level. But she's also teaching you about society. You get socialized by mom in every culture. So the fact that you might have some of mom's cells that have memory is very appealing to me.[59]

In these examples, it is easy to see how culturally and historically specific ideas—about motherhood, for example—become written into bodies in the language of science. It is also revealing, I think, that these scientists clearly conceive of cells as maintaining something of the person—you could call it spirit, biography, or self—to whom they can be genetically traced. In other words, a fetal cell (if it can be made visible) remains as such, regardless of whose body or petri dish it is in. When scientists propose that these cells are helping or hurting their "host" body, those insights are laden with sociocultural tropes, including a commitment to liberal individualism.

Those doctors and lab scientists who research chimerism and microchimerism agree that the phenomena are much more common than we think, possibly ubiquitous. The pediatrician told me:

It's pretty striking that we've all got Mom's cells and all fraternal twins will have their cells. It's just how many? Probably only one in eight twins conceived actually comes to term, and so you'd expect those to have cells in the surviving twin. And we know that Mom carries cells from her miscarriages, so there is a lot of that going on.[60]

One doctor, quoted in the *Nature* article, said: "I'm convinced that on the streets of London and Hamburg there are many undetected chimaeras."[61] David Bonthron, the geneticist who treated the chimeric boy in Britain, has suggested that the incidence is increasing because fertility treatments place many embryos in close proximity to each other in the womb, which increases the likelihood that they may fuse.[62]

This feature of genetic multiplicity—its immanence, or possible occurrence in any body—is what allows it to function more like multiple personality than it would seem to at first. Whereas personalities are invisible, DNA is ostensibly tangible. Scientists can look at it on autoradiographs and in karyotypes. Testing is, of course, what reveals chimerism in the first place. However, if you've never had occasion to be tested, how do you know whether you harbor multiple genomes? Even more mysteriously, the cell populations are completely unpredictable: any tissue could have both cell lines or just one or the other. A blood test showing a uniform blood type cannot rule out the possibility that one is chimeric. After Karen's story was publicized, a woman wrote on a Web post: "It's like she is their aunt and mother. But it doesn't cause a clinical problem, so I could be a chimera and not know it! How cool is that?!"[63] In this way, the existence of chimerism can have a looping effect, whereby it alters people's relation to and understanding of themselves, even if they are not one of the few people to be clinically described as chimeras.

[59] Interview with pediatrician, 15 Sept. 2004.
[60] Ibid.
[61] Pearson, "Human Genetics" (cit. n. 52), 11.
[62] L. Strain et al., "A True Hermaphrodite Chimera Resulting from Embryo Amalgamation after in Vitro Fertilization," *New England Journal of Medicine* 338(3) (1998): 166–9.
[63] Http://keribug.blogspot.com/2003_09_01_keribug_archive.html, posted 17 Sept. 2003 (accessed 24 June 2004).

LOOPING EFFECTS AND VANISHING TWINS

One of Hacking's recurrent themes in several decades of writing about "making up people" is that classifications have what he calls "looping effects" on the people so described. New descriptions of people provide new ways for people to act under that description. As Michael Lynch points out in his review of Hacking's work, this may be a variation on a long-studied sociological theme, often called "labeling," and also "looping," by interactional sociologists such as Erving Goffman.[64] I suggest that chimeric conditions are susceptible to looping and labeling effects, and it is this feature that makes chimeras the same "kind" of entities as multiples. In a way remarkably analogous to Hacking's multiples, members of a small but growing group of people, allied mostly through Internet communities, support groups, and particular psychotherapeutic approaches, conceive of themselves as multiple because of an early loss and/or absorption of their twins.

These people describe themselves as "wombtwin survivors" or "twinless twins." They are singletons who believe or know themselves to be surviving siblings of the in utero death of a twin. The Vanishing Twin Syndrome (VTS) is increasingly well known since the routinization of ultrasound, and the existence of early twin loss is accepted by obstetricians and gynecologists as a relatively common event.[65] The medical community is, however, reluctant to affirm that the loss of a twin in utero can have psychological and emotional sequelae. Survivors and sufferers have established their own support communities and are collecting anecdotes and observations to get the disorder medically recognized and to share strategies for healing.[66]

My evidence of this community comes from Web sites, booklets that survivors have published, and an archived email group called vanishingtwins@yahoogroups.com. In these fora, survivors describe, and share, emotional repercussions of their early loss:

> Due to a deep longing for some undefined, missing part of themselves that, it seems, no mate can quite fulfill, single twins may experience problems with relationships and/or even with their sexual identity. They often suffer from feelings of guilt. They may be haunted by feelings that they're "parasites."[67]

The first step in treatment, a self-proclaimed specialist in the field writes, "is to create a distinct entity in your mind that is completely separate from you. This tiny little person may be named. Giving a name is a very important step because it marks the fact that your wombtwin and you were separate little people."[68] As with multiples, naming creates a node around which a person can be made up. Another similarity is that knowledge of the disorder travels by semantic contagion. Those who identify as

[64] Michael Lynch, "The Contingencies of Social Construction," *Economy and Society* 30 (May 2001): 240–54.

[65] Ann L. Anderson-Berry and Terence Zach, "Vanishing Twin Syndrome," published online at http://www.emedicine.com/med/topic3411.htm, last updated 8 Aug. 2005 (accessed 17 April 2006).

[66] Their alliance around their (real or imagined) genetic condition has become a key to their personal and group identity, and they seek to shape biological knowledge about vanishing twins. This is an example of biosociality, as described by Paul Rabinow, "Artificiality and Enlightenment: From Sociobiology to Biosociality," in *Essays on the Anthropology of Reason* (Princeton, N.J., 1996), 91–111.

[67] Althea Hayton, *Wombtwin Survivors: An Introduction* (St. Albans, England, 2005), 12.

[68] Ibid., 15. The community's discourse and its commitments to prenatal experience resonate with "right to life" discourse, and there may be important overlaps.

wombtwins report that they began to suspect the cause of their psychological distress
after hearing of VTS in a psychology class or after a practitioner of a chiropractic therapy called neuro-emotional technique (NET) uncovered their twins.[69]

While the groups and the phenomenon are not formed around genetic chimerism
per se, the survivors embrace the genetic phenomenon as further proof of their predicament and include information about chimerism in basic overviews of the syndrome. Discussions of chimerism would inevitably surface on the Web forum following a news article or television show about chimerism, such as the CSI episode or an
article about Karen. For example, a member wrote:

> A team was invited in to view me, pictures were taken, and the team determined that the
> pigment is not vitiligo (as I had previously thought), but the skin of a completely differ-
> ent person, with its own DNA and cell properties. Basically, I am literally two people
> walking around as one.[70]

Another wrote:

> My conclusion, after many months of pondering, is that I am a chimera, and that in the
> womb I had a girl-twin with whom I fused so completely that we share physical as well
> as emotional and spiritual characteristics. Since admitting the possibility I have begun to
> feel her presence very strongly—in retrospect I have done so my whole life, but could
> never bring myself to admit it.[71]

The discovery of a second cell population legitimates sufferers' claims to the protracted
existence of their twins, both to themselves and to skeptics around them. Some participants of the group ask how they can get tested to determine if they are chimeras, too.
Hence the VTS community is appropriating chimerism—and a view of selfhood that
is genetically reductionist—to make sense of their lived experience of multiplicity.

While vanishing twin survivors readily extrapolate from cells of their twins to a
named second self of sorts, this literal projection is not inevitable. Karen, whose
chimerism has made her somewhat of a celebrity, does not seem to have internalized
a splitting of her self into two, although this is how her doctors and her interviewers
characterize her. In interviews, she appears articulate, composed, and somewhat
bemused by her novel situation. Her main concern is not about her own multiple
make-up but about the interruption in hereditary ties implied by her genetic mis-
match with her sons:

> Telling my sons about this was the hardest part because I felt that part of me hadn't passed
> on to them. I thought, "Oh, I wonder if they'll really feel that I'm not quite their real mother
> somehow because the genes that I should've given to them, I didn't give to them."[72]

While Karen seems to have adopted a genetic paradigm for talking about heredity; she
also seems to be rephrasing a much older worry about bloodlines and descent, though one
that traditionally has been allied with dubious paternity rather than suspect maternity.

---

[69] This is a peculiar variant of the recovered memory movement, also discussed by Hacking in
Rewriting the Soul (cit. n. 4).

[70] Posted on vanishingtwins@yahoogroups.com, 29 June 2005. (Incidentally, my grammarcheck
objects to this sentence because the subject does not agree with the verb.)

[71] Posted on vanishingtwins@yahoogroups.com, 22 Jan. 2006.

[72] "Karen," NPR, "Sophisticated DNA Testing" (cit. n. 42).

Lydia, too, claims to eschew looping effects that would render her multiple. In a television interview, she replies to an invisible interviewer:

> I don't think of myself as a freak, no. I don't think that I'm like this weird person or whatever. Because I'm still a person; I don't have different personalities. I'm a normal person. It's just that there's something wrong with the DNA.[73]

More interesting than her answer, I think, is the unheard question posed to her, which evidently suggested that she should or would alter her self-perception in light of the label "chimera." Indeed, she seems somewhat provoked by the experience in a way that may affect her identity: "I feel scared more than anything, like what is this, and why me, and I really want to know more about it to really get to know what this really is and how it came about."[74] While Lydia is somewhat distressed and confused by the self-knowledge, she is also relieved that it ended her difficulties in court.

CONCLUSION

Ian Hacking's history of multiple personality shows how, in the late 1800s, psychologists used cases of *dédoublement* to refute the existence of a transcendental self, soul, or ego: "For in those individuals, there was not one single self. Those individuals had two personalities, each connected by a continuous or normal chain of memories, aside from amnesic gaps. . . . Hence (it seemed) there were two persons, two souls in one body."[75] Studies of multiple personality disorder, including Hacking's own study, explore the existential confusion encountered by multiples in a world where social and political institutions, such as courts, require individuality to be singular.[76] Similarly, in her work on conjoined twins, Alice Dreger foregrounds the social imperative to contain only one body per person.[77] While almost all twins who have remained conjoined report that they prefer it that way, singleton doctors, judges, and parents presume that a life that so radically impinges on normal individuality would be unlivable.

These studies speak not just to a multiplication of selfhood in some rare or disordered people but also to a dissolution—or at least a historical, cultural, and *biological* fluidity—of the very entity called *the person*. Exceptions to the alignment of body and person could be interpreted as proof that this alignment is not, in fact, prescribed by nature. Chimerism is, I suggest, a potential exemplar of the kind of fragmentation of the self that was heralded by poststructuralist theory. However, unlike personalities, digital selves, social roles, or names, cells are in the body, of the body. Hacking writes: "Some thinkers find atomistic versions of human nature to be obviously false. Rather we are born into a society, educated by it, and our 'selves' are sculpted out of biological raw material by constant interaction with our fellow humans."[78] In chimerism, our "biological raw material" is itself sculpted by interactions with our fellow

---

[73] "*I Am My Own Twin*" (cit. n. 47).

[74] Ibid.

[75] Hacking, *Rewriting the Soul* (cit. n. 4), 208.

[76] Ibid.; Elyn R. Saks and Stephen H. Behnke, *Jekyll on Trial: Multiple Personality Disorder and Criminal Law* (New York, 1997).

[77] Alice Domurat Dreger, *One of Us: Conjoined Twins and the Future of Normal* (Cambridge, Mass., 2004).

[78] Hacking, *Rewriting the Soul* (cit. n. 4), 16.

humans (mothers and their children, for example). A further novelty of chimerism is that the challenge to intact and inviolable personhood is coming from the work of biologists rather than from critical social theorists. When researchers say that their mothers are always with them, and when vanishing twin survivors speak of their embodied sisters, they are quite literally (and not just figuratively) referring to the biological material as one and the same as the person. The cell scientists I have quoted in this paper seem to envision a radical reconceptualization of personhood that breaks down the individual in its last bastion: the material of the body.

Despite the potential that chimerism offers for a radical reevaluation of the boundaries of self and other, biomedical and political discourses most often police the "naturalness" of singular identity by portraying any exceptions (such as chimeras, multiples, and conjoined twins) as rare, pathological, or monstrous and therefore not disruptive to the fundamental norm of discrete individuality. The consistent reference to the monster—the Chimera—in most medical descriptions of this phenomenon attests to its marginalization. Although I have shown that people speak and write about chimeras as though they are "two people in one body," no one seriously proposes that chimeras be given more than one vote. Although the very existence of chimerism throws doubt on the fundamental premise of forensic testing (a point often noted by the public), so far forensic scientists discount the challenge on the basis of the (presumed) rarity of chimerism. If the court in Lydia's case subscribed to a thoroughgoing genetic reductionism, it may have ruled, even in light of the new evidence, that most of Lydia (the genetically inconsistent part) was not, in fact, the mother of her children.[79] Such a finding is unthinkable given the structure of the court and the phenomenological oneness, or indivisibility of the woman called Lydia. (Although a few decades ago it would have been unthinkable that a woman who knew herself to be the mother of her children—and whose conviction was supported by her obstetrician—would have been in this predicament).

In summary, it is clear that chimerism causes people to flirt with notions of multiplicity in a more than trivial way. Karen's interviewer in the NPR segment, for example, ominously posits: "the biggest impact of chimerism is not practical, but philosophical, existential, psychological. For Karen, who began life as twins, learning of her unusual past required some adjustment."[80] Between the lifetimes of Mrs. McK and Karen, something of the self has become bound up in cells, in response to a cultural rhetoric of genetic reductionism. I have suggested that this was facilitated by a broader political shift toward privatization and individual responsibility in the late twentieth century in America and in other advanced liberal states.[81] Coupled with this, and more "on the ground," genetic identification projects employed by states since the 1980s have contributed to a reductionist genetic imaginary. However, the ultimate failure of chimerism (so far) to spur a redistribution of rights and relationships along cellular lines illustrates the robustness of liberal individualism. Comparatively, genes are only skin deep.

---

[79] The cells that matched her children were only found in her cervical sample and might therefore be confined to her ovaries.

[80] David Baron, NPR host, "Sophisticated DNA Testing" (cit. no. 42).

[81] A similar case could be made for the historically and geographically idiosyncratic investment in embryos and stem cells by American conservatives, a phenomenon that is entwined with my case study but far too complex to address here.

Self-Development and Civic Virtue:
Mental Health and Citizenship in the Netherlands (1945–2005)

By Harry Oosterhuis[*]

ABSTRACT

This article is about the development of mental hygiene and mental health care in the Netherlands from the Second World War to the present, aiming to explore its relation to social and political modernization in general and the changing meanings of citizenship and civic virtue in particular. On the basis of three different ideals of individual self-development, my account is divided into three periods: 1945–1965 (guided self-development), 1965–1985 (spontaneous self-development), and 1985–2005 (autonomous self-development). In the conclusion, I will elaborate some more general characteristics of Dutch mental health care in its sociopolitical context.

INTRODUCTION

In the nineteenth and early twentieth centuries, the relationship between institutional psychiatry and citizenship was "negative" or "exclusive" in the sense that hospitalization in a mental asylum generally implied legal certification and therefore the loss of, and potential serious infringement on, basic civil rights. In the course of the twentieth century, however, in two ways a more "positive" or "inclusive" connection between psychiatry and liberal-democratic citizenship was established. First, the last three decades of the century saw increased attention to and recognition of the civil rights of the mentally ill. In many Western countries, the legislation on insanity was amended, reflecting a shift from values associated with maintaining law and order to values associated with mental patients' autonomy, responsibility, and consent, as well as their right to adequate care and treatment. Second, from the early twentieth century on, in psychiatry as well as in the broader field of mental hygiene and mental health care, psychological definitions of citizenship were advanced. Expressing views about the position of individuals in modern society and their possibilities for self-development, psychiatrists, psychohygienists, and other mental health workers connected mental health to ideals of democratic citizenship and civic virtue. Thus they were clearly involved in the modern liberal-democratic project of promoting not only virtuous, productive, responsible, and adaptive citizens but also autonomous, self-conscious, assertive, and emancipated individuals as members of an open society.

* Department of History, Faculty of Arts and Social Sciences, University of Maastricht, Postbus 616, 6200 MD Maastricht, The Netherlands; harry.oosterhuis@history.unimaas.nl.

This article is about the development of mental hygiene and mental health care in the Netherlands after the Second World War and explores its relation to social and political modernization, in general, and the changing meanings of citizenship and civic virtue, in particular.[1] In the course of the previous century, citizenship in the Netherlands took on a broad meaning, not just in terms of political rights and duties but also in the context of material, social, psychological, and moral conditions that individuals should meet to develop themselves and be able to act according to those rights and duties in a responsible way. Notions such as fairness, social justice, social responsibility, tolerance, emancipation, and personal development became elements of the definition of good citizenship. On the basis of the three different ideals of individual self-development that I identify, my account is divided into three periods: 1945–1965 (guided self-development), 1965–1985 (spontaneous self-development), and 1985–2005 (autonomous self-development).[2] Before turning to the postwar period, I will briefly sketch the rise of the mental movement in the Netherlands and its sociopolitical background during the first half of the twentieth century. In the conclusion, I will elaborate some more general characteristics of Dutch mental health care in its sociopolitical context.

MENTAL HEALTH AND CITIZENSHIP BEFORE THE SECOND WORLD WAR

From the late nineteenth century, Dutch psychiatrists had aligned themselves with social hygiene, in which the effort to prevent people from falling ill through a reform of their living conditions and way of life held center stage. The assumed danger of degeneration and the increase in the number of new clinical phenomena, such as neurasthenia, moral insanity, and criminal psychopathy—whereby less the rational powers than the emotional life and moral awareness were affected—provided psychiatrists with arguments for expanding their intervention domain from mental asylums to society at large. To counter the harmful influences of modern society that were supposedly undermining people's minds and nerves, psychiatrists pointed to the relevance of proper hygiene and also self-control, willpower, a sense of duty and responsibility, moral awareness, and moderation as ways of thwarting mental disorders.

Between the mid-1920s and the early 1940s, the groundwork was laid for the Dutch psychohygienic movement and a national network of social-psychiatric and public outpatient mental health care provisions, which developed mainly independently and at a distance from mental asylums. Most of these facilities were established by secu-

---

[1] The Dutch terms *burgerlijk* and *burger(schap),* just like their German equivalents *bürgerlich* and *Bürger(tum),* are not easily translatable. The Dutch and German terms combine at least two meanings for which in English, as well as in French, there are separate words. In this paper, I will use "bourgeois" or "middle class" to refer to a social group with specific socioeconomic and cultural features and "civil," "civic," and "citizen" in the sense of public domain, political rights and duties, and the political status of individuals.

[2] These models of self-development are borrowed from J. W. Duyvendak, *De planning van ontplooiing: Wetenschap, politiek en de maakbare samenleving* (The Hague, 1999); and E. Tonkens, *Het zelfontplooiingsregime: De actualiteit van Dennendal en de jaren zestig* (Amsterdam, 1999). These two studies focus on the development of the Dutch welfare state and on social work and mental health care, in particular. Their periodization is in line with more general political and cultural histories of the Netherlands in the twentieth century: the reconstruction after the war (1945–1965), the "sixties," which as a cultural period lasted from the mid-1960s to the early 1980s, and the last two decades, which have been characterized in terms of "no-nonsense" and witnessed a rejection of the heritage of the sixties.

lar as well as religious voluntary organizations, and they received support from local or provincial governments. The individuals involved—psychiatrists as well as other physicians, teachers, educational experts, psychologists, criminologists, lawyers, social workers, and clergymen—were concerned about the perceived increase in mental and nervous disorders in modern society. This growth could be contained, they argued, by taking preventive measures, such as treatment of the early stages of mental and behavioral problems to prevent them from becoming worse—an approach that had proven effective in the fight against epidemics and contagious diseases.

The professional domain claimed by psychohygienists was wide: it stretched from marriage, sexuality, procreation, and family life to education, work, leisure activities, alcoholism, crime, and the care for mentally ill, feebleminded, and psychopathic individuals. The psychohygienic ideal materialized in the establishment of pre- and aftercare services for the mentally ill and retarded, Child Guidance Clinics and other mental health facilities for problem children, centers for marriage and family problems, a public institute for psychotherapy, and a growing number of counseling centers for alcoholics. The regime of these facilities basically consisted of providing consultations, mobilizing social support, conducting surveillance, offering a form of moral re-education aimed at building self-discipline, and promoting social reintegration and rehabilitation.[3]

The underlying reasoning of psychohygienists was rooted in a more broadly shared cultural pessimism about the assumed harmful effects of the rapid changes in society as well as in the optimistic belief in the sheer potential of scientific knowledge to help solve those problems. Psychohygienists viewed modern society's pace of change and mounting complexity as major causes of the increase of mental and nervous problems. A rising number of people would have trouble keeping up with the rapid technological advances and the high-paced lifestyle of industrialized and urbanized society. During the period between the two world wars, such cultural pessimism was, in fact, widespread among Dutch intellectuals; it was intensified in the 1930s by anxieties about Americanization as well as the rise of totalitarianism in other European countries. Fearing cultural decay and social disintegration, intellectuals repeatedly stressed the significance of spiritual values and a sense of community.[4]

Mental health care developed against the backdrop of social and political modernization. The emergence of mass society and ongoing democratization—universal suffrage was introduced in 1919—caused mounting concerns in society's upper echelons regarding the dominance of irrational emotions and drives, which would only lead to more unruliness, mental slackening, and social disintegration. The question was whether all people had the necessary rational and moral qualities to meet the social

---

[3] T. E. D. van der Grinten, *De vorming van de ambulante geestelijke gezondheidszorg: Een historisch beleidsonderzoek* (Baarn, Netherlands, 1987); J. C. van der Stel, *Drinken, drank en dronkenschap: Vijf eeuwen drankbestrijding en alcoholhulpverlening in Nederland. Een historisch-sociologische studie* (Hilversum, Netherlands, 1995); L. de Goei, *De psychohygiënisten: Psychiatrie, cultuurkritiek en de beweging voor geestelijke volksgezondheid in Nederland, 1924–1970* (Nijmegen, Netherlands, 2001); H. Oosterhuis, "Insanity and Other Discomforts. A Century of Outpatient Psychiatry and Mental Health Care in the Netherlands, 1900–2000," in *Psychiatric Cultures Compared: Psychiatry and Mental Health Care in the Twentieth Century: Comparisons and Approaches,* ed. M. Gijswijt-Hofstra et al. (Amsterdam, 2005).

[4] R. van Ginkel, *Op zoek naar eigenheid: Denkbeelden en discussies over cultuur en identiteit in Nederland* (The Hague, 1999), 86–98; R. Schuursma, *Jaren van opgang: Nederland 1900–1930* (Amsterdam, 2000), 76–100.

responsibilities of an increasingly complex society and would be able to act as accountable political citizens. However, modernization had given rise not only to deeply felt worries but also to a social and moral activism aimed at tackling material and moral deprivation. Various behaviors, ranging from drinking, dancing, gambling, fair going, and other forms of "low entertainment" to idleness and money squandering, and from impulsive satisfaction of needs and sexual licentiousness to child abandonment and crime, became the targets of interference and intervention by both voluntary organizations and the state.[5] Resolving social wrongs and misfortunes, such as poverty, illness, backwardness, and exploitation, was not all that mattered; it was considered equally important to achieve a virtuous life and a sense of social responsibility for everybody.

The psychohygienic doctrine basically fit in with efforts to "civilize" the people, particularly the lower classes. In the nineteenth century, these activities had been promoted by the liberal bourgeoisie, but since the turn of the century they had become entangled with orthodox Protestant and Catholic as well as socialist politicians to further the social emancipation and national integration of their constituencies. These efforts indeed suggested an optimistic belief in the perfectibility of mankind, even though such a vision was frequently couched in a more or less conceited moral-didactic paternalism. In pleas for a national-level education of the common people, "character formation" was central. While classic liberalism had emphasized rational and autonomous thinking as the engine of social progress, the focus at this point was on teaching a sense of norms and duties, raising community spirit, and instilling willpower and self-discipline.[6] Although Dutch society and politics was divided and hierarchically organized along class as well as religious lines—the so-called pillarization[7]—the various social elites generally propagated an ideal of citizenship that stressed middle-class values. An industrious and productive existence, self-reliance, a sense of order and duty, thrift, and the family acted as cornerstones of the democratized bourgeois ideal of citizenship. Central notions were self-control and having a sense of responsibility: the curbing of erratic impulses and the postponement of instant gratification of needs aimed at a proper balance between individual independence and community spirit, as well as at long-term personal and collective well-being.[8]

In the interest of a well-ordered, democratic society, it was considered essential to elevate the people morally and to inculcate a civil sense of responsibility and decency

[5] A. de Regt, *Arbeidersgezinnen en beschavingsarbeid: Ontwikkelingen in Nederland, 1870–1940; een historisch-sociologische studie* (Amsterdam, 1984); P. Koenders, *Tussen christelijk réveil en seksuele revolutie: Bestrijding van zedeloosheid in Nederland, met nadruk op de repressie van homoseksualiteit* (Leiden, Netherlands, 1996); D. J. Noordam, "Getuigen, redden en bestrijden: De ontwikkeling van een ideologie op het terrein van de zedelijkheid, 1811–1911," *Theoretische Geschiedenis* 23 (1996): 494–518.

[6] H. te Velde, *Gemeenschapszin en Plichtsbesef: Liberalisme en Nationalisme in Nederland, 1870–1918* (The Hague, 1992).

[7] The three main pillars—networks of organizations in the fields of politics, economy, health, education, and culture—were those of orthodox Protestants, Catholics, and Social Democrats. The liberal bourgeoisie, which had dominated Dutch politics until the First World War, never organized itself into a pillar.

[8] H. te Velde, "How High Did the Dutch Fly? Remarks on Stereotypes of Burger Mentality," in *Images of the Nation: Different Meanings of Dutchness, 1870–1940,* ed. A. Galema, B. Henkes, and H. te Velde (Amsterdam, 1993), 59–79; R. Aerts and H. te Velde, eds., *De stijl van de burger: Over Nederlandse burgerlijke cultuur vanaf de middeleeuwen* (Kampen, Netherlands, 1998); J. Kloek and K. Tilmans, eds., *Burger: Een geschiedenis van het begrip "burger" in de Nederlanden van de Middeleeuwen tot de 21ᵉ eeuw* (Amsterdam, 2002).

in them. Apart from politicians, inspired social reformers, and moral entrepreneurs, the proponents of this social-moral activism were found especially among the professional groups that were gaining influence and self-awareness, such as physicians, teachers, educational specialists, youth leaders, civil servants, engineers, social workers, and from the 1920s on, psychohygienists and mental health workers.[9] With their particular understanding of public mental health, psychohygienists closely aligned themselves with the paradigm of an orderly mass society based on the unconditional adaptation of the individual to a collectively shared system of norms and values.

GUIDED SELF-DEVELOPMENT (1945–1965)

In the 1940s and 1950s, the Dutch outpatient mental health care facilities—the Child Guidance Clinics and Centers for Family and Marriage Problems, in particular—expanded rapidly. Worries about social disruption and moral decay in the wake of the German occupation and subsequent liberation by allied forces strongly promoted the growth of these facilities. Because the war and the atrocities of Nazism epitomized the cultural pessimism of the psychohygienists in quite concrete and dramatic ways, in the postwar years the psychohygienic doctrine won more support among politicians and social elites. Various forms of misconduct and shortcomings in ethical standards—including idleness, malingering, juvenile mischief, lack of respect for authority and ownership, along with family disruptions, growing divorce rates, greater autonomy of women, and sexual license—were considered serious threats to both the moral fiber and the mental health of the nation. The leitmotiv of this widespread anxiety was the observation that uncontrollable drives and urges had gained the upper hand, which seriously threatened the overall sense of community. It was widely felt that to rebuild the devastated country, create unity, and hold off the new threat of communism, people's moral resilience needed to be strengthened and broken-up families and individuals who had gone astray should be put back on track. Again, the insistence on self discipline and a sense of duty served to underline the importance of responsible citizenship in a democratic mass society as well as in the emerging welfare state.[10] Government officials and psychiatrists emphasized that a social security system would only be effective if its potential beneficiaries had a well-meaning attitude. Close monitoring and moral education were needed to cut off profiteers and those with malicious intentions.[11]

In their striving for a mental recovery of the Dutch people, psychohygienists displayed a great sense of mission while also claiming a broad professional domain.

---

[9] H. Nijenhuis, *Volksopvoeding tussen elite en massa: Een geschiedenis van de volwasseneneducatie in Nederland* (Amsterdam, 1981); De Regt, *Arbeidersgezinnen en beschavingsarbeid* (cit. n. 5); W. A. W. de Graaf, *De zaaitijd bij uitnemendheid: Jeugd en puberteit in Nederland, 1900–1940* (Leiden, Netherlands, 1989); S. Karsten, *Op het breukvlak van opvoeding en politiek: Een studie naar socialistische volksonderwijzers rond de eeuwwisseling* (Amsterdam, 1986); W. Krul, "Volksopvoeding, nationalisme en cultuur: Nederlandse denkbeelden over massa-educatie in het Interbellum," *Comenius* 36(9) (1989): 386–94.

[10] J. C. H. Blom, "Jaren van tucht en ascese: Enige beschouwingen over de stelling in Herrijzend Nederland 1945–1950," *Bijdragen en Mededelingen betreffende de Geschiedenis der Nederlanden* 96 (1981): 300–33; H. Galesloot and M. Schrevel, eds., *In fatsoen hersteld: Zedelijkheid en wederopbouw na de oorlog* (Amsterdam, 1986); Van Ginkel, *Op zoek naar eigenheid* (cit. n. 4), 177–205.

[11] I. de Haan, *Zelfbestuur en staatsbeheer: Het politieke debat over burgerschap en rechtsstaat in de twintigste eeuw* (Amsterdam, 1993), 92; F. S. Meijers, *Inleiding tot de sociale psychiatrie* (Rotterdam, 1947), 68–9.

Through the use of medical-biological metaphors—society viewed as body, the family as vital organ, the individual as cell, social wrongs as pathologies, and specific problem groups as nidi—social and moral problems were framed as issues of public mental health. Initially mental health workers, focusing on trouble children, uprooted juveniles, and "asocial families," continued to look for solutions in moral-pedagogical measures.[12] However, what in the late 1940s was still seen as lack of moral strength and willpower, in the 1950s was increasingly explained in psychological and relational terms. Personality defects, developmental disorders, and unconscious conflicts, brought about by a defective education and poorly functioning families, it was believed, constituted the underlying causes of deprivation and misbehavior. This meant that moral preaching and coercion needed to be replaced by treatment and cure. For instance, the psychiatrist S. P. J. Dercksen, who in Amsterdam headed a Dutch Reformed mental health institution, argued that a sense of responsibility could not be imposed through "authoritarian coercive advice" because people felt an inner aversion to such an approach. Instead, "subtle psychological work" was called for to make them accept mental health care.[13]

The results of preventive psychiatric treatment of allied soldiers during the war, the psychodynamic model, and new (American) psychosocial methods, such as social casework and counseling, raised expectations about the potential of psychiatry and the behavioral sciences to change and influence people's mental makeup. Even more than before the war, the psychohygienists linked a sustained cultural pessimism with an optimistic belief in the potential of scientific knowledge and professional expertise to avert doom. Inspired by the World Federation for Mental Health, they emphasized that it was not only important to prevent, treat, and cure mental disorders but also crucial to improve mental health in general, thereby ensuring maximal opportunities for all citizens to develop themselves in a wholesome way. Thus the distinction between normal and abnormal, or illness and health, was put into perspective. The notion of public mental health was turned into a comprehensive concept that was tied to the prevention of totalitarianism and the realization of a better world.

The development of mental health care was strongly influenced by the specific ways in which the experts in this field interpreted social transformations. When around 1950 the moral panic about the disruptive effects of the war had faded, the experts began to focus, in particular, on the potentially harmful influences of ongoing social and economic modernization. The Netherlands came out of the Second World War as a destroyed and impoverished nation, but the 1950s brought a new and vigorous economic dynamic, based on great confidence in science and technology. Large-scale urbanization, industrialization, and infrastructural innovation had far-reaching effects on people's social relationships and everyday life. The makeup of the working population changed drastically: as the agrarian sector declined, the industrial and services sectors saw great expansion. Spatial and social mobility rose sharply, allowing more individuals to evade the paternalism and social control of small communities, the church, and their families. In addition, the extension of the motorized traffic system, the growth of higher education, the increasingly international cultural orientation—

---

[12] A. Dercksen and L. Verplanke, *Geschiedenis van de onmaatschappelijkheidsbestrijding in Nederland, 1914–1970* (Meppel, Netherlands, 1987); F. W. van Wel, *Gezinnen onder toezicht: De stichting volkswoningen te Utrecht, 1924–1975* (Amsterdam, 1988).

[13] S. J. P. Dercksen, "Sociaal-psychiatrische ervaringen," *Folia psychiatrica, neurologica et neurochirurgica neerlandica* 59 (1956): 195–205, on 197.

geared toward America in particular—and the rise of new media such as television widened the horizons of many Dutch. A steadily increasing prosperity provided more material security, and class differences and other hierarchical relationships gradually lost their edge. Increasingly, the new dynamic of the everyday life of the Dutch was at odds with the still prevailing traditional middle class and Christian norms and values with their clearly defined dos and don'ts.[14]

In the views that Dutch psychohygienists articulated about these developments, a cultural pessimism reminiscent of the prewar years reverberated. Its essence seemed basically unchanged: the mental and moral development of man, if it had not been severely harmed by the ongoing economic and technological progress, had at least fallen out of step with it.[15] Like other intellectuals, they argued that the socioeconomic modernization caused society to be dominated by a one-sided, instrumental rationality that jeopardized moral and spiritual principles as sources of meaning. Their critique focused on modern man who was absorbed by mass culture (*de massamens*). This man, the embodiment of all evils that accompanied modernity, was lonely and uprooted, had no fixed norms and values, and no longer felt any ties with religion, tradition, and community. His mind was nihilistic, and he was swayed by the issues of the day; he let his life be dictated by his unconscious drives and emotions and showed no regard whatsoever for moral authority. His inner emptiness was shown by his flight into material consumption, popular entertainment, and sexual gratification. This rudderless man, critics argued, undermined social solidarity and democratic citizenship. They looked for a remedy in an activist cultural politics, as advanced by German sociologist Karl Mannheim before the war. Mannheim argued in favor of social planning and a normative education of the people directed by elites to prevent democratic mass society from degenerating into either anarchy or dictatorship. Although rationalization was regarded as one of the main causes of the cultural crisis, there was great confidence in the possibility of steering and controlling society with the help of the social and human sciences, which is why sociologists as well as psychohygienists believed they had a major task to fulfill.[16]

Initially, mental health workers stressed the significance of a fixed collective morality and the social adaptation of the individual to safeguard overall social stability, but in the 1950s the workers' defensive stance toward modernization gave way to an accommodating approach. More and more they acknowledged that moral restrictions and external coercion only affected the outer behavior of people while leaving their inner self untouched. The belief that socioeconomic progress was inevitable brought along a new perspective on their task: a striving for normalization and social integration, not only by offering support to people who did not manage to keep pace with the rapid developments but also by enhancing the mental attitude and psychological

---

[14] P. Luykx and P. Slot, eds., *Een stille revolutie? Cultuur en mentaliteit in de lange jaren vijftig* (Hilversum, Netherlands, 1997); K. Schuyt and E. Taverne, *1950: Welvaart in zwart-wit* (The Hague, 2000).

[15] T. de Vries, *Complexe consensus: Amerikaanse en Nederlandse intellectuelen in debat over politiek en cultuur, 1945–1960* (Hilversum, Netherlands, 1996); Van Ginkel, *Op zoek naar eigenheid* (cit. n. 4), 207–44.

[16] M. Gastelaars, *Een geregeld leven: Sociologie en sociale politiek in Nederland, 1925–1968* (Amsterdam, 1985); E. Jonker, *De sociologische verleiding: Sociologie, sociaal-democratie en de welvaartsstaat* (Groningen, Netherlands, 1988); De Goei, *De psychohygiënisten* (cit. n. 3); I. de Haan and J. W. Duyvendak, *In het hart van de verzorgingsstaat: Het ministerie van Maatschappelijk Werk en zijn opvolgers (CRM, WVC, VWS), 1952–2002* (Zutphen, Netherlands, 2002), 27, 76–83.

abilities individuals needed to function properly in a changing society. Thus the pursuit of more dynamic and flexible adaptation took the place of frantic attempts at restoring morality and community spirit. It was now believed that new social conditions required a redirection of norms and values and that individuals should be granted more responsibility for self-development.

Steering a middle course between tradition and renewal, paternalism and liberation, and spiritual values and psychological insight, leading psychohygienists began to present themselves as guides who prepared people for the particular dynamism of modern life. In their view, the main precondition for cultural improvement was a change in people's mentality. Inspired by phenomenological psychology and personalism—which stressed personality formation, spiritual reflection, and giving meaning to one's life in a self-conscious way—they now identified "maturity," "inner freedom," and "self-responsible self-determination" as the basis of mental health. Such mental qualities were the opposite of impulsive behavior; they entailed inner regulation, which would guarantee that people could do without external regulations to lead a responsible life. It became the individual's task to develop into a "personality" and to achieve a certain measure of inner autonomy regarding the outside world. What was crucial in this individualizing and psychologizing perspective was, in particular, the internalization of social norms and values in an autonomous self. The mentally healthy were not those who uncritically subjected themselves to rules and regulations but rather those who were independent, conscientious, and responsible—those who knew how to make decisions on their own, pursued optimal self-development, and thoughtfully adapted to social modernization.[17]

The psychohygienists backed up their argument for a mental reorientation not only with their psychological insights but also with a moral appeal—a form that gave their message a familiar ring to many in what was still largely a very Christian country. Invoking conscience and a sense of responsibility, they called upon people to identify with high moral values. Yet there was still concern about the harmful effects of social changes on people's mental balance. If individuals were to be able to decide on their own how to shape their lives, scrupulous self-examination was needed to assure that their intentions were conscientious and based on good grounds. Individuals were assumed to follow their own conviction, but they were also considered to do so in line with social expectations involving a morally responsible mode of life, as articulated by mental health workers and other expert leaders. Surely, this project of self-development was at odds with hedonism, extravagance, egoism, and egocentrism. People could only develop their personality in a meaningful way if they, of their own accords, were able to live up to high moral standards. For those who failed to realize their selves adequately, mental health supervision or treatment was the best solution. Constant reflection on individual conduct and motivation was called for in order to find the right balance between guidance and self-determination. By fostering such an

---

[17] F. J. J. Buytendijk, *De zin van de vrijheid in het menselijk bestaan* (Utrecht, 1958), 10; Buytendijk, *Gezondheid en vrijheid* (Utrecht, 1950); H. M. M. Fortman, *Een nieuwe opdracht: Poging tot historische plaatsbepaling en tot taakomschrijving van de geestelijke gezondheidszorg in het bijzonder voor het katholieke volksdeel in ons land* (Utrecht, 1955), 20; cf. De Goei, *De psychohygiënisten* (cit. n. 3), 154, 194–7; I. Weijers, *Terug naar het behouden huis: Romanschrijvers en wetenschappers in de jaren vijftig* (Amsterdam, 1991); H. Oosterhuis, *Homoseksualiteit in katholiek Nederland: Een sociale geschiedenis, 1900–1970* (Amsterdam, 1992).

attitude, mental health care would contribute to creating the conditions for participation in civil society and political involvement (which was a civic duty, after all) and thus for maintaining and deepening democracy.[18]

The ideal of citizenship promoted in the 1950s and early 1960s can be characterized as guided self-development. This model was geared toward socioeconomic modernization, a process that called for a functional individualization, meaning flexibility and mobility. Self-identity used to be a product of given and more or less stable social categories, such as class, religion, and family background, but it increasingly turned into a product of personal qualities and preferences. This individualization was understood as an inescapable effect of modernity, but in an effort to avoid social disintegration, psychohygienists considered it essential to offer moral guidance and add normative standards, as a counterbalance to the individual's growing freedom. Those who managed to internalize such standards successfully would be able to adapt to the constantly changing circumstances of modern society in flexible ways, while at the same time they would succeed in resisting its disintegrative forces on their own.

SPONTANEOUS SELF-DEVELOPMENT (1965–1985)

Dutch psychohygienists believed in controlled modernization and guided personal development through social and cultural planning under the supervision of a morally inspired and professionally trained elite. This patronizing approach was characteristic of the postwar period of reconstruction, but beginning in the mid-1960s it came under attack. In the ensuing decade, the Netherlands changed from a rather conservative and law-abiding nation into one of the most liberal and permissive countries of the Western world.[19] Secularization and depillarization, as well as growing prosperity and the expanding welfare state, caused more and more people to break away from established traditions and hierarchical relationships to enhance their independence and individuality. Since the 1950s, there had been a leveling of differences in income, a democratization of consumption, and widely available access to (higher) education, which increased the political awareness of many.[20] Various protest movements loudly voiced participants' concern for more openness, democratization, liberation, and self-determination. The control of emotions and the individual's adaptation to society were no longer considered signs of responsibility but rather examples of the repression of personal freedom and the authentic self. The ideal of spontaneous self-realization, extolling self-exploration and self-expression, superseded that of guided self-development. It paved the way for an assertive individualism that, together with

[18] G. Brillenburg Wurth et al., eds., *Geestelijke Volksgezondheid: Nederlands Gesprekcentrum Publicatie No. 17* (Kampen, Netherlands, 1959).

[19] J. Kennedy, *Nieuw Babylon in aanbouw: Nederland in de jaren zestig* (Amsterdam, 1995); H. Righart, *De eindeloze jaren zestig: Geschiedenis van een generatieconflict* (Amsterdam, 1995); S. Stuurman, "Terugblik op een *Ancien Régime:* Nederland in de twintigste eeuw," in *Sociaal Nederland: Contouren van de twintigste eeuw,* ed. C. van Eijl, L. Heerma van Voss, and P. de Rooy (Amsterdam, 2001), 201–16.

[20] G. van den Brink, C. Brinkgreve, and L. Heerma van Voss, "Verworven gelijkheid en gevoelde verschillen: Contouren van de sociale eeuw," in Van Eijl, van Voss, and De Rooy, *Sociaal Nederland* (cit. n. 19), 1–12; C. J. M. Schuyt, "Sociaal-culturele golfbewegingen in de twintigste eeuw," in ibid., 25–34, on 26; J. Luiten van Zanden, "De egalitaire revolutie van de twintigste eeuw: Nederland 1914–1993," in ibid., 187–200.

the democratization movement, rocked the foundations of Dutch society and its mental health care system. If beforehand individuals had been expected to comply with the social order, now society itself had to change to facilitate their optimal self-development and the ultimate fulfillment of democratic citizenship. After the liberal constitution (1848) had provided the Dutch people with basic civil rights, the introduction of universal suffrage (1919) had made them into citizens in the political sense, and the postwar welfare state had guaranteed their material security, now, as some psychohygienists argued, the time was ripe for taking the next step in this continuing process of emancipation: the settling of immaterial needs in order to advance personal well-being for everybody.[21] In the 1960s and 1970s, the welfare state, in general, and welfare work, in particular, received an aureole of moral dignity: they came to be seen as the touchstones of civilization and human solidarity.

Embracing some of the basic tenets of the protest movements and antipsychiatry, mental health workers increasingly voiced self-criticism and responded to clients who began to protest against what they saw as undemocratic relationships and a structural neglect of their own influence in the social services system. A growing number of professionals were trained in the behavioral sciences, sociology, and social work, and they demanded attention to the social causes of mental distress. Therapeutic treatment of individuals with the aim of adapting them to society became subject to debate. Instead, people needed to be liberated from the "social structures" that caused unlivable or intolerable situations and that restricted their spontaneous self-development. The realization of this objective seemed more dependent on welfare work and political activism than on psychiatry and mental health care.[22] However, whereas institutional and medical psychiatry were put on the defensive, in the 1970s the psychosocial and, especially, psychotherapeutic services more than ever increased in size and prestige. The critique of the 1960s protest movement and antipsychiatry was absorbed in a way that legitimized this expansion. The very dissatisfaction with medical psychiatry prompted new pleas for better and more alternative forms of mental health care, such as therapeutic communities in hospitals and outpatient facilities in society at large.[23] Their growth was facilitated by embedding mental health care in the welfare state: more and more collective social security and health care funds financed the costs. From an international perspective, welfare and mental health arrangements were generous and guaranteed their broad accessibility. Since about 1960, the growth of the expenditures for social services and government subsidies, in relation to the GNP, was nowhere more substantial than in the Netherlands.[24] The prevailing trend between 1965 and 1985 was, then, one of a substantial increase and scaling up of pub-

---

[21] J. A. Weijel, *De mensen hebben geen leven: Een psychosociale studie* (Haarlem, 1970); J. van den Bergh et al., *Verbeter de mensen, verander de wereld: Een verkenning van het welzijnsvraagstuk vanuit de geestelijke gezondheidszorg* (Deventer, Netherlands, 1970); G. van Beusekom-Fretz, *De demokratisering van het geluk* (Deventer, Netherlands, 1973).

[22] Weijel, *De mensen hebben geen leven;* Van den Bergh et al., *Verbeter de mensen, verander de wereld;* Van Beusekom-Fretz, *De demokratisering van het geluk.* (All cit. n. 21.)

[23] D. Ingleby, "The View from the North Sea," in *Cultures of Psychiatry and Mental Health Care in Postwar Britain and the Netherlands,* ed. M. Gijswijt-Hofstra and R. Porter (Amsterdam, 1998), 295–314; G. Blok, *Baas in eigen brein: "Antipsychiatrie" in Nederland, 1965–1985* (Amsterdam, 2004).

[24] G. Therborn, *European Modernity and Beyond: The Trajectory of European Societies, 1945–2000* (London, 1995), 93, 156; cf. P. Schnabel, *De weerbarstige geesteziekte: Naar een nieuwe sociologie van de geestelijke gezondheidszorg* (Nijmegen, Netherlands, 1995), 102; Schnabel, "Psychiatry after World War II: An Overview," in Gijswijt-Hofstra and Porter, *Cultures of Psychiatry and Mental Health Care* (cit. n. 23), 29–42; Oosterhuis, "Insanity and Other Discomforts" (cit. n. 3).

lic services, with steadily growing numbers of clients.[25] In the early 1980s, the various outpatient mental health facilities merged into Regional Institutes for Ambulatory Mental Health Care, the Dutch version of community mental health centers, which were aimed at a broad spectrum of psychosocial problems and psychiatric disorders.

It was striking how swiftly mental health workers, among them new professionals such as continuing education experts (*andragogen*) and (nonmedical) psychotherapists, restored their self-confidence and the belief in their own therapeutic effectiveness. While engaging in heated debates on the political implications of their work, they widened their professional domain to include welfare work, a sector that, in the 1970s, stimulated by a government dominated by Social Democrats and other leftists, experienced enormous growth. Together with social workers, psychotherapists undertook the task of supporting people to enable them to liberate themselves from the coercive social structures. While avoiding a patronizing stance at all costs, the practitioners were expected to encourage clients to become aware of their true needs and to "grow" as a way to develop their true selves and their assertiveness. As psychiatrist J. A. Weijel explained in his "psychosocial study" *De mensen hebben geen leven* (*People Have No Life*), personal unhappiness should not be viewed as an individual fate but as a social evil that can be remedied.[26] Mental health workers revealed themselves as inspired advocates of personal liberation in the areas of religion, morality, relationships, sexuality, education, work, and drugs, as well as advocates for the emancipation of women, youngsters, the lower classes, and other disadvantaged groups, such as the gay community and ethnic minorities. As some of these advocates emphasized, countering prejudice and advancing tolerance was part of the broader effort to improve the quality of social relations and "democratize happiness."[27]

There was much talk about "social action" among mental health workers, but it proved rather difficult to change society in the day-to-day practice of mental health care. Yet these years were the heyday of psychotherapy, which, apart from psychiatrists, was practiced more and more by nonmedical professionals, such as psychologists and social workers and which, in the popular view, was the *pars pro toto* of mental health care. By the late 1970s, the Netherlands had become one of the countries with the highest number of therapists in proportion to the size of the population, in part as a result of the rapid growth of the number and size of public psychotherapeutic institutes.[28] A growing number of people began to consider it more or less self-evident to seek psychotherapeutic help for all sorts of discomforts and personality flaws that bothered them, ones not previously regarded as mental problems. Both therapists and clients viewed themselves more or less as a cultural avant-garde: psychotherapy would liberate individuals from unnecessary inhibitions and limitations and provide

[25] Nationale Federatie voor de Geestelijke Volksgezondheid, *Gids voor de Geestelijke Gezondheidszorg in Nederland* (Amsterdam, 1965–1969), 11, 159, 223–4; Nationaal Centrum voor Geestelijke Volksgezondheid, *Gids Geestelijke Gezondheidszorg 1982* (Utrecht, 1981), 21, 43–241; C. T. Bakker and H. van der Velden, *Geld en gekte: Verkenningen in de financiering van de GGZ in de twintigste eeuw* (Amsterdam, 2004), 65.

[26] Weijel, *De mensen hebben geen leven* (cit. n. 21), 10.

[27] Van Beusekom-Fretz, *De demokratisering van het geluk* (cit. n. 21).

[28] F. M. J. Lemmens and P. Schnabel, "Vestiging en ontwikkeling van de psychotherapie," in *Oriëntatie in de psychotherapie,* ed. C. P. F. van der Staak, A. P. Cassee, and P. E. Boeke (Houten, Netherlands, 1994), 9–26, 15; W. J. de Waal, *De geschiedenis van de psychotherapie in Nederland* ('s-Hertogenbosch, Netherlands, 1992), 126; G. J. M. Hutschemaekers and H. Oosterhuis, "Psychotherapy in the Netherlands after the Second World War," *Medical History* 47 (2004): 429–48.

them with opportunities for self-discovery, personal growth, and improving the quality of their lives. The psychotherapeutic ethos was not without contradictions. Although critical mental health care workers blamed social evils for psychological problems, psychotherapy focused exclusively on the individual inner self. The ethos also tied in with optimistic expectations about the possibility of changing personal characteristics in a purposive and rational way, while at the same time, it focused on authenticity: spontaneous self-development implied that people had to discover and realize their hidden, preexisting natural cores and true selves.

Notwithstanding this turn back from social criticism to the inner self in mental health practice, psychiatrists and other psychohygienic experts played a crucial role in public debates, and some of them put controversial and sensitive issues on the social agenda. Already in the 1950s and early 1960s, psychohygienists such as the Catholic psychiatrist C. J. B. J. Trimbos were strongly contributing to changing the moral climate in the areas of family, marriage, and sexuality. They replaced the strained and suspicious attitude toward sexual matters and the predominant focus on reproduction with a more positive evaluation of satisfactory sexual relationships as the basis for affective bonds and individual well-being. By breaking down taboos about birth control and homosexuality, these practitioners laid the foundation for the sexual revolution.[29] From the late 1960s on, psychiatrists called attention to the suffering of war victims and other traumatized individuals. As a result of psychiatrists' concern about phenomena such as war traumas and concentration camp syndrome, politicians and the general public became aware of the mental suffering of war victims, which resulted in measures aimed at providing both material and psychological support. Touching on current controversies surrounding the war—the younger generations accusing the majority of the older ones of having failed to resist Nazism as well as of having ignored the suffering of its victims—the psychiatric logic proved especially effective in the effort to render the rights of war victims, and later those of sufferers from other "psychological traumas" as well, socially acceptable.[30] Obviously, it was neither the first nor the last time that disadvantaged groups and their spokespersons called attention to mental suffering in order to get public opinion on their side and see their interests and rights protected. Whoever in the Netherlands convincingly argued the case of an individual or group that suffered mentally on account of specific social wrongs could generally count on public attention and support from the government.

Psychiatrists also stood up for the self-determination of patients and the decriminalization of euthanasia, abortion, and drugs.[31] In this way, they contributed to a new

[29] D. A. M. van Berkel, *Moederschap tussen zielzorg en psychohygiëne: Katholieke deskundigen over voortplanting en opvoeding 1945–1970* (Assen, Netherlands, 1990); Oosterhuis, *Homoseksualiteit in katholiek Nederland* (cit. n. 17); Oosterhuis, "The Netherlands: Neither Prudish nor Hedonistic," in *Sexual Cultures in Europe: National Histories,* ed. F. X. Eder, L. A. Hall, and G. Hekma (Manchester, UK, 1999), 71–90.

[30] I. de Haan, *Na de ondergang: De herinnering aan de Jodenvervolging in Nederland, 1945–1995* (The Hague, 1997); J. Withuis, *Erkenning: Van oorlogstrauma naar traumacultuur* (Amsterdam, 2002).

[31] J. Kennedy, *Een weloverwogen dood: Euthanasie in Nederland* (Amsterdam, 2002); E. Ketting, *Van misdrijf tot hulpverlening: Een analyse van de maatschappelijke betekenis van abortus provocatus in Nederland* (Alphen aan den Rijn, Netherlands, 1978), 82–3; J. V. Outshoorn, *De politieke strijd rondom de abortuswetgeving in Nederland, 1964–1984* (Amsterdam, 1986), 123, 139, 179–80; M. de Kort, *Tussen patiënt en delinquent: Geschiedenis van het Nederlandse drugsbeleid* (Rotterdam, 1995).

public morality and the implementation of practices that were quite liberal, certainly when considered from an international perspective. In so doing, they drew on the 1960s culture of liberation and democratization; but they also followed in the footsteps of the reform-minded psychohygienists from the 1950s.[32] By raising issues that earlier were largely silenced, they sought to break taboos and put an end to hypocrisy, thus paving the way for more openness, understanding, tolerance, and liberation. To achieve all this, so they explained, a sense of responsibility, conscientious positioning, a sincere exchange of arguments, and the willingness of people to listen to each other was required. As psychiatrist R. H. van den Hoofdakker wrote in his book on medical power and medical ethics, "[I]n a world of emancipated and independent human beings" there was only one way to overcome outmoded ideas and habits, and that was "talking, talking, talking."[33] Rules and laws should not be rigidly applied but discussed and sensibly interpreted. Emphasizing an issue's "debatability" (*bespreekbaarheid*)—which in the Netherlands became a major norm that served as the basis for policies of controlled toleration (*gedogen*)—was essentially the opposite of being noncommittal or outright permissive.[34] What mattered was countering the invisible abuse of specific liberties and channeling and controlling them carefully, in good faith and in open-minded deliberations with all parties. Making sensitive issues debatable was inextricably bound up with the belief in an open, egalitarian, and fully democratized society. Only mature, self-reflective, socially involved citizens empathized with others, did not shy away from unpleasant truths, regulated their emotions, and were capable of making rational considerations and—through negotiation and mutual understanding—arriving at balanced decisions. This psychohygienic ideal of citizenship made great demands on people's psychological competence.

AUTONOMOUS SELF-DEVELOPMENT (1985–2005)

Until the late 1970s, there was great faith in the Netherlands in social planning as a way to change society in directions that would allow for individual self-development.[35] However, the practice of rational planning, which was self-evident during the reconstruction era's directed economy and guided democracy, as well as in the context of the Social Democratic reform policies of the 1970s, conflicted with society's increasing individualization. As there was progressively more emphasis on personal emotional life and individual self-realization, the socially critical dimension of the self-development ideal eroded: the pursuit of social reform was replaced with the values of the "me-generation," stressing an inner-directed, independent self. At the same time, the welfare state was under attack, mainly because its costs had gone up tremendously but also because critics argued that collective services nullified people's sense of responsibility and self-reliance. Around 1980, welfare work, in particular, was singled out as a target. Rather than enlarging people's self-autonomy, it was seen

---

[32] I. Weijers, "De slag om Dennendal: Een terugblik op de jaren vijftig vanuit de jaren zeventig," in Luykx and Slot, *Een stille revolutie?* (cit. n. 14), 45–65; Weijers, "The Dennendal Experiment, 1969–1974: The Legacy of a Tolerant Educative Culture," in Gijswijt-Hofstra and Porter, *Cultures of Psychiatry and Mental Health Care* (cit. n. 23), 169–84.

[33] R. H. van den Hoofdakker, *Het bolwerk van de beterweters: Over de medische ethiek en de status quo* (Amsterdam, 1971), 50.

[34] Kennedy, *Een weloverwogen dood* (cit. n. 31); cf. Kennedy, *Nieuw Babylon in aanbouw* (cit. n. 19).

[35] Duyvendak, *De planning van ontplooiing* (cit. n. 2).

as making them dependent.[36] In addition, the generous public funding of psychotherapy drew criticism: psychotherapists, who appeared as the elite among mental
health workers, made good money by serving a privileged YAVIS-clientele (young attractive verbal intelligent successful), while neglecting psychiatric patients with serious mental and behavioral disorders.[37] With their politics of deregulation and privatization, conservative liberals and Christian Democratic politicians began to shift the
emphasis from the state-organized collective care facilities to the self-reliance of citizens in the community and on the market. Autonomous self-development of responsible and independent individuals on the basis of their talents and efforts, with a minimum of interference from government and social bureaucracy, came to be the new
standard of good citizenship. Self-development was considered merely a personal
matter and no longer a social issue, let alone a political one (as leftist activists, welfare workers, and many mental health experts had argued).

However, the crisis of the welfare state, which led to a downsizing of welfare work,
hardly affected mental health care; on the contrary, the latter underwent more expansion in subsequent years, although its focus changed. Further growth of the outpatient
sector, in particular, was stimulated by the effort to push back institutional psychiatry
and to develop community care for psychiatric patients, which became a governmental priority. Mental health care also adapted better than welfare work to the changing
social climate, notably the depoliticization of social issues coupled with ongoing
individualization. Professionalism, efficiency, rationalization, standardization, and a
partial remedicalization of psychiatry as well as the issue of costs and benefits took
the place of the lofty ideals of the sixties movement. Increased attention to elements
of the free market and people's own sense of responsibility went hand in hand with
the development of a more formal, legally based relationship between clients and care
providers: rights and responsibilities were fixed into laws, rules, and procedures.[38]

The government and some psychiatrists repeatedly argued that the main outpatient
facilities, the Regional Institutes for Ambulatory Mental Health Care, were geared
one-sidedly to deal with clients with minor psychological afflictions, causing an endless increase in the demand for mental health care. The treatment and care of acute
and chronic psychiatric patients was now to become a priority, along with keeping the
number of admissions to mental hospitals as low as possible. Only those patients unable to get by in society without hurting themselves or others were considered to be
eligible for hospitalization. Others were to be cared for in halfway and outpatient facilities to allow them to be, as much as possible, regular members of society. In the
late 1990s, in order to improve cooperation between psychiatric hospitals and outpatient services, the government pressured both to merge into comprehensive mental

---

[36] H. Achterhuis, *De markt van welzijn en geluk: Een kritiek van de andragogie* (Baarn, Netherlands,
1980): De Haan and Duyvendak, *In het hart van de verzorgingsstaat* (cit. n. 16), 121–2, 182, 352–3.

[37] Among the clientele of psychotherapy, certain social groups were indeed over-represented. Workers with little or no education, for one, were hardly found; most clients had a middle-class background
and were familiar with the notions and thinking of psychotherapists. More specifically, clients tended
to be young, well educated, nonchurchgoing, and still studying or professionally active in sectors such
as health care, social work, and education. C. Brinkgreve, J. H. Onland, and A. de Swaan, *Sociologie
van de psychotherapie 1: De opkomst van het psychotherapeutisch bedrijf* (Utrecht, 1979), 97, 104,
124; A. de Swaan, R. van Gelderen, and V. Kense, *Sociologie van de psychotherapie 2: Het spreekuur
als opgave* (Utrecht, 1979), 37, 50, 84–6.

[38] J. Legemaate, "De juridisering van de psychiatrie," in *De Januskop van de psychiatrie: Waarden
en wetenschap*, ed. C. F. A. Milders et al. (Assen, Netherlands, 1996), 131–40; P. Schnabel, "Het jonge
en het oude gezicht van de psychiatrie," in Milders et al., *De Januskop van de psychiatrie*, 151–9.

health facilities that offered intramural as well as extramural care. The psychotherapeutic treatment of minor psychosocial problems was increasingly relegated to private practices. All of this marked a break with the historically developed constellation of Dutch public mental health care, which since the 1930s had been divided between clinical psychiatry for serious mental disorders and an outpatient sector with a strong psychosocial orientation for a wide spectrum of milder problems.

The "socialization" of psychiatry, as this policy was termed, echoed some of the democratic ideals of the 1960s and the 1970s, such as the need to counter the social isolation of psychiatric patients, improve their self-autonomy, and respect their civil rights. In 1970, the paternalistic, for-your-own-good criterion in the Insanity Act of 1884, which until then had justified involuntary institutionalization, was now replaced by the criterion of danger.[39] A new mental health law enacted in 1994 set down strict criteria for forced hospitalization against the will of patients, insisting on commitment only if someone posed a threat to himself or others. The law—which brought an end to the possibility of certification and the loss of full citizenship simply because it was considered in the best interest of patients—was a judicial stamp of approval for the increased recognition of the individual autonomy, freedom, integrity, and responsibility of the mentally ill. The mentally ill regained, so to speak, their status as citizens, an aim that since the 1970s had been championed by the critical patient's movement.[40]

One of the basic motivations for the policy of socialization was to assure that, although psychiatric patients—just like other groups in need of care—were limited in their autonomy, judgment, and decision-power, they should not be excluded in advance from exercising both their rights and duties as citizens. The degree to which they would be able to realize themselves as more or less independent members of society relied, in part, on social conditions that could be shaped: a mental health care that was organized in a way that made sure such individuals were not isolated from the rest of society and would receive sufficient social support to bring about their integration into society. This reasoning, which was largely rooted in the ideas of the 1960s and the 1970s, was quite similar to the way in which, one century before, social intervention was promoted to develop members of disadvantaged groups into full citizens. The underlying idea was that achieving citizenship largely depended on the degree to which the social structure actually enabled people's self-development, including the support and the encouragement they needed.

In practice, however, the citizenship of psychiatric patients met with obstacles time and again and was directly challenged in the 1990s. Critics pointed out that the principle of autonomous self-determination, on which the modern ideal of citizenship was grounded, entirely ignored what, in effect, constituted the essence of mental illness: the limited power for self-determination and self-reliance and the loss of the basic and taken-for-granted patterns of social interaction. Furthermore, critics argued that the emancipation of psychiatric patients as citizens was quite paradoxical as their representatives, more than they themselves, were the ones insisting that they should be able to take care of themselves, allowed to or even compelled to make decisions on their

[39] F. A. M. Kortmann, "Bemoeizorg en de WGBO," in Milders et al., *De Januskop van de psychiatrie* (cit. n. 38), 141–50, on 145.

[40] A. J. Heerma van Voss, "De geschiedenis van de gekkenbeweging: Belangenbehartiging en beeldvorming voor en door psychiatrische patiënten (1965–1978)," *Maanblad Geestelijke volksgezondheid* 33 (1978): 398–428; R. van der Kroef, *25 jaar en nog steeds geen normaal mens ontmoet: Pandora, psychiatrie en beeldvorming* (Baarn, Netherlands, 1990).

own, and participate in society. With the emphasis on autonomy and self-reliance as meaningful modes of existence, other needs of psychiatric patients moved to the background: safety, security, protection, rest, the longing for an orderly and quiet life shielded from society, and recognition of their own experiences and fragility.[41] As long as the defining qualities of citizenship were autonomy, agency, and active social participation (especially by having regular work), the mentally ill and disabled were in fact consigned, at best, to the category of marginal citizens.

Care providers and the government inspectors of public mental health, as well as patient organizations and their families, also questioned the positive evaluation of self-determination because it allowed the mentally ill with serious behavioral problems to refuse psychiatric treatment, even if they were unable to take care of themselves, caused social trouble, or were potentially aggressive. From this perspective, the socialization of psychiatry soon ran up against its limits. The striving for the social integration and employment rehabilitation of psychiatric patients was complicated by increasing pressure on the social cohesion in (sub)urban neighborhoods and the ever higher demands of the labor market (proper training, social skills, performance, assertiveness, competition, flexibility, and being immune to stress), which many psychiatric patients were certainly unable to meet. It also became clear that since the late 1980s, the tolerance of the Dutch population toward those with psychiatric disorders, particularly when accompanied by disturbing conduct, had begun to wane the more they were directly confronted with the mentally ill in everyday life.[42]

The policy of socialization had its downsides: isolation, abandonment, and impoverishment of some patients, a lack of daytime activities for many others, an overburdening of social care facilities and the general environment, and the rise of a variety of social problems caused by, among other things, homelessness, alcohol abuse, and drug addiction. "They keep coming back in and leave again (*draaideuren*), roam around, do drugs and move elsewhere," as one psychiatrist summarized the fate of many afflicted.[43] Neither society nor those "depraved and impoverished" individuals were seen as benefiting from the legally sanctioned reticence of care providers; hence the for-your-own-good criterion, it was believed, needed to be reconsidered.[44] Pleas for more pressure and coercion in socio-psychiatric care and for a vigorous public mental health policy under the government's authority, as well as new experiments in outreach care for mental patients who were unwilling to cooperate or hard to reach, put earlier ideals of emancipation and self-determination into perspective. Basically,

[41] A. K. Oderwald and J. Rolies, "De psychiatrie als morele onderneming," *Tijdschrift voor Psychiatrie* 32 (1990): 601–15; P. Schnabel, *Het recht om niet gestoord te worden: Naar een nieuwe sociologie van de psychiatrie* (Utrecht, 1992); G. van Loenen, "Van chronisch psychiatrische patiënt naar brave burger: Over de moraal van psychiatrische rehabilitatie," *Maandblad Geestelijke volksgezondheid* 52 (1997): 751–61; G. A. M. Widdershoven, R. I. P. Berghmans, and A. C. Molewijk, "Autonomie in de Psychiatrie," *Tijdschrift voor Psychiatrie* 6 (2000): 389–98; J. Rasmussen, "Bij zinnen: De betekenis van het lijden in de psychiatrie," *Maandblad Geestelijke volksgezondheid* 56 (2001): 833–41.

[42] M. H. Kwekkeboom, "Sociaal draagvlak voor de vermaatschappelijking in de geestelijke gezondheidszorg: Ontwikkelingen tussen 1976 en 1997," *Tijdschrift voor Gezondheidswetenschappen* 78(3) (2000): 165–71.

[43] J. Droës, "De metamorfose van de GGZ," *Maandblad Geestelijke volksgezondheid* 57 (2002): 143–5, on 143.

[44] Geneeskundige Inspectie voor de Geestelijke Volksgezondheid, *Jaarverslag 1994* (Rijswijk, Netherlands, 1995), 13; Geneeskundige Inspectie voor de Geestelijke Volksgezondheid, *Jaarverslag 1995* (Rijswijk, Netherlands, 1996), 9: E. Borst-Eilers, *Brief Geestelijke Gezondheidszorg aan de Tweede Kamer der Staten-Generaal* (Rijswijk, Netherlands, 1997), 9.

these were hardly relevant for those who suffered from serious psychiatric disorders, were incapable of living on their own, or could not assert their needs and lacked the ability to reflect on their possibilities and limitations. For them, social reintegration was no real option, and they were living proof that mental illness and full citizenship were hard to reconcile.

With respect to patients who suffered from serious mental disorders, psychiatry still largely proved to be the science of unsolved riddles and despair.[45] The optimism that had since the 1950s prevailed in the psychohygienic movement and large segments of the outpatient mental health community about the possibility of stimulating individuals' self-development and fashioning them into self-aware citizens had, in part, been facilitated precisely because there was a strong tendency to keep patients with serious psychiatric disorders out of the system. The psychotherapeutic institutes, as well as the Centers for Family and Marriage Problems and the Child Guidance Clinics, had distanced themselves from care provision for psychiatric patients in institutions, emphasizing their identity as welfare facilities with a psychotherapeutic orientation. In these facilities, mental health workers catered to a clientele with a variety of psychosocial and existential problems, and they focused on the improvement of people's psychosocial welfare, self-development opportunities, social participation, and assertiveness. A psychological perspective and various talking cures had increasingly set the tone in these facilities. Clients were expected to have some capacity for introspection, verbal talent, initiative, and a willingness to change, and this automatically excluded the mentally ill.

However, when, in the Regional Institutes for Ambulatory Mental Health Care in the 1990s, social psychiatry was prioritized and ever more of these services merged with psychiatric hospitals, the emphasis shifted toward people with more serious and unmanageable mental disorders, those who did not meet the ideal of voluntary perfectibility and malleability. The high expectations regarding people's potential for change and liberation were replaced by the more modest objective of trying to limit or alleviate mental suffering and control its symptoms as much as possible. Notwithstanding the dominance of biological psychiatry and the increasing use of psychopharmaceuticals, the various social, psychological, and behavioral therapies remained in use in public mental health facilities, but they were directed less at self-discovery, self-reflection, and personal growth than at acquiring social and practical skills to cope with life, in good times and bad.

Yet in another way, the ideology of individual liberation and emancipation of the 1960s and 1970s was called into question. Under the influence of the ongoing expansion of mental health care consumption, epidemiological research showing a high frequency of psychological disorders among the population, and prognostic data suggesting a further rise, the social dimension of mental disorders and their possible prevention were highlighted again in the 1990s.[46] The evidence motivated the government

[45] P. Schnabel, "Maakbaar en plooibaar," *Maandblad Geestelijke volksgezondheid* 42 (1987): 490–1; Schnabel, *Het recht om niet gestoord te worden* (cit. n. 41); Schnabel, "Het jonge en het oude gezicht van de psychiatrie" (cit. n. 38), 154–5.

[46] R.V. Bijl, G. van Zessen, and A. Ravelli, "Psychiatrische morbiditeit onder volwassenen in Nederland: het NEMESIS-onderzoek. II. Prevalentie van psychiatrische stoornissen," *Nederlands Tijdschrift voor Geneeskunde* 141 (1997): 2453–60; R. V. Bijl and A. Ravelli, "Psychiatrische morbiditeit, zorggebruik en zorgbehoefte: Resultaten van de Netherlands Mental Health Survey and Incidence Study (NEMESIS)," *Tijdschrift voor Gezondheidswetenschappen* 76 (1998): 446–57.

to conduct a number of studies on the perceived rise in the number of mental problems and the measures needed to address the problem.[47] The message put forward in subsequent reports and policy recommendations was ambiguous. On the one hand, experts explained that the rising demand for professional care was not necessarily a sign of deteriorating public mental health because the growing care consumption would be, in part, a result of the broadened supply of services, the public's greater familiarity with it, a declining tolerance toward all sorts of inconveniences and misfortunes (often articulated as psychological complaints), and the increased trust in the possibility of treating these groups professionally. On the other hand, the tone of the reports betrayed the resurfacing of a familiar cultural pessimism. They kept pointing to an array of social developments that were likely to trigger psychical problems: the high pace and intensity of social changes; the atomization of society, in part as a consequence of the high degree of social and geographic mobility and the weakening or loss of family ties and other social networks; the loss of normative and meaning-providing frames; mounting job pressures and the (too) high demands placed on people's social skills and mental elasticity; (immanent) unemployment; an information avalanche with which many could barely cope; the social disadvantage and discrimination experienced by ethnic minorities; and the diminishing sense of social security and safety. There was a great deal of emphasis, in particular, on the assumed loss of shared norms and values. For example, one of the reports suggested that in the densely populated and urbanized Netherlands, individual freedom and tolerance could not flourish without social responsibility and cohesion.[48] In another policy suggestion, reference was made to the disappearance of "traditional social bonds," "new risks of dropouts," and the "disintegration of the social-pedagogical infrastructure" that made "the systematic passing on of values" less self-evident.[49]

Inasmuch as policy advisers issued proposals for the improvement of public mental health, they reverted to remedies from the past: the recommendation to not limit care for mental patients and psychological problems to professional care alone but to give the afflicted a place in other social sectors and to involve laypersons as much as possible—an approach that right after the Second World War was recommended as well but hardly realized. In addition, there were pleas for the stimulation of, as it was described, "a *new form of civil society*" and "the articulation and teaching . . . of the values and norms that society wishes to defend."[50] In less shrouded terms, such recommendations were also echoed in the Manifest, in which, on the eve of the 1998 parliamentary elections, the National Fund for Public Mental Health (Nationaal Fonds Geestelijke Volksgezondheid) called on the government to pursue a more active policy to improve public mental health: "In a society like ours—with many disinte-

---

[47] Scenariocommissie Geestelijke Volksgezondheid en Geestelijke Gezondheidszorg en Onderzoeksteam van het Nederlands centrum Geestelijke volksgezondheid, *Zorgen voor geestelijke volksgezondheid in de toekomst: Toekomstscenario's geestelijke volksgezondheid en geestelijke gezondheidszorg, 1990–2010* (Utrecht, 1990); M. Gastelaars et al., *Vier gevaarlijke kruispunten: Een voorzet voor een geestelijk volksgezondheidsbeleid* (Utrecht, 1991); P. Schnabel, R. Bijl, and G. Hutschemaekers, *Geestelijke volksgezondheid in de jaren '90: Van ideaal tot concrete opgave* (Utrecht, 1992); Landelijke Commissie Geestelijke Volksgezondheid, *Zorg van velen: Eindrapport van de Landelijke Commissie Geestelijke Volksgezondheid* (The Hague, 2002).

[48] Schnabel, Bijl, and Hutschemaekers, *Geestelijke volksgezondheid in de jaren '90* (cit. n. 47).

[49] Landelijke Commissie, *Zorg van velen* (cit. n. 47), 58, 62, 64, 79.

[50] Schnabel, Bijl, and Hutschemaekers, *Geestelijke volksgezondheid in de jaren '90*, 38; Landelijke Commissie, *Zorg van velen,* 62 (emphasis in original). (Both cit. n. 47.)

grating families, aggression, violence, alcohol abuse, ever growing job pressure, much social fear, much stress, and a collective loss of norms and values—the chance of . . . mental problems or disorders increases." The authors argued that to promote the cohesion of society and mental health, the government had to "make rules, set limits, and articulate norms and values."[51]

Evidently, in mental health care the optimistic view of the 1960s and the 1970s, in which emancipated and motivated people tried to solve problems together in mutual interaction, had been replaced with concern about the loss of community spirit and public morality. In fact, this was in line with a broader criticism of the legacy of the sixties movement since the 1990s. The antiauthoritarian movement and the celebration of individual freedom, politicians and intellectuals argued, had degenerated into egoism, erosion of the personal sense of responsibility, an exaggerated assertiveness that was exclusively based on rights rather than duties, a coarsening of social interactions, and an increase in violent behavior and other forms of crime. The welfare state had resulted in calculating behavior and improper use of benefits. The balance between communal and individual interests was entirely disrupted: spontaneous self-development and assertiveness had led to a colonizing of the public sphere by all sorts of personal claims and preferences. The overall toleration policy and the new taboos of political correctness had led to a lack of self-restraint, a degradation of the public domain, and social disintegration. These developments would have to be countered by the restoration and revitalization of a sense of community and civic virtue.[52]

In the 1980s, the Christian Democrats, in particular, with their ideal of the "caring society," pointed to the significance of community spirit and social participation. But in the 1990s, Social Democrats and liberals also became convinced of the need to re-gauge collective and individual responsibilities and cultivate a sense of civic virtue with an emphasis on adjustment, integration, and moral regeneration. The policies of the "purple" government coalition (Social Democrats and liberals) foregrounded the reinforcement of social cohesion and the promotion of good citizenship. Exactly at that moment when neoliberalism could develop unchecked and the economy flowered, problem groups that were socially lagging, notably ethnic minorities and the longtime and poorly educated unemployed, became more visible. The taboo on coercion and duties began to recede, particularly in regard to efforts aimed at the reactivation of the unemployed and those previously declared unfit to work as well as at the integration of migrants.[53]

At the start of the twenty-first century, the concern for social disintegration and the degradation of the public domain mingled with fear of the loss of national identity on account of the rising ethnic diversity, continuing European integration, and globaliza-

[51] Nationaal Fonds Geestelijke Volksgezondheid, *Manifest van het Nationaal Fonds Geestelijke Volksgezondheid: Verontrustende ontwikkelingen* (Utrecht, 1998), 3, 8.

[52] H. Wigbold, *Bezwaren tegen de ondergang van Nederland* (Amsterdam,1995); H. Vuijsje, *Correct: Weldenkend Nederland sinds de jaren zestig* (Amsterdam,1997); H. Beunders, *Publieke tranen: De drijfveren van de emotiecultuur* (Amsterdam, 2002); D. Pessers, *Big Mother: Over de personalisering van de publieke sfeer* (The Hague, 2003); R. Diekstra, M. van den Berg, and J. Rigter, eds., *Waardenvolle of waardenloze samenleving? Over waarden, normen en gedrag in samenleving, opvoeding en onderwijs* (The Hague, 2004); G. van den Brink, *Schets van een beschavingsoffensief: Over normen, normaliteit en normalisatie in Nederland* (Amsterdam, 2004).

[53] H. R. van Gunsteren, *Eigentijds burgerschap* (The Hague, 1992); H. R. van Gunsteren and P. den Hoed, eds., *Burgerschap in praktijken* (The Hague, 1992); S. Koenis, *Het verlangen naar gemeenschap: Politiek en moraal in Nederland na de verzuiling* (Amsterdam, 1997); Duyvendak, *De planning van ontplooiing* (cit. n. 2); De Haan and Duyvendak, *In het hart van de verzorgingsstaat* (cit. n. 16).

tion. After 2002, the threat of terrorism and two political murders caused a polarization that centered on multicultural values and the role of Muslims in Dutch society. The government has been pushing for a restoration of norms and values—with a prime minister who is inspired by communitarianism—and pursues policies that emphasize a further downsizing of the welfare state, the responsibility of citizens, the activation of the unemployed, the mandatory enculturation (*inburgering*) of migrants, a repressive approach of previously tolerated (mis)behavior, and a toughening of criminal law.

The last decade or so saw a change—a hardening, to be more specific—in the social and political climate, one that has called into question the optimistic view of humankind and the citizenship ideal that since the 1960s had been promoted in mental health care. As to citizenship, the mental health workers seem to have been forced on to the defensive: they mingle in public debates much less than they did in the decades between 1950 and 1980, less inclined to promote a specific public morality than had been the case earlier. The psychiatrist A. van Dantzig, former director of the psychotherapeutic institute in Amsterdam, was one of the few who continued to advocate a socially engaged mental health care. He considered attention to mental suffering and its professional treatment a touchstone of humanitarian and democratic progress and social justice. Mental health care (and psychotherapy, in particular), he claimed, is a valuable product of secularization and growing scientific understanding, and it has, in part, enabled the emancipation of the individual. It must be as comprehensive as somatic health care, so that, in principle, everyone is granted the opportunity to raise their quality of life and achieve maximal happiness with the help of psychotherapy. To avoid mental disorders becoming "privatized," he insisted, mental health care also has the task of exposing the social wrongs that are harmful to individual well-being. If Van Dantzig still embodied the inspiration that had marked many of his colleagues in the 1960s and 1970s, in the 1990s, he could hardly count on support in the world of mental health care, let alone outside of the community.[54]

CONCLUSION

The link between the democratization and psychologization of citizenship—illustrated here by following the development of mental health care in the Netherlands—is, of course, part of a more general historical process in the Western world. In traditional systems of social control and political domination, which subjected people by external coercion, no matter whether they accepted it or not, their inner selves were relatively irrelevant. The need to form individuals and to make them internalize certain values and behavior patterns became greater the more society was democratized. It was in democratic societies, which rejected force and coercion and presupposed that the social and political orders were basically founded on the autonomous consent of individual citizens, where inner motivation was considered of crucial importance for the quality of the public domain. A democratic social order can only be main-

---

[54] A. van Dantzig, "Persoonlijk lijden als publieke zorg," *Maandblad Geestelijke volksgezondheid* 46 (1991): 635–48; Van Dantzig, *Is alles geoorloofd als God niet bestaat? Over geestelijke gezondheidszorg en maatschappij* (Amsterdam, 1995); Van Dantzig, "Psychologisering en geestelijke gezondheidszorg," in *Het verlangen naar openheid: Over de psychologisering van het alledaagse*, ed. R. Abma et al. (Amsterdam, 1995), 69–74; Van Dantzig, "Geestelijke volksgezondheid," *Maandblad Geestelijke volksgezondheid* 57 (2002): 557–63.

tained, it has been thought, if individuals use their basic liberties in a responsible way. Ironically, the pursuit of individual autonomy and self-determination went hand in hand with gentle, but persistent, pressure on people to open their inner selves for scrutiny by others and account for their urges and motivations (for example, before mental health experts). Where it could no longer be assumed that the individual's conformity was something natural, in theory each member of society acquired an interest in what went on in the minds of others. If, in the nineteenth century, citizens were largely judged on external aspects (such as property ownership, financial autonomy, sex, tax duty), in the twentieth century—the era of general suffrage and the welfare state's softening of the contradiction between formal political rights and socioeconomic inequality—the formation of a proper mentality gained prominence. This psychologization, which drew attention to the major role of drives and emotions in both individual and collective life, called for an "inner mission."

As said, against this backdrop, the Dutch developments are hardly unique. In Britain for example, from the 1920s on, mental health provided a paradigm to articulate in psychological terms a secular ideal for self-development as the groundwork for responsible democratic citizenship. In the United States, the mental hygiene movement displayed a strong impulse to formulate a diagnosis of modern American society from the perspective of psychiatry and psychoanalysis. The ills of modern society and the malaise in individuals were linked and mental health experts used theories of personality development to show how they could contribute to the formation of robust and self-reliant democratic subjects. In Germany, critical reflection on and the search for fundamental reforms in psychiatry took place in the 1960s and 1970s, whereby the Nazi past was explicitly used as specter, giving mental health care a strong political dimension. Against the complicity of psychiatry in the atrocities of the Third Reich, a democratic and emancipatory countervision of mental health care emerged, based on a concept of citizenship that stressed political awareness, independence of mind, liberalization, and social rights of, and solidarity with, the infirm and indigent.[55]

However, what was often missing in these countries was an extensive network of public outpatient mental health facilities to tie the rhetoric about mental health and citizenship with concrete care-providing practices. In the Netherlands, models of psychological self-development and citizenship were not mere abstract theories: in the practice of outpatient provisions these ideals materialized. From the 1950s on, clients were encouraged to be self-reflective about their conduct and motivations within their private lives as well as in the public sphere.[56] The Dutch psychohygienic movement and the outpatient services were more lasting and broader in the Netherlands

[55] M. Thomson, "Before Anti-Psychiatry: 'Mental Health' in Wartime Britain," in Gijswijt-Hofstra and Porter, *Cultures of Psychiatry and Mental Health Care* (cit. n. 23), 43–59; Thomson, "Constituting Citizenship: Mental Deficiency, Mental Health, and Human Rights in Inter-war Britain," in *Regenerating England: Science, Medicine, and Culture in Inter-war Britain,* ed. Chr. Lawrence and A.-K. Mayer (Amsterdam, 2000), 231–50; J. C. Pols, *Managing the Mind: The Culture of American Hygiene, 1910–1950* (PhD diss., Univ. of Pennsylvania, 1997); F.-W. Kersting, ed., *Psychiatrie als Gesellschaftsreform: Die Hypothek des Nationalsozialismus und der Aufbruch der sechziger Jahre* (Paderborn, Germany, 2004); Kersting, "Between the National Socialist 'Euthanasia Programme' and Reform: Asylum Psychiatry in West Germany, 1940–1970," in Gijswijt-Hofstra et al., *Psychiatric Cultures Compared* (cit. n. 3).

[56] P. van Lieshout and D. de Ridder, eds., *Symptomen van de tijd: De dossiers van het Amsterdamse Instituut voor Medische Psychotherapie (IMP), 1968–1977* (Nijmegen, Netherlands, 1991); Oosterhuis, *Homoseksualiteit in katholiek Nederland* (cit. n. 17).

than in Britain, Germany, or the United States. Already in the 1940s, these were well-established parts of the mental health sector, and this would continue to be the case until the early 1980s, when they merged into one comprehensive system, the Regional Institutes for Ambulatory Mental Health Care.

The large degree of continuity, a distinctive feature of the Dutch outpatient mental health sector, was perhaps partly caused by the influence of the Dutch (pillarized) social system and the major role played by private initiative, facilitating more or less stable organizational structures even before the government became an active player in this area. That confessional groups had their own mental health facilities lowered the threshold for them to ask for professional care, while it also caused psychohygienic views to be spread more widely than would have been possible in a situation in which only generic services were offered. The pillarized system raised the chances of religious people coming into contact with a more psychological approach toward normative issues.[57] The Dutch government's interference in mental health care only began around 1970, but from then on it greatly contributed to the fact that this sector, compared with such sectors in other countries, prospered. This was an immediate effect of the generous collective funding that since the late 1960s had officially been set aside for mental health care. The Dutch welfare state—one as comprehensive as the Scandinavian welfare states and geared not only toward material security but also toward enhancing immaterial qualities of life—guaranteed that public mental health care facilities were available and accessible to all Dutch citizens and that they functioned properly.

Another striking element of the outpatient sector in the Netherlands was its broad orientation: it not only consisted of social psychiatric for patients but, from the 1930s and the 1940s, also included various counseling centers for problem children, existential problems, marriage and family-related issues, psychotherapy, and alcohol and drug addiction. This broad orientation is, in part, accounted for by the fairly early differentiation between institutional psychiatry and the outpatient sector as well as by the strong psychosocial (rather than biomedical) focus of extramural facilities (at least until the 1990s). In other European countries, the institutional and public mental health sectors were more exclusively geared toward psychiatric patients, while there was also a closer link with the domain of (poly)clinical psychiatry.[58]

In the twentieth century, the mental hygiene movement and the outpatient mental health sector successfully established themselves in the Netherlands. The notion of mental health, which heaped together a host of problems in and between people, caught on precisely because its vagueness served a major strategic function in linking various social domains and appealing to a variety of groups. Mental health applied to both the individual and society, establishing a connection between the private and public spheres. The notion of health care evoked associations with medicine and hygiene, while "mental"—the Dutch *geestelijk* also means "spiritual"—referred to psychical features as well as to religious, moral, cultural, and political values. Thus it was possible to establish an explicit connection with the strong charitable tradition in the

[57] P. J. van Strien, *Nederlandse psychologen en hun publiek: Een contextuele geschiedenis* (Assen, Netherlands, 1993), 88–9.

[58] H. Oosterhuis, "Outpatient Psychiatry and Mental Health Care in the Twentieth Century: International Perspectives," in Gijswijt-Hofstra et al., *Psychiatric Cultures Compared* (cit. n. 3).

Netherlands and the bourgeois civilization offensive, which, in the form of a moral-didactic ethos, was adopted by both confessionals and socialists. The ideal of mental health tied in with the need to articulate public morals and a certain utopian message, not only among liberals and Christians, but also, especially in the 1960s and 1970s, among socialist and other leftist groups that believed strongly in the perfectibility of society. Once the establishment of the welfare state had guaranteed material security, mental and social well-being became the standard for the good life.

The modernization of Dutch society and the evolving views of democratic citizenship provided a sociopolitical context for the pursuit of mental health, whereby either a cultural pessimism or an optimistic belief in society's progress prevailed. In this light, it is possible to identify a turning point in the mid-1950s. Around this time, the defensive response to the modernization process and the emphasis on Christian and traditional middle-class values were exchanged for a much more accommodating stance. At the same time, in reflections about citizenship, there was a shift from unconditional adaptation to the existing system of values and norms ("character") to individual self-development ("personality"). People's personal lives and experiences and their inner motivations came to be center-stage, and therapeutic treatment and social integration were definitively prioritized over external coercion and social exclusion. In the years between 1950 and 1965, by building on the ideal of guided self-development, mental health care hooked up with socioeconomic modernization: individuals had to shape their personalities, develop their autonomy and flexibility, be open for renewal, and in a responsible way achieve self-realization. In the late 1960s and the 1970s, mental health workers embraced spontaneous self-development as a core value, thus legitimizing the need for assertiveness, democratization, and personal liberation. Subsequently, in the last two decades of the twentieth century, they approached their clients as autonomous, mature, and self-responsible citizens, whose freedom to make choices as members of a pluralist market society was perceived as self-evident. At the close of the twentieth century, however, a cultural pessimism reappeared, and the emphasis on self-determination and autonomy as more or less absolute values was brought up for discussion.

The Dutch national government generally kept a low profile regarding the organization and implementation of mental health care and the articulation of civic virtues. At least until the late 1960s, when it began to play a more active role, the state mainly left these issues to voluntary organizations or lower governments, in part because it did not want to intervene in the activities of the various ideological pillars, but also because, in the Netherlands, there has always been a strong aversion to state compulsion.[59] Models for self-development and citizenship were hardly imposed from above by the state, but they were developed and enunciated by leading groups in pillarized civil society itself. Mental health care played a major part in the articulation of the psychic dimension of personal as well as public life, but the spread of a psychological habitus among the Dutch population also took place as an effect of more general social developments. Psychologization, a change of mentality characterized by a combination of growing individualization and internalization, was connected with the democratization of social relationships, the change in manners and authority struc-

[59] R. Aerts et al., *Land van kleine gebaren: Een politieke geschiedenis van Nederland, 1780–1990* (Nijmegen, Netherlands, 1999).

tures, the shift from external coercion to self-control, the transition from a command order to an order based on negotiation, and the increasingly subjective way of fashioning personal identity.[60]

From the 1950s on, people's behavioral orientation shifted from a submissiveness to fixed, unambiguous norms and guidelines, dictated by given social positions and hierarchies, religious dos and don'ts, a general sense of decency, and authority figures to a valuation of personal autonomy and individual consideration. Yet at the same time, the increased equality forced one to reckon more and more with others and, paradoxically perhaps, show more restraint in social interactions. As explicit rules and formal conventions lost some of their relevance, and individual social conduct became less predictable, the significance of self-regulation, subtle negotiation, and mutual consent grew accordingly. To find the proper balance between assertiveness and compliance, though, one needed social skills, empathy, self-knowledge, and an inner, self-directed regulation of emotions and actions. What mattered in a democratized social dynamic was a strongly developed sense of self-identity and mental resilience as well as insight into, and understanding of, the drives and motivations of others. Thus the interactions between people and the ways in which they evaluated each other became determined more and more by psychological insight. Tensions and conflicts between them had ramifications for their inner lives, potentially leading to mounting mental pressures and increasing the chance of their suffering from serious doubts, fears, and uncertainties.

In the Netherlands, which in social and cultural terms used to be quite conservative and Christian, the cultural revolution of the 1960s was more sweeping than in other countries because it coincided with rapid secularization and depillarization. After the stable and familiar moral frame began to be discussed publicly, it soon lost its relevance for many. In few other countries were control and coercion from above and others so radically excised as in the Netherlands.[61] The moral and spiritual vacuum was partially filled by a psychological ethos; from the 1960s on, mental health care, psychotherapy in particular, expanded at an unprecedented rate. The strongly developed democratization of public and everyday life replaced hierarchy, (group) coercion, and formal power relations with self-development, emancipation, and informal manners. This subsequently required subtle social regulation and psychological insight from individuals. It was more and more common for people to talk about themselves or others in psychological terms and to refer to their moods or feelings as ways to legitimate their behavior. Promoted in mass media and self-help books and by all sorts of therapists, trainers, advisers, and consultants, psychotherapeutic jargon and knowledge have basically—albeit in a watered-down version—penetrated all social

---

[60] C. Brinkgreve and M. Korzec, *"Margriet weet raad": Gevoel, gedrag, moraal in Nederland, 1938–1978* (Utrecht, 1978); W. Zeegers, *Andere tijden, andere mensen: De sociale representatie van identiteit* (Amsterdam, 1988); C. Wouters, *Van minnen en sterven: Informalisering van omgangsvormen rond seks en dood* (Amsterdam, 1990); Abma et al., *Het verlangen naar openheid* (cit. n. 54).

[61] That external control and coercion were replaced by a high degree of self-control is demonstrated by, among other things, the fact that the Dutch, despite their aversion to authority, are much less inclined than other nationalities toward civil disobedience and unconventional forms of protest, such as boycotts, spontaneous strikes, demonstrations, and occupations. Instead they show greater confidence in interaction and deliberations as means of solving conflicts. L. Halman et al., *Traditie, secularisatie en individualisering: Een studie naar de waarden van Nederlanders in een Europese context* (Tilburg, Netherlands, 1987).

and cultural domains, ranging from education and religion to sports, advertising, politics, business, public happenings, and politics.[62]

With their emphasis on self-reflection and raising sensitive issues, mental health experts articulated new values and offered a clear alternative for outdated norms. They not only adapted their views to the continuously changing social circumstances, but, especially in the 1950s, 1960s, and 1970s, also functioned as major agents of sociocultural renewal, which, if anything, won them overall public respect. Talking was their preferred strategy for solving problems, linking them not only with the Dutch culture of negotiation and consensus (holding meetings is, after all, a favorite Dutch pastime) but also with the practices of everyday life of many people.[63] Already in the 1930s, the largest segment of the working population had been active in the services sector, where social interaction and communications have increasingly grown central.[64] The strong inclination toward psychologization is also tied to the specific ways in which social and ethical issues are addressed in the Dutch political culture of consensus. It is a culture in which experts figure prominently. Their expertise is frequently called in because their supposedly objective professional stance neutralizes social conflicts associated with sensitive issues. In the articulation of policies involving euthanasia, sexuality, birth control, abortion and drugs, for example, experts such as physicians, psychiatrists, psychologists, and social workers have had a large say. They generally contributed to formulating practical solutions that are both pragmatic and well considered and that focus on individual conditions and motivations.

However, in the past two decades, confidence in the possibility of motivating individuals through considerate, soft psychosocial support toward self-guidance and socializing them in such way that they automatically integrate into an egalitarian and democratic society as full citizens has lost its taken-for-granted status. This approach has proved unsuitable in a society in which neoliberalism gained a foothold, where cultural diversity and polarization have become stronger, and where a large part of the population have viewed crime and safety as the major social problems. As a result of the emphasis on the market, individualization became increasingly couched as competition and the need to perform, rather than as liberation and well-being.[65] The freedom to develop seemed to benefit self-reliant and thick-skinned individuals in particular. They embodied an ideal of citizenship in which (economic) autonomy was elevated to the highest good. Those who wanted to back out of the hectic dynamic of the stress society, or had no choice but to do so, were quickly seen as problem cases. Those lagging behind, many of whom depended on the shrinking welfare state or belonged to ethnic minorities, were increasingly met with force or coercion so as to activate them toward social participation and self-reliance. For them, the emphasis shifted from rights to duties. Dependence on the welfare state and a lack of social integration—because of unemployment, educational disadvantages, insufficient

[62] Abma et al., *Het verlangen naar openheid* (cit. n. 54); J. C. van der Stel, "Individualisering, zelfbeheersing en sociale integratie," in *Individualisering en sociale integratie,* ed. P. Schnabel (Nijmegen, Netherlands, 1999), 126–58; Beunders, *Publieke tranen* (cit. n. 52).

[63] W. van Vree, *Nederland als vergaderland: Opkomst en verbreiding van een vergaderregime* (Groningen, Netherlands, 1994).

[64] H. Knippenberg and B. de Pater, *De eenwording van Nederland: Schaalvergroting en integratie sinds 1800* (Nijmegen, Netherlands, 1990), 128–30; Schuyt, "Sociaal-culturele golfbewegingen in de twintigste eeuw" (cit. n. 20), 223; Beunders, *Publieke tranen* (cit. n. 52), 61, 125–6.

[65] H. Wansink, *De opmars van de stressmaatschappij* (Amsterdam, 1994).

language skills, or certain religious (that is, Islamic) values—came to be more or less at odds with full citizenship. The view that citizenship had to be earned began to make headway, but along with this, more was expected from educational, employment, entrepreneurial, and criminal law circles than from the psychological subtleties of mental health care. Apart from being an issue of social participation, citizenship is still a matter of the proper mentality, yet the psychologizing angle has largely been replaced with a resurgent inclination toward moralizing paternalism and didactic instruction, on the one hand, and political polarization and juridical correction and repression, on the other.

Notes on Contributors

Greg Eghigian is Associate Professor of Modern European History at Penn State University. He is the author of *Making Security Social: Disability, Insurance, and the Birth of the Social Entitlement State in Germany* (2000) and co-editor of *Pain and Prosperity: Reconsidering Twentieth-Century German History* (2003) and *Sacrifice and National Belonging in Twentieth-Century Germany* (2002). He is presently writing a book about the science and politics of correctional incarceration in Nazi, West, and East Germany as well as editing a collection of primary sources on the history of madness and mental illness in western society.

Andreas Killen is Assistant Professor of History at the City College of New York/CUNY. He is the author of *Berlin Electropolis: Shock, Nerves, and German Modernity* and *1973 Nervous Breakdown: Watergate, Warhol, and the Birth of Post-Sixties America*. His current work concerns the medical, scientific, and cultural discourse about film spectatorship in Germany in the early twentieth century.

Christine Leuenberger teaches in the Department of Science & Technology Studies at Cornell University. She received her PhD in Sociology/Social Sciences in 1995 from the University of Konstanz, Germany. She has held Visiting Appointments at the Max-Planck-Institute for Human Development (Berlin), the University of California San Diego, and Korea University (Seoul). Her research interests are at the intersection of sociology, Science & Technology Studies, and the social and historical studies of the human sciences. Her work has been published in various edited collections as well as in sociological, philosophical, and historical journals, including *Social Problems, Theory & Society, Journal of the History of the Behavioral Sciences,* and *Human Studies: A Journal for Philosophy and the Social Sciences.*

Daniel Beer is a lecturer in Modern European History at Royal Holloway, University of London. His research interests are in the intellectual and cultural history of nineteenth- and early twentieth-century Russia. His forthcoming monograph, *Renovating Russia: The Human Sciences and the Fate of Liberal Modernity, 1880–1930,* is currently under review. He is embarking on a new research project exploring the cultural history of accusations of ritual murder made against Jews in the Russian Empire between 1800 and 1917.

Geoffrey Cocks is Julian S. Rammelkamp Professor of History at Albion College. He is the author of *Psychotherapy in the Third Reich: The Goering Institute* (1985, 1997), *Treating Mind and Body* (1998), and *The Wolf at the Door: Stanley Kubrick, History, and the Holocaust* (2004). He is presently working on a second edition of the Kubrick book and on a social history of illness in Nazi Germany.

Slava Gerovitch is a lecturer in the Science, Technology and Society Program at MIT. His scholarly interests include the history of Soviet cosmonautics, cybernetics, and computing. He has published *From Newspeak to Cyberspeak: A History of Soviet Cybernetics* (2002), several articles in the journals *Science in Context, Social Studies of Science, The Russian Review,* and *Technology and Culture,* and chapters in the collections *Universities and Empire* (1998), *Cultures of Control* (2000), *Science and Ideology* (2003), and *Critical Issues in the History of Spaceflight* (2006). He is currently working on a book on the technopolitics of automation in the Soviet space program.

Ellen Herman is Associate Professor of History at the University of Oregon. She is the author of *The Romance of American Psychology: Political Culture in the Age of Experts* and is currently completing a book on child adoption in the twentieth-century United States, *Kinship by Design.* She maintains a public service website, The Adoption History Project: www.uoregon.edu/~adoption

Volker Janssen is Assistant Professor of History at California State University, Fullerton. He completed his dissertation, *Convict Labor, Civic Welfare: Rehabilitation in California's Prisons, 1941–1971,* at the University of California, San Diego, in 2005. He is currently exploring the history of California women prisoners but is also venturing into the history of postwar politics of technology.

Aryn Martin is Assistant Professor of Sociology at York University in Toronto. She returned to Canada after completing her doctorate in Science & Technology Studies at Cornell University. In addition to her project on human chimeras, she is engaged in research on the contingencies of counting with Michael Lynch and, with Soraya de Chadarevian, a sociohistorical study of the visual standardization of the normal human karyotype.

Harry Oosterhuis studied history at the University of Groningen and took his doctoral degree in social science at the University of Amsterdam. Since 1992, he has been teaching history at the Faculty of Arts and Culture of the University of Maastricht, but he is also affiliated with the Huizinga Graduate School for Cultural History in Amsterdam. His current research focuses on the cultural and social history of mental disorders and psychiatry as well as the development of scientific knowledge about sexuality and gender. At present, he is codirecting a research project on the twentieth-century history of psychiatry and mental health care in the Netherlands. His main publications include *Homoseksualiteit in katholiek Nederland: Een sociale geschiedenis 1900–1970* (1992), *Homosexuality and Male Bonding in Pre-Nazi Germany: The Youth Movement, the Gay Movement and Male Bonding Before Hitler's Rise* (1992), and *Stepchildren of Nature: Krafft-Ebing, Psychiatry, and the Making of Sexual Identity* (2000).

Hans Pols is director of the Unit for History and Philosophy of Science at the University of Sydney. He is interested in the history of psychiatry, mental hygiene, and the history of medicine in the former Dutch East Indies and modern Indonesia.

Index

SUGGESTIONS FOR CONTRIBUTORS TO OSIRIS

OSIRIS is devoted to thematic issues, conceived and compiled by guest editors who submit volume proposals for review by the OSIRIS Editorial Board in advance of the annual meeting of the History of Science Society in November. For information on proposal submission, please write to the Editor at Osiris@georgetown.edu.

1. Manuscripts should be submitted electronically in Rich Text Format using Times New Roman font, 12 point, and double-spaced throughout, including quotations and notes. Notes should be in the form of footnotes, also in 12 point and double-spaced. The manuscript style should follow *The Chicago Manual of Style*, 15th ed.

2. Bibliographic information should be given in the footnotes (not parenthetically in the text), numbered using Arabic numerals. The footnote number should appear as superscript. "Pp." and "p." are not used for page references.

 a. References to books should include the author's full name; complete title of book in *italics*; place of publication; date of publication, including the original date when a reprint is being cited; and, if required, number of the particular page cited (if a direct quote is used, the word "on" should precede the page number). *Example*:

   [1] Mary Lindemann, *Medicine and Society in Early Modern Europe* (Cambridge, 1999), 119.

 b. References to articles in periodicals or edited volumes should include the author's name; title of article in quotes; title of periodical or volume in *italics*; volume number in Arabic numerals; year in parentheses; page numbers of article; and, if required, number of the particular page cited. Journal titles are spelled out in full on the first citation and abbreviated subsequently according to the journal abbreviations listed in *Isis Current Bibliography*. *Example*:

   [2] Lynn K. Nyhart, "Civic and Economic Zoology in Nineteenth-Century Germany: The 'Living Communities' of Karl Möbius," *Isis* 89 (1999): 605–30, on 611.

 c. Journal articles are given in full in the first reference. For succeeding citations, use an abbreviated version of the title with the author's last name. *Example*:

   [3] Nyhart, "Civic and Economic Zoology" (cit. n. 2), 612.

3. Special characters and mathematical and scientific symbols should be entered electronically.

4. A small number of illustrations, including graphs and tables, may be used in each volume. Hard copies should accompany electronic images. Images must meet the specifications of The University of Chicago Press "Artwork General Guidelines" available from the Editor.

5. Manuscripts are submitted to OSIRIS with the understanding that upon publication copyright will be transferred to the History of Science Society. That understanding precludes consideration of material that has been previously published or submitted or accepted for publication elsewhere, in whole or in part. OSIRIS is a journal of first publication.

OSIRIS (SSN 0369-7827) is published once a year.

Single copies are $33.00.

Address subscriptions, single issue orders, claims for missing issues, and advertising inquiries to *Osiris*, The University of Chicago Press, Journals Division, PO Box 37005, Chicago, IL 60637.

Postmaster: Send address changes to *Osiris*, The University of Chicago Press, Journals Division, PO Box 37005, Chicago, IL 60637.

OSIRIS is indexed in major scientific and historical indexing services, including *Biological Abstracts*, *Current Contexts*, *Historical Abstracts*, and *America: History and Life*.

Paperback edition, ISBN 978-0-226-19087-7

Osiris

A RESEARCH JOURNAL DEVOTED TO THE HISTORY OF SCIENCE AND ITS CULTURAL INFLUENCES

A PUBLICATION OF THE HISTORY OF SCIENCE SOCIETY

EDITORIAL OFFICE
BMW CENTER FOR GERMAN & EUROPEAN STUDIES
SUITE 501 ICC
GEORGETOWN UNIVERSITY
WASHINGTON, D.C. 20057-1022 USA
osiris@georgetown.edu